Teaching and Learning for a Sustainable Future:

Innovative Strategies and Best Practices

Chai Lee Goi
Curtin University, Malaysia

IGI Global
PUBLISHER of TIMELY KNOWLEDGE

A volume in the Practice, Progress, and Proficiency in Sustainability (PPPS) Book Series

Published in the United States of America by
IGI Global
Information Science Reference (an imprint of IGI Global)
701 E. Chocolate Avenue
Hershey PA, USA 17033
Tel: 717-533-8845
Fax: 717-533-8661
E-mail: cust@igi-global.com
Web site: http://www.igi-global.com

Library of Congress Cataloging-in-Publication Data

Names: Goi, Chai Lee, 1976- editor.
Title: Teaching and learning for a sustainable future : innovative strategies and best practices / edited by Chai Lee Goi.
Description: Hershey PA : Information Science Reference, [2024] | Includes bibliographical references and index. | Summary: "The book seeks to contribute to the ongoing discourse on sustainability education by showcasing innovative pedagogical approaches and strategies that can be adapted and applied across different contexts and disciplines. The book aims to inspire educators to incorporate sustainable practices into their teaching and to equip learners with the knowledge, skills, and values necessary for a sustainable future"-- Provided by publisher.
Identifiers: LCCN 2023037139 (print) | LCCN 2023037140 (ebook) | ISBN 9781668498590 (hardcover) | ISBN 9781668498606 (ebook)
Subjects: LCSH: Sustainability--Study and teaching (Higher) | Environmental education.
Classification: LCC GE196 .T43 2024 (print) | LCC GE196 (ebook) | DDC 304.2071/1--dc23/eng/20231016
LC record available at https://lccn.loc.gov/2023037139
LC ebook record available at https://lccn.loc.gov/2023037140

This book is published in the IGI Global book series Practice, Progress, and Proficiency in Sustainability (PPPS) (ISSN: 2330-3271; eISSN: 2330-328X)

British Cataloguing in Publication Data
A Cataloguing in Publication record for this book is available from the British Library.

Practice, Progress, and Proficiency in Sustainability (PPPS) Book Series

Ayman Batisha
International Sustainability Institute, Egypt

ISSN:2330-3271
EISSN:2330-328X

Mission

In a world where traditional business practices are reconsidered and economic activity is performed in a global context, new areas of economic developments are recognized as the key enablers of wealth and income production. This knowledge of information technologies provides infrastructures, systems, and services towards sustainable development.

The **Practices, Progress, and Proficiency in Sustainability (PPPS) Book Series** focuses on the local and global challenges, business opportunities, and societal needs surrounding international collaboration and sustainable development of technology. This series brings together academics, researchers, entrepreneurs, policy makers and government officers aiming to contribute to the progress and proficiency in sustainability.

Coverage

- E-Development
- Environmental informatics
- Intellectual Capital
- Sustainable Development
- Technological learning
- Outsourcing
- Eco-Innovation
- Green Technology
- Socio-Economic
- Global Content and Knowledge Repositories

IGI Global is currently accepting manuscripts for publication within this series. To submit a proposal for a volume in this series, please contact our Acquisition Editors at Acquisitions@igi-global.com or visit: http://www.igi-global.com/publish/.

The Practice, Progress, and Proficiency in Sustainability (PPPS) Book Series (ISSN 2330-3271) is published by IGI Global, 701 E. Chocolate Avenue, Hershey, PA 17033-1240, USA, www.igi-global.com. This series is composed of titles available for purchase individually; each title is edited to be contextually exclusive from any other title within the series. For pricing and ordering information please visit http://www.igi-global.com/book-series/practice-progress-proficiency-sustainability/73810. Postmaster: Send all address changes to above address.

Titles in this Series

For a list of additional titles in this series, please visit:
www.igi-global.com/book-series/practice-progress-proficiency-sustainability/73810

Water-Soil-Plant-Animal Nexus in the Era of Climate Change
Ahmed Karmaoui (Moulay Ismail University, Meknès, Morocco & Moroccan Center for Culture and Science, Morocco)
Engineering Science Reference • copyright 2024 • 478pp • H/C (ISBN: 9781668498385) • US $240.00 (our price)

Operational Research for Renewable Energy and Sustainable Environments
Joshua Thomas (UOW Malaysia KDU, Malaysia) Gerhard Wilhelm Weber (Poznan University of Technology, Poland) Roman Rodríguez Aguilar (Universidad Panamericana, Mexico) Elias Munapo (North West University, South Africa) and Pandian Vasant (Modeling Evolutionary Algorithms Simulation and Artificial Intelligence (MERLIN), TDTU, Vietnam)
Engineering Science Reference • copyright 2024 • 300pp • H/C (ISBN: 9781668491300) • US $240.00 (our price)

The Role of Women in Cultivating Sustainable Societies Through Millets
Raghvendra Kumar (GIET University, India) and Ishaani Priyadarshini (University of California, Berkeley, USA)
Engineering Science Reference • copyright 2024 • 293pp • H/C (ISBN: 9781668498194) • US $265.00 (our price)

Opportunities and Challenges for Women Leaders in Environmental Management
Perfecto Gatbonton Aquino, Jr. (The University of Cambodia, Cambodia) Revenio Cabanilla Jalagat, Jr. (Al-Zahra College for Women, Oman) and Mercia Selva Malar Justin (Xavier Institute of Management and Entrepreneurship, India)
Engineering Science Reference • copyright 2024 • 259pp • H/C (ISBN: 9781668459867) • US $250.00 (our price)

Quality of Life and Climate Change Impacts, Sustainable Adaptation, and Social-Ecological Resilience
Kasturi Shukla (Symbiosis International University) Yogesh B. Patil (Symbiosis International University, India) Ronald C. Estoque (Forestry and Forest Products Research Institute (FFPRI), Japan) and Pedro Antonio López de Haro (Universidad Autónoma Indígena de México, Mexico)
Engineering Science Reference • copyright 2024 • 300pp • H/C (ISBN: 9781668498637) • US $250.00 (our price)

Sustainable Development in AI, Blockchain, and E-Governance Applications
Rajeev Kumar (Moradabad Institute of Technology, India) Abu Bakar Abdul Hamid (Infrastructure University, Kuala Lumpur, Malaysia) Noor Inayah Binti Ya'akub (Infrastructure University, Kuala Lumpur, Malaysia) Hari Om Sharan (Amity University, Kolkata, India) and Sandeep Kumar (Rightzone Technologies Pvt. Ltd., India)
Engineering Science Reference • copyright 2024 • 350pp • H/C (ISBN: 9798369317228) • US $285.00 (our price)

IGI Global
PUBLISHER of TIMELY KNOWLEDGE

701 East Chocolate Avenue, Hershey, PA 17033, USA
Tel: 717-533-8845 x100 • Fax: 717-533-8661
E-Mail: cust@igi-global.com • www.igi-global.com

Table of Contents

Detailed Table of Contents

Growing concerns about the planet and society have led to the evolution of the concept of sustainable development. This concept gained popularity when the World Commission on Environment and Development released its report Our Common Future in 1987. In 1975, United Nations Educational, Scientific and Cultural Organization (UNESCO) brought sustainability as the International Environmental Education Program focusing on environmental education. This gave birth to the idea that Higher Educational Institutions (HEIs) can play a significant role in promoting the sustainability agenda. Over the years, this was done by implementing sustainability initiatives on the campus. These initiatives mostly focused on the environment and ignored the social and economic dimensions of sustainability. Given the paucity of adequate knowledge in this field, the chapter aims to explore the challenges in implementing sustainable initiatives and suggest a framework that will guide HEIs to act as a catalyst for sustainability development.

Multidisciplinary collaboration is crucial to sustainable education. It places a strong emphasis on combining many academic disciplines to address complex environmental issues and promote sustainable lifestyles. Students can put their knowledge into practice and raise awareness of the world through community-based learning, corporate partnerships, and international collaborations. The chapter emphasizes effective methods for evaluating the effectiveness and impact of collaborative activities. It highlights how crucial innovation, knowledge sharing, and international cooperation are to building a sustainable future. Social justice, action-oriented learning, and lifetime professional growth are all components of sustainability education. Institutional barriers, cultural differences, and resource constraints impact these cooperative initiatives. Future generations' perceptions of the Earth can be affected via interdisciplinary collaboration and partnerships.

The 2030 Agenda for Sustainable Development calls on all countries to adopt a series of measures to promote sustainability. The circular economy emerges as a new paradigm of production and consumption aligned with sustainable development represents a major challenge for all the agents involved. At the corporate level, it represents the inclusion of sustainable practices which put marketing in the spotlight of business strategy. In this context, education plays a decisive role in contributing to the sustainable development of societies, and this educational process can be promoted in different areas. The companies, for their potential to booster responsible demand from consumers, can be a powerful sustainability tool that contributes to reducing the gap between attitude and sustainable behavior. Based on a desk research methodology, this chapter pursues a twofold objective: to analyze whether there is a gap between consumers' stated attitude and sustainable behavior and to identify whether an appropriate marketing strategy can help overcome it.

The United Nations' Sustainable Development Goals (SDGs) provide a collaborative framework for sectors worldwide to pursue a just, prosperous, and environmentally sustainable future. This chapter investigates the application of design thinking as a pedagogical strategy to equip students with the skills necessary to meet sustainability challenges. Rooted in a human-centered approach, design thinking seeks to intimately comprehend complex challenges to create solutions that bolster the welfare of both individuals and communities, thus enabling students to act as agents of societal change. Introducing academic design thinking pedagogy (ADTP), this chapter is tailored to the context of higher education and articulates its phases to provide educators with actionable insights on its implementation.

There have been trends in experiential learning (EL) to implement self-directed learning (SDL) in higher education institutions in the 21st century. Due to the necessity of a real-life, interactive environment in the lecturing approaches in higher education, the problem is to ensure that the learning becomes interactive and meaningful to the students. Students commit to learning when their goals and objectives are considered realistic and important, though they may face socio-economic challenges. Therefore, applying knowledge in the real world is important and relevant to the student's needs. This chapter explores the implications of experiential learning theory to gain work-related skills and knowledge for the work environment. The implication of the experiential learning theory (ELT) to higher education is that lecturers should design course materials that empower students to acquire knowledge and skills to solve real-life problems. Thus, ELT points to teaching practices that aim at enabling students to perform activities that are close to those found in the workplace.

This chapter explores integrating pedagogical approaches for sustainable teaching and learning, emphasizing the capacity to meet present requirements without compromising the future. It highlights the merits of experiential, inquiry-based, problem-based, and project-based methods in sustainable practices in education. Experiential learning emphasizes practical application and reflection, whereas inquiry-based learning promotes inquiry and exploration. Problem-based learning immerses students in real-world sustainability challenges and interdisciplinary collaboration, whereas project-based learning enables students to take on leadership roles. Integrating these techniques offers a variety of options for addressing complex environmental issues. Future obstacles include integrating technology and ensuring equitable access. Integrated educational practices require learner-centred approaches, collaboration, and continuous feedback, empowering students to become proactive sustainability advocates and promoting positive change for a sustainable future.

Environmental education (EE) has become an essential component of the curriculum from primary to higher education. It connects the students to the world around us, making them aware about both natural and built environments. It heaves awareness of issues impacting the environment upon which we all depend, as well as actions we can take to improve and sustain it. Environmental education has a lot of advantages to students, teachers, schools, and the society as a whole. The curriculum attempts to incorporate the components of environmental education either in language, science, mathematics, or a complete subject as well. The teachers can deliver such knowledge by bringing nature into the classroom, take students outside to learn, or find impromptu teachable moments on an eco-trail. The present study explores the awareness level of students using new ecological paradigm (NEP) scale after and before implementation of an EE program. Further, it differentiates their environmental awareness with respect to their demography.

Social emotional learning (SEL) can be defined as a process in which individuals make responsible decisions and choices by managing their emotions, setting goals, establishing positive relationships with empathy towards others. This chapter aims to discuss the current SEL literature through a transformative lens of social justice and equity for a sustainable educational future. With this purpose in mind, this chapter reviews the following issues: fundamentals of SEL along with core components, the crucial role of SEL in celebrating diversity, promoting equity in EFL classes through SEL, and benefits of SEL integrated foreign language learning and teaching. Moreover, applications of transformative SEL integration methods and strategies are addressed. Based on the current review, some implications for educators on how to integrate SEL key components into their teaching contexts and suggestions for further research to enhance equity and sustainability in the field of foreign language education are yielded.

The role of teachers in raising awareness and equipping students with the necessary skills to contribute to a sustainable future is crucial. This study aims to identify the descriptions of the characteristics of effective teachers made by pre-service teachers who have just started university. The research was conducted with the participation of pre-service teachers attending a state university. The data were collected through semi-structured interview forms prepared by the researchers. The interview form was comprised of questions about what the effective and qualified teachers' characteristics are and about what characteristics of effective and qualified teachers would have positive reflections into teaching environments. Each interview lasted approximately 15-20 minutes. The data collected were then put to content analysis. The analyses demonstrated that the participants usually describe effective teachers as individuals who have comprehensive knowledge of their branch and who continuously make self-improvement.

The study aimed at investigating the attitudes of secondary school teachers and students towards STEM approach. The study is a follow up of the social initiative of SASTRA University, India, where STEM labs were set up in rural schools to promote STEM education. Data for this chapter came from 10 schools where the labs were set up in the first phase of the STEM project. The tool of the study was two questionnaires which were validity and reliability verified. Thirty-three secondary school teachers and 136 students were administered the questionnaire after one year of exposure to the lab. Descriptive techniques were used to analyze the data. Results of the study showed positive student and teacher attitude toward STEM education. In light of results of the study, the researcher highly recommends using STEM approach in teaching subjects in schools as being crucial to personal development and citizenship in the 21st century. The study also recommends the 'adoption' of a school, or a cluster of schools, by a local STEM industry/enterprise, university, NGOs, and other public, private bodies.

Whilst individual efforts to achieve social sustainability are positive, the challenges it poses are such that without a collective, proactive, and collective approach, intergenerational success for the most significant critical global issues will become a lost and irretrievable opportunity. Gamification is one pedagogical approach to engaging the collective in issues pertaining to social sustainability, where collective learning and engagement will be central to the creativity and innovation that can ensue. Capacity for critical reflexivity, rather than a reflective stance on the past, is needed to drive the perspective lens of transformation forward

Preface

In the face of critical global challenges, the pursuit of a sustainable future has become an imperative that transcends borders, disciplines, and generations. The complexities of sustainability demand collective action and a concerted commitment to nurturing a future that safeguards the well-being of our planet and its inhabitants. Education emerges as a linchpin in this transformative journey, wielding the power to shape values, instill knowledge, and cultivate skills essential for sustainable practices.

Teaching and Learning for a Sustainable Future: Innovative Strategies and Best Practices stands as a testament to the urgency of this educational mandate. In the following pages, we delve into the interdisciplinary and multi-sectoral landscape of teaching and learning for sustainability. Our aim is to provide a comprehensive collection of innovative strategies and best practices that resonate across diverse disciplines, educational levels, and contextual frameworks.

At the heart of sustainability education lies a transformative process that empowers learners to become catalysts for change in their communities, workplaces, and beyond. From climate change and biodiversity loss to social justice and poverty reduction, the breadth of themes covered underscores the holistic nature of this educational endeavor. This book seeks to inspire educators to incorporate sustainable practices into their teaching methodologies, equipping learners with the knowledge, skills, and values indispensable for navigating the challenges of the future.

ORGANIZATION OF BOOK

Chapter 1: This chapter delves into the evolution of the concept of "Sustainable Development" and its integration into Higher Educational Institutions (HEIs). It explores the historical context, focusing on the environmental initiatives within HEIs. Recognizing the need for a comprehensive approach, the chapter identifies challenges in implementing sustainable initiatives and proposes a framework to guide HEIs in becoming catalysts for sustainability development.

Chapter 2: Emphasizing the pivotal role of multidisciplinary collaboration in sustainable education, this chapter explores how combining academic disciplines can address complex environmental issues. It underscores the importance of community-based learning, corporate partnerships, and international collaborations. The chapter also provides insights into evaluating the effectiveness and impact of collaborative activities, acknowledging challenges and highlighting the transformative potential of interdisciplinary collaboration.

Chapter 3: Focusing on the corporate dimension of sustainability, this chapter delves into the circular economy as a paradigm aligned with sustainable development. It examines the role of marketing in promoting sustainable practices at the corporate level. The chapter employs a desk research methodology to analyze the gap between consumers' stated attitude and sustainable behavior, exploring how marketing strategies can bridge this gap and contribute to sustainability.

Chapter 4: This chapter investigates the application of design thinking as a pedagogical strategy to equip students with the skills needed to address sustainability challenges. Introducing Academic Design Thinking Pedagogy (ADTP), the chapter articulates its phases, offering educators actionable insights for implementation within the context of higher education. The goal is to foster human-centered solutions that contribute to societal change and align with the United Nations' Sustainable Development Goals.

Chapter 5: Focusing on the trends in experiential learning, this chapter explores the implementation of Self-Directed Learning (SDL) in higher education. It addresses the need for interactive and meaningful learning experiences, emphasizing the relevance of applying knowledge in real-world scenarios. The chapter advocates for course design that empowers students to acquire work-related skills and knowledge through experiential learning theory (ELT).

Chapter 6: This chapter offers a comprehensive exploration of pedagogical approaches for sustainable teaching and learning. It highlights the merits of experiential, inquiry-based, problem-based, and project-based methods, emphasizing their capacity to meet present needs without compromising the future. The chapter addresses challenges such as technology integration and equitable access, advocating for learner-centered approaches that empower students as proactive sustainability advocates.

Chapter 7: Exploring the essential role of Environmental Education (EE), this chapter examines its incorporation into the curriculum from primary to higher education. It emphasizes the advantages of EE in connecting students to the natural and built environments, raising awareness, and fostering sustainability actions. The chapter utilizes the New Ecological Paradigm (NEP) Scale to assess students' environmental awareness before and after an EE program, differentiating results based on demographics.

Chapter 8: This chapter delves into Social Emotional Learning (SEL) through a transformative lens of social justice and equity. It discusses the fundamentals and core components of SEL, highlighting its crucial role in celebrating diversity and promoting equity in English as a Foreign Language (EFL) classes. The chapter explores transformative SEL integration methods, offering implications for educators and suggestions for further research in the field of foreign language education.

Chapter 9: Focusing on the pivotal role of teachers, this chapter investigates pre-service teachers' descriptions of effective teachers. Through semi-structured interviews, the study identifies characteristics associated with effective teaching and their positive reflections on teaching environments. The findings contribute insights into the importance of comprehensive knowledge and continuous self-improvement in effective teaching for a sustainable future.

Chapter 10: This chapter explores the attitudes of secondary school teachers and students toward the STEM approach. Building on a social initiative, the study investigates the impact of STEM labs set up in rural schools. The results reveal positive attitudes toward STEM education, leading to recommendations for wider adoption of the STEM approach in schools to foster personal development and citizenship in the 21st century.

Chapter 11: Addressing the challenges of social sustainability, this chapter explores gamification as a pedagogical approach to engage collective learning. Emphasizing the need for a proactive and collective approach, the chapter discusses how gamification can foster creativity and innovation. It highlights the importance of empowering individuals across educational contexts to address key global issues through a holistic and humanitarian approach.

Chapter 12: Investigating the application of virtual reality (VR) in higher education, this chapter covers types of VR, implementation in academic settings, and effects on student engagement and learning outcomes. The comprehensive review explores how VR refines cognitive skills, practical knowledge, and personalized learning. The chapter provides insights into the implementation process and highlights the positive impact of effective VR content and prepared instructors on engagement and academic performance.

Chapter 13: This chapter draws inspiration from rituals and beliefs rooted in Anatolian societies to facilitate students' understanding of climate change's impact on local and global communities. It explores activities like readings, storytelling, and portfolio work to develop pro-environmental values and discuss the collective consciousness of past and present civilizations. The researchers aim to preserve socio-cultural identities and promote intergenerational knowledge dissemination.

Chapter 14: Tracing the infusion of sustainability education in business schools, this chapter provides an exploratory case study approach to two Australian university branch campuses in Malaysia. Thematic analysis of unit course details reveals the presence of sustainability education across a three-year bachelor's degree. The chapter emphasizes the importance of building sustainability fundamentals from the first year to produce future-ready graduates.

Chapter 15: This chapter addresses the challenges in preservation and conservation education within Library and Information Sciences in the Eastern and Southern Africa Regional Branch of the International Council of Archives (ESARBICA) member states. It weaves together literature, explores observed challenges, and proposes a sustainable framework for integrating conservation into the preservation curriculum. The chapter concludes with recommendations for improvement in conservation education.

Chapter 16: Highlighting the role of Science, Technology, Engineering, and Mathematics (STEM) in Africa's economic development, this chapter advocates for teaching STEM in local languages. It contends that teaching STEM in a foreign language hinders sustainable development. The chapter recommends the use of translanguaging and multilingualism to improve student performance in STEM, emphasizing the importance of acquiring scientific knowledge in mother tongues for innovation and development.

This compendium is more than a mere compilation of pedagogical approaches—it is a call to action. Through a variety of innovative teaching tools and technologies, curriculum development strategies, and cross-disciplinary collaborations, we aim to spark a dialogue among educators, academics, researchers, policy makers, and students. This platform is designed to facilitate the exchange of experiences, insights, and best practices in teaching and learning for sustainability.

As you embark on this intellectual journey, we invite you to explore the diverse topics encapsulated within these pages. From experiential learning to the integration of sustainability in different educational levels, from cutting-edge technologies to cross-disciplinary partnerships, the breadth of coverage is intended to resonate with a wide audience. This book is not just a resource; it is a catalyst for change.

We extend our gratitude to the contributors who have enriched this volume with their expertise and experiences. Their commitment to advancing sustainability education is evident in the richness of the content presented here. We hope that Teaching and Learning for a Sustainable Future serves as a source of inspiration and a practical guide for educators and stakeholders committed to shaping a sustainable and equitable world for generations to come.

Chai Lee Goi
Curtin University, Malaysia

Chapter 1
Higher Education Institutions as a Catalyst for Sustainability Development

Shilpa H. Shetty
https://orcid.org/0000-0002-0048-4645
Christ University, India

ABSTRACT

Growing concerns about the planet and society have led to the evolution of the concept of sustainable development. This concept gained popularity when the World Commission on Environment and Development released its report Our Common Future in 1987. In 1975, United Nations Educational, Scientific and Cultural Organization (UNESCO) brought sustainability as the International Environmental Education Program focusing on environmental education. This gave birth to the idea that Higher Educational Institutions (HEIs) can play a significant role in promoting the sustainability agenda. Over the years, this was done by implementing sustainability initiatives on the campus. These initiatives mostly focused on the environment and ignored the social and economic dimensions of sustainability. Given the paucity of adequate knowledge in this field, the chapter aims to explore the challenges in implementing sustainable initiatives and suggest a framework that will guide HEIs to act as a catalyst for sustainability development.

INTRODUCTION

The modern world has witnessed rapid economic development. Immense economic growth came at the cost of environmental and social resources. According to world data, the world carbon dioxide emission was 6 billion tonnes in 1950, almost quadrupled to 22 billion tonnes in 1990 and is over 35 billion tonnes by 2020 (Ritchie & Roser, 2020). On the other hand, according to the world inequality report the richest 10% of the global population currently takes 52% of global income, whereas the poorest half of the population earns 8.5% of it (Chancel, Piketty, Saez, & Zucman, 2022). This imbalance in the ecosystem and inequality in the income began with the onset of industrialisation. With rapid industrialisation, this has been increasing in an alarming rate.

DOI: 10.4018/978-1-6684-9859-0.ch001

In the recent years growing population and resource use has increased the pressure on the environment. This has caused an increased concern about climate change and environment degradation in every corner of the world. The intensifying human interference with the earth's ecosystem has threatened the sustained availability of natural resources. Under this circumstance, the need to stimulate the coordinated growth of economy, environment and society was felt. Post World War II, many strains of thoughts have emerged in this context. It began with the realisation that the available resources are inadequate to expand industrial activities. Eventually, the quality of environment surpassed the concern for the quantity of resources. This has led to the emergence and worldwide recognition of the concept of 'sustainability'. From a concept, today sustainability has become a widely accepted idea across the world.

BACKGROUND

The term sustainability was first used in 1972 in a British book, Blueprint for Survival. In the book, the industrial way of life was criticized as non-sustainable and a need for a sustainable society was stressed upon. Following this The International Union for the Conservation of Nature (IUCN), found one of the first instances of the use of the word "sustainable" in the context of the environment. In 1974, the Woodlands conference report for the first time used the word sustainability. The subsequently held woodlands conferences led to books that were the first to use the word "sustainable" in their titles: Dennis Meadows- Alternatives to Growth I: A Search for Sustainable Futures (1977), James C. Coomer: Quest for a Sustainable Society (1979) and Harlan Cleveland: The Management of Sustainable Growth (1979). The term sustainability also appeared in the 1976 fisheries act, The Magnuson Act, 1976. In defining optimum yield, the act stated, "the amount of fish which is prescribed as such on the basis of the maximum sustainable yield from such fishery". Further, in 1977, Professor Ignacy Sachs of the Centre International de Recherche sur invented the term "ecodevelopment" and defined it as "an approach to development aimed at: harmonizing social and economic objectives with ecologically sound management, in a spirit of solidarity with future generations." However, he did not use the word "sustainability." But the contribution of Sachs is significant because it provided the basic rationale for the 1978 UNEP document that marked the first use of the term sustainability in a U.N. document.

In the UN Documents the idea of sustainability emerged slowly between 1968 and 1976. During this period there was growing concerns towards environmental degradation and deterioration in the quality-of-life due to industrialisation. In 1978, UNEP used the word sustainability in its report, Review of the Areas: Environment and Development, and Environment Management. The term gained popularity between 1980 and 1988 when sustainability was used by government and non-government agencies in their publications. In 1981, the new President of the World Bank, A.W. Clausen, chose Sustainability as the theme for his Fairfield Osborn Memorial Lecture, Sustainable Development: The Global Imperative. He developed the central idea that "Sustainable development and wise conservation are, in the end, mutually reinforcing and absolutely inseparable goods:' In 1987, the World Bank also held a conference on sustainability and released Report of the World Bank Workshop on Sustainability issues in Agriculture. Sustainability was the central theme in Our Common Future released by The World Commission on Environment and Development 1987. This is popularly called as the Brundtland report, named after its chairman Mrs. Gro Harlem Brundtland (Kidd, The Evolution of Sustainability, 1992). According to the report, sustainability development was defined as the as "development that meets the needs of the present without compromising the ability of future generations to meet their own needs."

Sustainability began as a strategy to protect environment but, the idea has extended way beyond over the years. To accomplish sustainable development, in 2000, leaders of 189 countries signed Millennium declaration committing to attain eight measurable goals by 2015. It includes Eradicate extreme poverty and hunger, achieve universal primary education, promote gender equality and empower women, reduce child mortality, improve maternal health, combating HIV/AIDs, malaria, and other diseases, ensure environmental sustainability and develop a global partnership for development. With the help of eight Millennium Development Goals (MDG) substantial progress was made. But the crises and the agenda still continue.

Sustainable Development Goals were constituted in 2015 to maintain the momentum created by MDG's. 17 Sustainable Development goals were adopted by 193 countries. It includes no poverty, zero hunger, good health and wellbeing, quality education, gender equality, clean water and sanitation, affordable and clean energy, decent work and economic growth, reduced inequalities, sustainable cities and communities, responsible consumption and production, climate action, life below water, peace justice and strong institution and partnership for the goals. Thus, sustainability has changed from a concept to a global agenda today.

SUSTAINABILITY IN EDUCATION

In 1975, United Nations Educational, Scientific and Cultural Organization (UNESCO) brought sustainability as the International Environmental Education Program focusing on environmental education. Later it changed to Education for Sustainability (EfS) which years later culminated in the UN Decade of Education for Sustainable Development (DESD) from 2005 to 2014 (Moura, Frankenberger, & Tortato, 2019). HEI's are identified as a critical player in the move towards sustainable development. They play a very important role in promoting sustainability because they educate the future leaders. In the recent years HEI's have begun to focus on sustainability especially after the launch of UN Sustainability goals. Today, along with the government and the business sector, the education sector is also taking initiatives to address environmental, social, and economic challenges. This can bring about a long-lasting change in the form of improved quality of life, reduced ecological footprint and enhanced economic vitality. Currently, HEI's directly contribute to SDG 4- Quality education. By integrating sustainability into education, other SDG's like no poverty, good health and wellbeing, reduced inequalities, responsible consumption and production can also be attained.

HEI involvement in sustainability development begun following the policy initiatives that were undertaken globally mostly in the form of declarations. These declaration on sustainability encourage HEI's to support and contribute to the sustainability movement. Some of the significant mentionable sustainability policy initiatives in the field of education are discussed in this section.

Talloires Declaration

The Talloires Declaration was one of the first attempt to promote sustainability among HEI's. This declaration made in October 1990 at Talloires, France by the University leaders of different countries spelled out the key actions to be undertaken to create sustainable future. The 10-point action plan suggested by the Talloires declaration are summarised below-

1. Increase awareness of environmentally sustainable Development
2. Create an institutional culture of sustainability.
3. Establish programs that educate for environmentally responsible citizenship.
4. Foster environmental literacy for all
5. Practice institutional ecology like recycling, waste reduction etc.
6. Involve all stakeholders to assist in exploring solution to environmental problems.
7. Collaborate with environmental practitioners to develop interdisciplinary approaches in curriculum, research, operations, and outreach activities.
8. Enhance capacity of primary and secondary schools' with capacity for interdisciplinary teaching about population, environment and sustainable development.
9. Broaden service and outreach nationally and internationally by collaborating with national and international organisations.
10. Maintain the movement by establishing steering committees.

Since the launch of The Talloires Declaration, it has successfully created awareness across the world about the significance of sustainable development in education. Following the Talloires Declaration, many other universities and organisations were inspired, and few other such declarations were made in different parts of the word. It includes Halifax declaration, Swansea declaration and so on (The Talloires Declaration 10 Point Action Plan, 1990).

Halifax Declaration

The head of universities from Canada, Brazil, Zimbave, Indonesia and others along with senior representatives of the International Association of Universities, the United Nations University and the Association of Universities met in Halifax, Canada to discuss the role of universities in sustainable development in 1991. The result of the meeting was the Halifax declaration that was signed by 16 Canadian Universities committing to become more sustainable. Specific activities suggested by this declaration includes-

1. Creating environment education programs at local, regional, national, and international level
2. Sponsoring prizes for students, faculty, and staff to encourage sustainable development.
3. With the help of national media identifying ways that can help university contribute to sustainable development
4. Associating with a national university network that focuses on sustainable development.
5. Encouraging faculty and students to link their research to sustainability goals (Wright, 2003).

Swansea Declaration

Following The Talloires declaration and Halifax declaration, the Association of Commonwealth Universities (ACU) 15th Quinquennial Congress drawn from over 400 universities in 47 different countries met at Swansea, Wales in 1993 to address the challenge of "People and the Environment - Preserving the Balance". The Swansea declaration suggested the following actions by all ACU.

1. Establish and disseminate a clearer understanding of sustainable development.
2. Utilise university resources to spread understanding about the dangers facing the planet.

3. Emphasise on the ethical obligation of the current generation to overcome the practices that lie at the root of environmental unsustainability.
4. Enhance the university capacity to teach and undertake research that promote environmental sustainability.
5. Cooperate with one another to achieve sustainable development.
6. Encourage universities to review their own operations to reflect best sustainable development practices.
7. Urgently identify ways and means to implement this declaration (Association of Commonwealth Universities, 1993)

The other declarations that followed are Kyoto Declaration on sustainable development (1993), Thessaloniki Declaration (1997), The Luneburg Declaration on higher education for sustainable development (2001), Ubuntu Declaration (2002), Declaration of Barcelona, education in sustainable development (2004), Bergen Declaration (2005), Lucerne Declaration (2007), Sapporo Sustainability Declaration (2008), Turin Declaration on education and research for sustainable and responsible development (2009) and so on. All these declarations and many more suggested how HEI's can play a role in addressing sustainability issues (Jnr, 2021).

CASE STUDIES

In the recent years, increased number of HEI's around the world are taking up new and innovative activities that favours sustainable development. From undertaking green initiatives in managing the campus to designing curriculum on sustainability, the HEI's are making a sincere attempt in this direction. Institutions are also voluntarily preparing sustainability reports, participating in sustainability ranking, signing up for international sustainability declarations and so on. According to (Ralph & Stubbs, 2013), HEI that has integrated sustainability will have the following characteristics.

- Leadership and vision that expresses commitment to, and promotes, sustainability.
- Concepts and practices of sustainability incorporated into the teaching and research of all academic disciplines.
- An emphasis on fostering the inter- and trans-disciplinary teaching and research needed to provide solutions to sustainability challenges.
- Recognition of the ecological footprint of the institution, along with sustainable policies and practices that minimise this footprint and
- Engagement in community outreach that enhances environmental sustainability.

In this section, case studies of HEIs from Malysia, Bazil and USA that have successfully incorporated sustainability in their activities are discussed.

Malaysia

There is a growing inclination towards sustainable development in Malaysia. Malaysia had committed to decrease carbon dioxide emission by 40% in the 1992 Earth Summit held in Rio. Malaysian Universities have already begun to adopt sustainability practices. In a research article authored by (Jnr, 2021), 16 Malaysian HEI's were studied to understand the green practices and initiatives undertaken to foster environment, economic and social development. A summary of the initiatives is listed below.

- Use of eco-friendly materials in day-to-day operations
- Promoting energy saving, waste management and water conservation
- Developing green and energy efficient buildings.
- Encouraging the practice of recycling and reusing.
- Promoting awareness on the dangers of using plastic bags and substituting it with banana leaves.
- Encouraging the use of bicycles for commute in the campus
- Promoting the use of public transport and providing planned buses to decrease carbon emission.
- Deploying computer power control setting and use of power strip or surge protector for monitors, printers and other accessories.
- Printing only when needed initiatives and printing double-sided to save paper.
- Deployment of rainwater harvesting system.
- Endorsing landscape, watershed, and habitat management.
- Undertaking community outreach programmes to promote sustainability.
- Organising seminars, workshops, and lectures on sustainability.
- Developing courses and modules on sustainable development.

Brazil

A study conducted by (Moura, Frankenberger, & Tortato, 2019) discussed how Brazilien HEI's have adopted sustainability practices by analysing 92 Brazilian educational institutions. The sustainability activities identified by the authors are summarised below:

- Organisation structure designed with a position dedicated to sustainability.
- Developing courses and lectures on sustainability especially in the areas of engineering, architecture, and business.
- Forming research groups that deliver lectures and involve in debates on sustainability.
- Publishing booklets and Newsletter on sustainability
- Designing technical courses and sustainability centres
- Instituting academic journals that encourage publications on sustainability.
- Launching environment improvement activities with respect to water, building, energy, waste management, food, and nature.
- Undertaking social activities like educating different sections of the society, offering healthcare programmes like blood glucose test, breast examination etc; financial incentives for income generation and infrastructure support to execute social projects, initiates to improve wellbeing of local community; providing guidance like legal advice, psychological care etc
- Provide training to university staff to be multipliers of good practices of use and reuse of raw materials used in research.

USA

A Study undertaken by (Purcell, Henriksen, & Spengler, 2019), explores the different ways in which Harvard university strategizes sustainability in attaining the SDGs. Harvard University has set up Office of Sustainability (OFS) through which healthy and sustainable campus community is promoted through pan university sustainability plan. Some of the activities undertaken through OFS are discussed below:

- Advise and mentor staff and students to experiment with sustainable solutions to tackle real life challenges.
- Campus Sustainability Innovation Fund and Green Revolving Fund are set up to co-fund innovation that can address problems of people and planet.
- Through its convening power, OFS empowers people to make informed choices.
- Facilitates hands- on research by creating a trans-disciplinary climate change learning course that provides a space for students from across the university to come together with faculty and work on projects that provide solution to real life sustainability challenges.
- In association with various civic organisations, OFS works towards developing strategies for fighting climate change.

Existing work on sustainability in HEI's have mostly focussed on sustainable practices adopted by institutions. However, the is no adequate and dedicated literature uncovering the challenges faced by HEI's in promoting and smoothly implementing sustainable practices. Further, in the case studies discussed, it is observed that though the HEIs in collaboration with the staff, students and the community are trying to come together to promote sustainable development in and around the campus, the initiatives discussed are mostly aimed at reducing the carbon footprints thus, addressing only the environment dimension of sustainability development. This shows that participants are mostly engaged in the environment issues thus, ignoring the broader aspects of sustainability. In other words, the economic and social dimensions of sustainability are overlooked by the HEI's. According to Segura and Zamar (2021), the economic dimension of sustainability focuses on reducing poverty by meeting basic needs and increasing economic equality without impairing natural resources. On the other hand, social dimension covers cultural diversity, social justice etc.

There is a need to firstly identify and put together the major challenges faced by HEI's in promoting and executing sustainable practices. In addition, there is a need to develop a framework that can guide the HEIs in being a catalyst for sustainable development. As HEI's educate the future leaders, adopting this framework will provide the right exposure to the students who will create a more sustainable world in the future.

CHALLENGES

The barriers that deter the successful planning and execution of sustainability policies and practices include-

1. Lack of clarity in the definition of the term 'sustainability' and absence of deeper understanding of sustainable HEI and its practices (Mahameed, 2022).

2. Disconnect between policy development and implementation of sustainability initiatives (Macheridis & Paulsson, 2021).
3. Lack of support from the management like training, teaching material etc to the staff (Macheridis & Paulsson, 2021)
4. Due to disciplinary barriers, the extent of inclusion of sustainability in curriculum and research is limited.
5. Inadequate financial support to implement sustainability initiatives due to multiple projects competing for limited resources.
6. Poor understanding and awareness about sustainability issues and the need to address them.
7. Achieving a coherent institutional approach where operations, teaching, research, and outreach are synergised is the major challenge of sustainability development (Ralph & Stubbs, 2013).
8. Lack of wholehearted cooperation and participation by the stakeholders in implementing sustainability practises
9. Absence of monitoring and reporting of sustainability activities as the number of HEI's tracking and reporting sustainability activities is very limited compared to business corporations.
10. Weak acknowledgement of sustainability at the national level and limited policies promoting sustainability in education (Dagiliute & Liobikiene, 2015)
11. Acknowledgement and celebration of successful sustainability projects is a rare event.

HEIs as Catalyst for Sustainable Development

This section presents a framework that addresses the challenges and touches all dimensions of sustainability- Environmental, Social and Economic. The suggested framework is basically derived and categorised from the activities that commonly take place in an academic set up. The vision and mission of the HEI lies at the core of the framework. The other elements of the framework comprise of operations, extension, curriculum, and research. Figure 1 presents the framework. With all these elements coming together, its implications will be synergised and amplified. Further, constantly monitoring and periodically reporting the activities will build trust and credibility with the stakeholders.

VISION AND MISSION STATEMENT

The activity of any HEI is largely guided by its vision and mission statement. Incorporating the sustainability agenda in the vision and mission statement shows that the university or institutions activities are aligned with the sustainability goals 2030. The greatest challenge of getting support from the management in terms of adequate finances, requisite training and celebrating successful implementation of sustainable projects will all be addressed once the sustainability agenda becomes part of the vision and mission statement.

Consider the Vision and Mission Statement of Carnegie Mellon University-

Vision

Carnegie Mellon University will have a transformative impact on society through continual innovation in education, research, creativity, and entrepreneurship.

Figure 1. Framework for HEIs to become catalyst of sustainable development

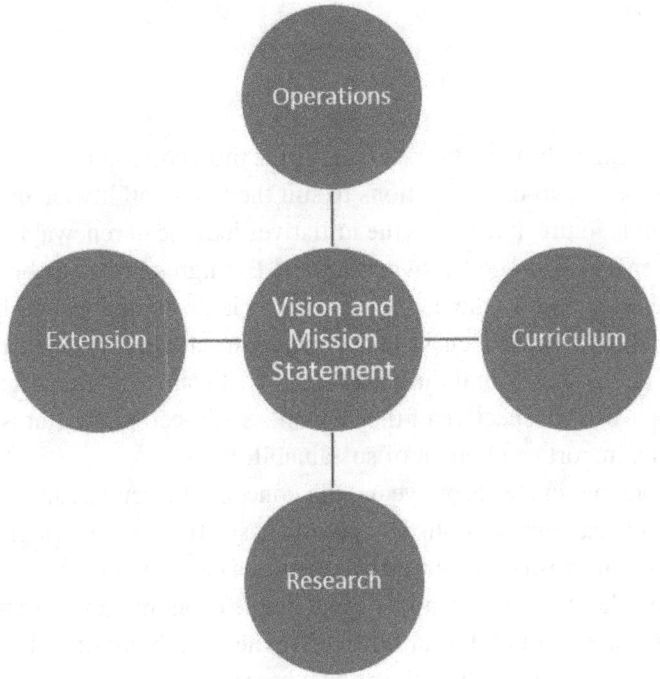

Mission

To create a transformative educational experience for students focused on deep disciplinary knowledge; problem solving; leadership, communication, and interpersonal skills; and personal health and well-being.

To cultivate a transformative university community committed to (a) attracting and retaining diverse, world-class talent; (b) creating a collaborative environment open to the free exchange of ideas, where research, creativity, innovation, and entrepreneurship can flourish; and (c) ensuring individuals can achieve their full potential.

To impact society in a transformative way — regionally, nationally, and globally — by engaging with partners outside the traditional borders of the university campus (Carnegie Melon University, n.d.).

Once the vision and mission statement are in alignment with the sustainability agenda, all other initiatives and activities undertaken by HEI's finds its way to promote sustainability.

Both vision and mission statement of the Carnegie Mellon University emphasise on the importance of having a transformational impact on the society through education, research, creativity, and entrepreneurship. Thus, by clearly setting the vision and mission statement in the direction of sustainability, the HEI can-

- Encourage its management and staff to integrate sustainability into its day-to-day operations.
- Introduce the essence of sustainability in its curriculum.
- Inspire the faculty, students, and other stakeholders to join hands with the university in its sustainability journey through their research and consultancy initiatives.

- Build trust and get support from the wider community resulting in more impactful extension and outreach by the HEI.

Operations

Introducing sustainable culture through its operations is the most common way used by HEI's to promote sustainability. Altering the day-to-day operations to suit the sustainability agenda can go a long way in creating a more sustainable future. It includes the initiatives like use of renewable resources by installing solar panels, reducing energy wastage by switching to LED lights, using techniques to recycle water, composting food waste, encouraging the use of bicycles in large campuses etc. In the case studies section, the instances of HEI that were discussed have mostly used this path to sustainability. While most of the initiatives focus on environmental dimension, the HEI's are also encouraged to make education more accessible and inclusive irrespective of the student's socio-economic status, gender, and ethnicity which also constitutes an important element of sustainability.

It is interesting to note that in the recent years, the concept of circular campus is slowly gaining the attention of HEI's. Circular campus is a holistic, systems-based framework designed to reduce emissions and waste, reduce costs, transform consumption patterns on campus, increase access and affordability for students, and support the transition to a just, sustainable economy. In a circular campus, resources like books, food etc are shared within the campus and in the neighbourhood. This way resources easily move from the point of excess availability to the point where it is needed. In other words, circular campus encourages sharing, reusing, recycling, refurbishing, repairing etc thus, increasing the utility and the lifecycle of the item (Goldmark, Raucher, & Cardenas, 2023).

Curriculum

Curriculum constitutes a very significant part of teaching learning process in HEI's. Integrating sustainability as a part of curriculum can be a very good way to inculcate sustainability thinking in young minds and this has gained momentum over the years. HEI's generally offer multitude programmes under each discipline and under each programme, there are several courses, and each course consists of multiple modules. For instance, a HEI may offer graduate and post graduate programmes in Sustainability under the disciplines of Life Science, Physical Science, Social Science and humanities. Under each of the programme, there many courses (major and minor) that a student should choose to acquire adequate credits to attain the desired qualification. These courses consist of numerous modules and projects.

Firstly, graduate and/or post graduate programmes on sustainability can be offered by HEI's. For example, Monash University, Australia offers Master's degree in Environment and sustainability. This programme offers interdisciplinary and specialised knowledge and skills in the field of sustainability through lectures, internships, projects, and research thesis. Currently, there are many other universities offering similar programs on sustainability that aims at creating environment consciousness and social responsibility like Oxford University offers masters in Sustainability, Deakin University has a programme titled Master of Sustainability Enterprise and the Environment, University of Pennsylvania has Masters of Environmental Studies, University of Cambridge offers Master of studies in Sustainability Leadership, National University of Singapore provides Master of Science in Sustainable and Green finance, Franklin University, Switzerland and Connecticut College, USA offers Bachelor of Arts in Social Justice and Sustainability, The London School of Economics and Political Science offers online certificate course on Sustainability: Environment, Economy and Society and many more such offering by different University across the world..

Further, HEI's can also introduce credit-based course on sustainability. As we know, in a formal academic setup, every student goes through multiple courses (major and minor) with clearly defined learning outcomes to attain the desired educational qualification. It is recommended to offer a credit-based course on sustainability in every programme thus covering all dimensions of sustainability since the enrolment till graduation. With learning outcomes clearly defined in a course, the expectations are set at the very outset. This helps the students to learn and practice their learning on sustainability effectively.

Introducing modules on sustainability in a course can also be a simple beginning to promoting sustainability via curriculum. HEI's can also explore the possibility of integrating the concept of sustainability into all courses in a discipline. Besides module on sustainability, project work or assignments that promote sustainable behaviour can kindle a sense of responsibility amongst the student community. Also, as a part of curriculum, workshops and seminars are often held to provide added information to the students. Hold sustainability themed workshops or seminars facilitated by a sustainability practitioner or expert can offer better insights to the students. Organising field trips to sites or organisations that focus on sustainability can be a fun way to introduce sustainability into the curriculum.

Let's take the example of the New Education Policy (NEP) proposed by the Ministry of Human Resource Development, Government of India. In the proposed NEP, HEI's are encouraged to have flexible and innovative curricula by including credit-based courses and projects in the areas of community engagement, environment education, waste management, management of biological resources and diversity, sustainable development and living and so on. Infact, in the proposed curricular components, a mandatory 3 credit course on Environment science/ education is proposed. Besides this, HEI's are given the autonomy to include sustainability and its areas as major or minor course under the programmes offered. The core objective of this move is to form Global citizenship and education for sustainable development as an integral part of the curriculum to empower learners to become aware of and understand global and sustainable development issues and thereby become active promoters of more peaceful, tolerant, inclusive, secure, and sustainable society (UGC, 2020).

Research

Today Research has become a significant part of higher education. Academicians are encouraged to adopt Research Based Pedagogical Tools (RBPTs) to persuade the students to look beyond classroom teaching and textbooks. This helps the student community to get a concrete understanding of the Theories and concepts, advance their knowledge and induce creative and innovative thinking. Besides, RBPT initiatives can also spark new ideas among the faculty to explore and solve practical problems in their areas of expertise.

HEI's have a major role to play in creating a research culture in the campus and therefore, drive the sustainability agenda through this culture. A major step in this direction can be establishing sustainability centre where faculty, students and experts come together to discuss latest developments in the field and explore opportunities for collaborative research that can address an ongoing sustainability issue. Inter disciplinary research can go a long way in attaining more success in this area. Further, the HEI through the sustainability centre can fund these research initiatives in the form of Major and Minor Research Projects. It can also identify sponsors through its associations to fund the research initiated under the sustainability centre banner. The institution may seek alumni support also in this initiative. Alumni can provide financial support, in-kind support, or volunteer to provide their time to back the research initiatives of the centre. The centre can also provide the requisite support to the faculty and students to apply for external research projects and secure the funds.

Many HEI's have constituted the sustainability centre to advocate and promote sustainability practices and research. Some of the examples include University of Sunshine Coast, Australia; University of Leeds, UK; Cornell University, New York; Stockholm School of Economics, Sweden; Azim Premji University, India and so on who have set out on a mission to make a difference in the world through research and innovation. The University of sunshine Coast established its sustainability research centre in 2007 with an aim to provide evidence-based solution through policy, planning and decision making for a socially just and thriving world. Some of its initiatives include research on coastal governance to protect from environmental change and urbanisation supported by Australian governments, in association with the Sunshine Coast council the centre is working on a research to develop a performance measurement framework and community report to help transform the sunshine coast into Australias's most sustainable region, a transdisciplinary research funded by MISTRA, Swedish Foundation for Strategic Environmental Research to mainstream an advanced and inclusive understanding of environment by developing new insights and approaches and translate into communication practices (University of the Sunshine Coast, n.d.). These are just very few examples of sustainability centres and some of the initiatives taken but, there are many other HEI's who have initiated sustainability centres to study, research and innovate in all spheres of sustainability.

Extension

HEI's with their extension activities and outreach programs can bring about major changes in and around their surroundings. There are several instances of HEIs that promote sustainability through its extension and outreach programs. Centre for Social Action, initiated by Christ University, India is one such example.

Centre for Social Action (CSA) was initiated by Christ University in its Bengaluru campus in 1999 to sensitizes the students on plethora of issues that affect the society and thereby encourages staff and students to extend support to the surrounding communities through various extension activities. With the help of international universities, other educational institutions in India, local institutions, NGO's, funding partners and rural networking organisations, Christ University have successfully undertaken and completed several development initiatives. Some of the initiatives include community development with respect to health, education, land development etc, promoting education to bring about equality and address poverty, transforming trash to treasure and others. Through these initiatives, over 400 students engage in volunteering activities each year, more than 2500 women have benefitted by self-help groups, 4,500 plus children benefitted through activity centre, reached over 10,000 families in villages and urban slums, more than 3,500 students sponsored under "Child sponsorship program" and more than 4500 children benefitted through children care centres (Christ University, n.d.). This initiative of Christ University largely addresses social and economic dimensions of sustainability besides the environmental dimension.

While Christ University is recognised for its exemplary contribution towards sustainability agenda, the following are few common ways through which every university can contribute to sustainability development through its extension activities in the surrounding communities. HEI's can hold workshops on sustainability related topics for the public where it can share ways to reduce energy consumption, methods to compost food waste, promoting health, environmental education, and awareness etc. HEI's can also publish or give away materials promoting sustainability. In addition, sustainability themed events like street plays, community clean up drives, exhibitions etc to attract the attention of the masses can be held. Use of social media, blogs and podcasts can also be a powerful way to promote sustainable activities in the present day. Institutions or Universities can collaborate with corporate especially under their CSR wing in their attempt to promote sustainability. HEI's can also partner with government bodies in not only promoting sustainable activities but also in bringing about policy changes that will have a greater and farther impact. Thus, HEI's through their extension activities and outreach programs can make a significant contribution towards the achievement of sustainable goals agenda.

CONCLUSION

Global awareness on sustainability was developed post The World Commission on Environment and Development released the report Our Common Future in 1987. Following this, with the launch of Millenium Development Goals in 2000 and Sustainable development Goals in 2015, the need for sustainable development and popularity of this idea was seen in all spheres of life. However, initially the idea was limited to environment as sustainable practices were adopted to address the problems of climate change, pollution, depletion of natural resources and others. Overall, these initiatives aimed at reducing the adverse impact of human activities on the environment. Eventually, sustainability became a holistic approach to include not just environment but also social and economic impact of today's decisions and actions.

HEI's comprise of educational institutions that impart learning beyond secondary education like universities, colleges, professional institutions, vocational training schools and others. HEI's are a place where future decision makers, entrepreneurs, policy makers and other professionals are prepared. Thus, they play a very important role in promoting sustainability agenda. Analysis of cases of sustainability practices among HEI's across the world shows that HEI's are mostly engaged in tackling the environmental issues, the social and economic aspect of sustainability is often overlooked. This might be due to the lack of deeper understanding of sustainability. HEI's also face challenges with respect to support from management and stakeholders, availability of adequate financial resources, absence of sustainability reporting, disciplinary barriers and so on.

To address all these challenges a framework is suggested which can be adopted by HEI's actively to promote sustainability in all its activities. While the vision and mission statement of the University/institution forms core to this framework, the other crucial elements include operations, extension activities, curriculum and research and consulting. The framework recommends how HEI's can incorporate the sustainability agenda in its vision and mission statement and thus attain support from management and stakeholders. This framework touches not only on the environmental aspect of sustainability but also throws light on the social and economic aspects of sustainability through cases, examples and suggestions which is often ignored by HEI's.

The suggested framework when used can bring about transformation in the way HEI's plan and execute their activities. Adopting sustainable ways to run the day-to-day operations in the campus can help in efficiently managing the resources and practicing inclusivity. Extension and outreach programs designed in this direction can have a great influence on the people in and around the campus. Incorporating sustainability in the curriculum either in the form of a dedicated programme or course or module or project on sustainability can create an awakening awareness amongst the learners and thus make them active promoters of sustainable production and consumption. Finally, all the research and consulting activities of the HEI can be brought under a single umbrella by constituting sustainability centre. This centre can encourage research in the areas of sustainability by funding faculty and student research, constantly looking for sponsors for funds to back research, inviting alumni support, providing requisite encouragement in the form of training and guidance to the staff and students to apply for external funding and so on. By establishing these sustainability centres, HEI can go a long way in conducting cutting-edge, interdisciplinary research from different dimensions of sustainability. Incorporating sustainability in curriculum and encouraging research and innovation in the areas of sustainability will help the young minds to look at sustainability from all three dimensions- environment, social and economic. The process of sustainability is complete when HEI's demonstrate their commitment by monitoring and voluntarily report all its sustainability activities at regular intervals.

REFERENCES

Association of Commonwealth Universities. (1993, August). Retrieved from https://www.iau-hesd.net/sites/default/files/documents/the_swansea_declaration.pdf

Carnegie Melon University. (n.d.). Retrieved from https://www.cmu.edu/about/mission.html

Chancel, L., Piketty, T., Saez, G., & Zucman, G. (2022). *World Inequality Report 2022*. Academic Press.

Christ University. (n.d.). *Centre for Soccial Action*. Retrieved from Christ University: https://christuniversity.in/center/C/Centre-for-Social-Action

Dagiliute, R., & Liobikiene, G. (2015). University contributions to environmental sustainability: Challenges and opportunities from the Lithuanian case. *Journal of Cleaner Production*, *108*, 891–899. Advance online publication. doi:10.1016/j.jclepro.2015.07.015

Florida State University. (n.d.). https://sustainablecampus.fsu.edu/about/mission

Goldmark, S., Raucher, L., & Cardenas, A. (2023). Building a Circular Campus: Consumption, Net Zero Emissions, and Environmental Justice at Barnard College. In M. R. Maulicci, S. Pfirman, & H. Callahan (Eds.), *Transforming education for sustainability* (pp. 347–376). Springer. doi:10.1007/978-3-031-13536-1_20

Jnr, B. A. (2021). Green campus paradigms for sustainability attainment in higher education institutions – A comparative study. *Journal of Science and Technology Policy Management*.

Kidd, C. V. (1992). *The Evolution of Sustainability*. doi:10.1007/BF01965413

Kidd, C. V. (1992). The Evolution of Sustainability. *Journal of Agricultural & Environmental Ethics*, *5*(1), 1–26. doi:10.1007/BF01965413

Macheridis, N., & Paulsson, A. (2021). Greening higher education? From responsibilization to accountabilization in the incorporation of sustainability in higher education. *International Journal of Sustainability in Higher Education*, *22*(8), 208–222. doi:10.1108/IJSHE-09-2020-0338

Mahameed, M. A. (2022). The implementation of sustainability practices in Arab higher education institutions. *Journal of Financial Reporting and Accounting*. doi:10.1108/JFRA-11-2022-0415

Moura, M. M., Frankenberger, F., & Tortato, U. (2019). Sustainability in Brazilian HEI: Practices Overview. *International Journal of Sustainability in Higher Education*, *20*(5), 832–841. Advance online publication. doi:10.1108/IJSHE-01-2019-0021

Purcell, W. M., Henriksen, H., & Spengler, J. (2019). Universities as the engine of transformational sustainability toward delivering the sustainable development goals. *International Journal of Sustainability in Higher Education*, *20*(8), 1343–1357. Advance online publication. doi:10.1108/IJSHE-02-2019-0103

Ralph, M., & Stubbs, W. (2013). Integrating environmental sustainability into universities. *Higher Education*. Advance online publication. doi:10.1007/s10734-013-9641-9

Ritchie, H., & Roser, M. (2020). *CO2 emissions*. Our World in Data.

Segura, E., & Zamar, M. D. (2021). Sustainable economic development in higher education institutions: A global analysis within the SDGs framework. *Journal of Cleaner Production*, *294*, 126133. Advance online publication. doi:10.1016/j.jclepro.2021.126133

The Talloires Declaration 10 Point Action Plan. (1990). Association of University Leaders for Sustainable Future.

UGC. (2020). *Curriculum Framework and Credit Sustem for Four-Year Undergraduate Programme*. UGC.

University of the Sunshine Coast. (n.d.). Retrieved from usc.edu.in: https://www.usc.edu.au/about/structure/schools/school-of-law-and-society/sustainability-research-centre

Wright, T. S. (2003). A Tenth Year Anniversary Retrospect: The Effect of the Halifax Declaration on Canadian Signatory Universities. *Canadian Journal of Environmental Education*.

KEY TERMS AND DEFINITIONS

Catalyst: An action or event or factor that brings about a change. In this context, Higher Education Institutions (HEI's) is the factor that is referred to which can bring about a change in the production and consumption habits of the humans.

Course: A subject that is taught to the students in HEI during an academic term say semester. Multiple modules on related areas are brought together to make a course.

Curriculum: A structured plan that clearly describes the concepts and skills that the students will learn, the teaching materials and pedagogy that the teacher will use and the assessments that will be applied to measure the learning.

Declaration: A formal announcement made by an individual or group. In this context, declaration refers to the formal announcement made by an HEI or group of HEI's clearly stating their actions to create a sustainable future.

Discipline: A field of study constituted by concepts, theories, principles, methods and techniques, principles. The choice of discipline in HEI's determines the domain of specialisation, course of study and the area of research.

Extension: Activities beyond curriculum designed to give back to the community and bring about a difference in the life of others along with developing empathy, compassion, communication, leadership, teamwork, and other valuable qualities among students.

Framework: A structure for application. In this context, the chapter provides a structure that exhibits the key components that can be adopted by HEIs in promoting sustainable development.

HEI: Higher Education Institutions (HEI) are educational institutions that provide education beyond secondary education i.e., high school. It includes Universities, colleges, Technical and vocational institutions. These institutions offer undergraduate degrees, post-graduation degrees, doctoral degrees, and professional degrees.

Module: An academic course in an HEI is built on many related modules. In a course, modules are arranged in a way to create a flow of content.

Operation: It refers to the day-to-day activities in HEIs which includes managing water, food, electricity, and other maintenance activities in the campus.

Pedagogy: Method of teaching.

Programme: Students pursuing higher education, register for programme, successfully complete the corresponding curriculum and are awarded appropriate degree for example Bachelor of Engineering or Master of Science, etc.

Research: Investigation undertaken by faculty and students in their respective field of knowledge to solve an ongoing problem.

Sustainability Development: Promoting all those ways and means to live on earth without compromising with the quality of environment and life of people living in the society.

Chapter 2
Cross-Disciplinary Collaborations and Partnerships for Sustainability Education:
Including Community-Based Learning, Industry Partnerships, and International Collaborations

Surjit Singha
https://orcid.org/0000-0002-5730-8677
Kristu Jayanti College (Autonomous), India

Ranjit Singha
https://orcid.org/0000-0002-3541-8752
Christ University, India

ABSTRACT

Multidisciplinary collaboration is crucial to sustainable education. It places a strong emphasis on combining many academic disciplines to address complex environmental issues and promote sustainable lifestyles. Students can put their knowledge into practice and raise awareness of the world through community-based learning, corporate partnerships, and international collaborations. The chapter emphasizes effective methods for evaluating the effectiveness and impact of collaborative activities. It highlights how crucial innovation, knowledge sharing, and international cooperation are to building a sustainable future. Social justice, action-oriented learning, and lifetime professional growth are all components of sustainability education. Institutional barriers, cultural differences, and resource constraints impact these cooperative initiatives. Future generations' perceptions of the Earth can be affected via interdisciplinary collaboration and partnerships.

DOI: 10.4018/978-1-6684-9859-0.ch002

1. INTRODUCTION

It is now more important than ever to address global environmental challenges, as evidenced by recent events. In order to achieve a sustainable future, the conventional strategy of discrete disciplinary activities needs to consider the intricate relationships and diverse character of sustainability concerns. In this environment, cross-disciplinary partnerships and collaborations become essential cornerstones of sustainable education. Goi & Ng (2008) state that all evaluated elements must be considered crucial success factors in e-learning for sustainability education. Designing effective and high-quality e-learning programs that effectively address sustainability challenges and engage learners in interdisciplinary collaborations and partnerships for a sustainable future requires a commitment from the institution, practical methodologies, and the learner's perspective. Additionally, companies are learning from their successes and failures in implementing E-learning projects, and the transformative influence of mixed and entirely online delivery techniques on learning is becoming increasingly apparent. The usage of E-learning in a variety of educational contexts is growing as efficient methods for providing online teaching to audiences around the world are still being created and improved (Yahiaoui et al., 2022; Zalat et al., 2021; Nikou & Maslov, 2021; Giannakos et al., 2021). Cross-disciplinary collaborations bring together professionals from several disciplines, including environmental science, engineering, economics, and social sciences, to enable a thorough and holistic understanding of our difficulties. By combining knowledge from several disciplines, educators and students can create novel and workable solutions to pressing problems, including resource depletion, climate change, and environmental degradation (Brodin & Avery, 2020; Klein, 2020; Wisz et al., 2020).

In order to address global sustainability concerns in a more integrated, dynamic, and flexible manner, this chapter examines the significance of fostering interdisciplinary cooperation in sustainability education. Sustainable education can be effectively imparted outside of the traditional classroom. The gap between theoretical and practical application must be closed to give students the abilities and information to make a discernible difference in the world. International collaborations, industrial partnerships, and community-based learning all depend on this goal. Using a community-based learning approach, students can participate in real-world sustainability concerns and projects. Students have invaluable possibilities for experiential learning by taking part in community projects. These exchanges foster cultural sensitivity, empathy, and a greater understanding of the contextual complexity of sustainable development. Through industry-education collaboration, students can have access to the resources and expertise of companies that are actively adopting sustainable practices. Students learn about cutting-edge technologies, environmentally friendly business models, and creative solutions through industry partnerships. Students' impact on sustainability projects is increased by these kinds of interactions, which allow them to use their knowledge directly in the corporate sector (Nyemba et al., 2021; Awasthy et al., 2020; Yu et al., 2020). Since environmental problems transcend national boundaries, tackling global sustainability requires international cooperation. Collaborations across international borders enable the sharing of diverse viewpoints, optimal methodologies, and local and cultural expertise. It trains pupils to participate in international efforts to protect the planet's future by instilling a feeling of global citizenship (UN, 2023; Mariani, 2023).

This chapter has examined how industry alliances, international partnerships, and community-based learning may improve sustainability education and provide students with the tools to actively engage in their local and global communities as change agents. In order to create a shared understanding and lay the groundwork for future conversations, this chapter has identified vital terms and concepts related to sustainability education. Readers will have a strong basis for exploring the rest of the chapter's material by clearly defining words like sustainability, cross-disciplinary collaboration, community-based learning, industry partnerships, and international collaborations.

2. COMMUNITY-BASED LEARNING

A combination of classroom instruction and active community interaction is known as community-based learning (CBL) (Rocha, 2023). It motivates students to work in teams with nearby communities to solve problems in the real world and support efforts for sustainable development. Within sustainability education, CBL promotes empathy, responsible citizenship, and a deeper comprehension of environmental concerns (Hart & Akhurst, 2017). In sustainability education, community-based learning augments theoretical understanding with real-world student experiences (Bruce & Bloch, 2013; Wilhelmsson et al., 2009). Students learn about the social, economic, and environmental aspects that impact sustainable practices and initiatives by interacting with community members. Students can apply their knowledge, imagination, and critical thinking to real-world problems by taking part in community projects, which fosters a sense of empowerment and agency.

Through community-based learning, students can witness firsthand how their activities affect the environment and the community. Outside of the classroom, this practical learning fosters a commitment to sustainable practices and a sense of responsibility. Students gain a more sophisticated understanding of cultural diversity and local knowledge in the context of addressing global concerns through their engagement with local communities. Through the Sustainable Agriculture Project, students participate in community-based learning by working with nearby farmers and agricultural groups to advance sustainable farming methods. The program covers various topics related to agricultural sustainability, such as social justice, environmental protection, and economic viability. Students are taught organic agriculture techniques, focusing on natural inputs and procedures that minimize environmental impacts. They might work on farms that use biological insect management, crop rotation, and decomposition to lessen the usage of artificial chemicals and support healthier ecosystems. Students who participate in the Crop Succession and Biodiversity program learn about crop succession planning, which involves growing different crops one after the other in order to maintain the fertility and health of the soil. Students gain an understanding of the value of biodiversity and how different plant species may support the resilience and food security of an ecosystem.

A variety of crucial components that support sustainable practices make up the Sustainable Agriculture Project. It relies on efficient water management; to protect this vital resource, students learn effective irrigation methods, precipitation harvesting, and water recycling. The goal of the Soil Health and Conservation program is to increase agricultural yields, sequester carbon, and benefit ecosystems by promoting healthy soils through practices like cover cropping, less tilling, and soil nutrient management (Rodríguez et al., 2022). Students participating in local food systems are taught the value of patronizing local businesses and the advantages of doing so, including lower carbon emissions and financial support for local communities. This project relies on community involvement, building links between the educational institution and the community, and expanding understanding of the opportunities and problems associated with sustainable agriculture. Through practical farm experience, the Experiential Learning program helps students better comprehend sustainable agriculture and its broader implications. In order to support evidence-based decision-making, Impact Assessment assigns students the responsibility of assessing the effectiveness of sustainable farming methods and their effects on the environment and local community (Graça et al., 2022; Huambachano et al., 2022; Weißhuhn et al., 2017). Overall, students get an understanding of the intricate connections that exist between society, the environment, and agriculture through the Sustainable Agriculture Project, which equips them to advocate for just and resilient communities and contribute to the development of sustainable agricultural systems. Through this community-based learning program, children are exposed to the direct consequences of their activities on nearby communities and the environment, fostering a sense of responsibility and civic engagement.

Urban Renewal Initiative

Through the Urban Renewal Initiative, students engage in community-based learning while working with local leaders, architects, and planners to regenerate urban areas through sustainable development projects. Numerous issues related to urban sustainability are covered in this program, such as energy conservation, transportation, and ecological areas. Students work in teams with experts to develop and execute sustainable transit plans, encourage the use of public transportation, construct pedestrian-friendly streets, and apply cutting-edge technology to lower carbon emissions and traffic jams. They also work with community members and landscape architects to create urban gardens and natural areas that enhance the well-being of locals and urban biodiversity. Furthermore, to reduce energy consumption, students collaborate with engineers and architects to develop energy-efficient buildings and adapt already-existing structures (Chen et al., 2023; Cerreta & La Rocca, 2021).

The Urban Renewal Initiative places a strong emphasis on collaborative decision-making and community engagement. Students interact with leaders and residents to learn about their needs and preferences, then match activities with the community's vision for sustainable urban development. Pupils learn about important urban planning ideas that balance social justice, economic prosperity, and environmental preservation, such as walkable neighbourhoods and mixed-use developments. This practical experience improves my understanding of challenges related to urban sustainability. Impact assessments quantify urban renewal projects' effects on the environment, society, and economy to inform future decisions. Students are encouraged to think about the long-term viability of their suggested projects as part of the curriculum, which emphasizes long-lasting and sustainable effects in urban settings (Fransen et al., 2021; Daher et al., 2021; Xie et al., 2021; Glasson & Wood, 2009). Students who take part in this project help to revitalize urban areas, gain crucial knowledge and skills for sustainable urban development, and build a sense of ownership and responsibility in the process.

Environmental Restoration Project

As part of the Environmental Restoration Project, students work with regional environmental organizations to protect biodiversity, clean up contaminated areas, and restore ecosystems. Students enrolled in this program are taught environmental stewardship and ecological restoration. They actively participate in the cleanup of hazardous areas while observing how their efforts are affecting the surrounding ecosystem immediately. Tree planting, habitat restoration, and eradicating invasive species are all parts of ecosystem restoration, which aims to improve biodiversity and ecological resilience. Students work with professionals to discover and conserve endangered animals and their habitats by participating in wildlife monitoring and conservation initiatives. As part of the community education and outreach component of this initiative, students plan lectures, awareness campaigns, and educational activities to encourage sustainable practices (Clewell & Aronson, 2013).

In order to make judgments based on empirical evidence, students undertake assessments, water quality testing, and biodiversity surveys to acquire information on environmental conditions and the effectiveness of restoration. Ensuring congruence with ecological and community requirements is ensured by collaboration with stakeholders, including government agencies, environmental specialists, and people. By integrating students from different academic areas, this strategy adds specialized knowledge and multiple viewpoints to the learning process. The Environmental Restoration Project teaches students the value of sustainability in preserving healthy ecosystems and highlights the long-term consequences of restoration efforts (Wei et al., 2022; Greening et al., 2022; Lindenmayer, 2020). Students who take part in this program actively contribute to the preservation of their local environment and grow to have a strong sense of environmental responsibility. They learn more about local environmental issues and develop the abilities needed to solve them as they work with community members and environmental organizations. Students who complete the program are equipped to continue fighting for sustainability and ethical behaviour. Strong partnerships with local communities and organizations are crucial for the success of community-based learning in sustainability education. Other essential factors include matching learning objectives with community aspirations, promoting reflection, offering continuous support, and handling logistics and evaluation (Holst, 2022; Joseph & Said, 2020). By bridging the gap between theory and practice, this method empowers students to effect change and work toward a sustainable future actively.

3. INDUSTRY PARTNERSHIPS

Establishing connections between academic institutions and businesses that participate in industry partnerships in sustainability education is essential. These collaborations link academia and industry, giving students practical experience and giving companies access to fresh viewpoints, creative ideas, and skilled future labour. A dynamic learning environment that addresses pressing environmental challenges and encourages ethical business practices is created by integrating industry collaborations into sustainability education (Watson et al., 2010). Through these collaborations, students can participate in actual sustainability projects that improve their understanding of real-world problems and solutions, making their education more applicable. Educational institutions can also get access to industry professionals, cutting-edge technologies, and priceless resources through these relationships. Experts in sustainability from various industries participate, which improves the learning process by introducing students to the latest developments in the field.

Furthermore, by fusing academic rigour with industrial understanding, collaborative research initiatives between academia and industry allow the creation of creative solutions to sustainability concerns. These collaborations enable research that addresses urgent and valuable needs, guaranteeing that the information produced is practical and applicable. Through industry partnerships, students can get a deeper grasp of industry-specific environmental issues and help to develop sustainable practices within those industries by concentrating on sustainability difficulties that are specific to those businesses.

As examples, a few case studies of fruitful industry collaborations are provided. For example, a prestigious institution in an area where interest in renewable energy was growing aimed to increase the scope of its solar energy research, aligning with the goals of top renewable energy firms to increase the sustainability and efficiency of solar panels. This alignment of objectives led to the formation of a collaboration aimed at developing solar panel technology and tackling environmental concerns. Utilizing its proficiency in physics, engineering, and materials science, the institution formed research groups focused on various facets of solar panel technology. These groups collaborated closely with experts from renewable energy companies, who gave them access to advanced solar panel prototypes and production facilities as well as financial support. The collaboration also allowed students to work on research projects, which aided in practical learning and created a talent pool for the renewable energy sector. By working together, the two sides shared information, ideas, and discoveries that improved materials science and photovoltaic cell design, ultimately leading to the production of solar panels with higher efficiency. By implementing these study findings into their production processes, the renewable energy companies improved their competitiveness and helped solar energy become widely used. This cooperative effort promoted sustainable practices and stimulated regional economic growth in addition to strengthening research skills and having a significant impact on the renewable energy sector.

A reciprocal opportunity to develop sustainable supply chain management evolved in a substantial collaboration between multinational firms interested in implementing sustainable supply chain practices and a top business school known for its commitment to sustainability. The main goals were as follows: the multinational corporations wanted to optimize their supply chains, lessen their environmental impact, and improve responsible procurement practices; the business school wanted to lead the way in developing innovative programs in sustainable supply chain management, satisfying the growing demand for sustainability expertise. In order to accomplish these objectives, the collaboration involved the creation of an extensive curriculum with feedback from business leaders, giving students a well-rounded education that combined academic theory with practical knowledge. Additionally, it allowed students to work with executives from the business on actual supply chain initiatives, promoting practical application and hands-on learning. In order to find creative solutions and best practices, faculty members and business professionals also worked together on cooperative research projects. This enhanced the curriculum and added to industry knowledge. Frequent seminars and workshops, held in collaboration with the firms, provided a forum for knowledge sharing where students, teachers, and business leaders could discuss new developments and difficulties in sustainable supply chain management. The business school was able to successfully implement cutting-edge curricula that prepared a diverse set of students for success in sustainable supply chain management, and the results were revolutionary. The adoption of ethical procurement procedures, more environmentally friendly supply chains, and smaller environmental footprints by multinational firms resulted in a shift towards supply chain activities that are more sustainable. Both academia and industry were elevated by the cooperative interchange of ideas and research discoveries; students gained real-world experience and were highly sought-after candidates for roles in sustainable supply chain management. This case study demonstrates how this partnership significantly increased sustainable supply chain management, promoting learning, propelling industry innovation, and eventually helping to create a corporate environment that is more ethical and sustainable.

A progressive community college understood the need for sustainable solutions in a town facing urgent waste management and recycling issues. Concurrently, a nearby recycling business steadfastly dedicated to ethical waste handling looked for creative ways to support recycling and waste minimization initiatives. The community college and the recycling company collaborated on a project due to their shared commitment to sustainability. There were two main goals for this relationship. The primary goals are to provide students with practical experience in trash reduction and recycling, as well as practical skills and knowledge. Secondly, the program aims to use students' novel and creative viewpoints to improve recycling procedures, reduce waste, and advance sustainability. This partnership materialized in several ways, such as the incorporation of recycling and waste reduction elements into the college's curriculum, the student participation in actual recycling and waste reduction projects in conjunction with the recycling company, the holding of innovation workshops to foster innovative problem-solving, and cooperative research projects for the gathering and analysis of data. The student's practical experience and comprehensive understanding of appropriate trash management were noteworthy outcomes. New perspectives were welcomed by the recycling company, which resulted in the adoption of creative recycling techniques and heightened community involvement. Increased community involvement due to the campaign led to higher recycling rates and a cleaner, more sustainable local environment. The local environment benefited from lower waste, less landfill usage, and higher recycling rates, all of which significantly lessened the impact on the environment. Students also gained valuable skills in waste reduction and recycling, improving their employability in sustainable waste management roles. This case study shows how a community college and a nearby recycling business worked together to enhance education and develop creative solutions for ethical trash management. In addition to giving students practical learning opportunities and advancing recycling techniques, the cooperation made the community cleaner and more sustainable by reducing waste and boosting recycling efforts, which had a long-lasting positive effect on the local economy and environment.

It can be challenging to bring the objectives of industry and educational institutions related to sustainability into alignment. It usually necessitates ongoing negotiations since it demands a delicate balance between industry interests and academic aspirations. Lack of funding and human resources might make it challenging to establish and maintain solid industrial partnerships. Furthermore, during joint research projects, issues about ownership, rights, and preservation of intellectual property may surface, requiring careful thought and agreement.

Industry collaborations can put short-term financial gains ahead of long-term sustainability goals, so it takes constant communication and commitment to find a balance. These collaborations are essential to sustainable education because they give students practical experience, exposure to business knowledge, and chances for creativity. They create a lively learning atmosphere that tackles contemporary problems and encourages ecologically conscious corporate operations. In order to fully reap the rewards of these kinds of partnerships and create long-lasting effects for both industry and educational institutions, challenges, including resource shortages and misaligned goals, must be overcome.

4. INTERNATIONAL COLLABORATIONS

In order to address global sustainability concerns, international collaborations in sustainability education relate to alliances and exchanges between educational institutions, groups, and students from different countries. Through these partnerships, participants can better understand how sustainability issues are interconnected globally by exchanging knowledge, best practices, and varied viewpoints (Annan-Diab & Molinari, 2017; Laurie et al., 2016; Lawson, 2005). International collaborations help to provide a more inclusive, thorough, and valuable approach to sustainable development by fostering cross-border cooperation.

Students have a rare opportunity to be exposed to various regional cultural viewpoints and sustainable practices through international collaborations. By expanding their awareness of global challenges, these encounters help people build a sense of global citizenship. Collaborating with professionals and students from different nations encourages the creation of fresh approaches to common sustainability problems. The varied cultural backgrounds of the individuals engaged in these partnerships offer unique viewpoints and methods for addressing environmental issues, which improves the process of group problem-solving (Finger et al., 2021; Yarime et al., 2012). Furthermore, by introducing participants to one another's cultures, customs, and sustainable living practices, these initiatives promote cross-cultural understanding and respect. Crucially, international partnerships can give developing countries priceless resources, know-how, and technological advancements to quicken their pursuit of sustainable development (Berglund et al., 2019; Payne & O'Neil, 2019). Through this exchange, the world community is encouraged to collaborate for a more sustainable future.

A few instances of this kind of cooperation are given, such as the worldwide Water Management Project, which colleges started on several continents in response to the urgent worldwide issues of pollution and water scarcity. This cooperative project brought together scholars and students from many backgrounds to explore approaches to sustainable water management, share research results, and implement plans to address water-related problems in their local communities. The Global Water Management Project's main goals were to conduct a thorough investigation into water scarcity and pollution, disseminate research results and best practices to project members, and establish locally applicable sustainable water management techniques. Global collaboration between researchers and students examined pollution, water scarcity, and sustainable water management techniques while considering the particular environmental, social, and economic elements that exist in each location. To enable the sharing of information and research findings, regular conferences, seminars, and online forums were set up. This allowed participants to modify effective tactics to fit their specific local settings. In order to use sustainable water management principles, each participating university started local projects. These projects ranged from community-driven water-saving programs to effective irrigation techniques. Community participation was essential in addition to academic research and activities, as academics and students worked with local communities to promote awareness and the adoption of sustainable water practices. The Global Water Management Project produced significant results, including creative approaches to pollution and water scarcity, raised public awareness of water-related concerns, enhanced water availability and quality, and established a global network of academic institutions, students, and researchers committed to solving water-related problems. This project is a prime example of how international cooperation and knowledge exchange can successfully address complicated global problems, contributing significantly to the global endeavour to guarantee that everyone has access to clean water supplies.

The annual Climate Change Youth Summit was founded by an international youth-led group, bringing together students from various countries to discuss climate change. Beyond dialogue, the occasion promotes doable, locally-driven climate action. Its goals include bringing attention to the problems associated with climate change on a global scale, encouraging youth collaboration across borders, fostering conversations and debates about creative solutions to the problem, and inspiring participants to start climate action projects in their local areas. The summit will include panel discussions, workshops, and cross-cultural interactions to enhance attendees' comprehension of climate change from local and global viewpoints. The knowledgeable direction of environmental experts gives these young leaders even more authority. The summit has a significant impact because it gives young people the tools to combat climate change actively, fosters global perspectives, helps local communities by implementing practical projects, and creates a robust global network of young climate activists. The Climate Change Youth Summit combines awareness-raising with practical action for a more resilient and sustainable society, showcasing the potential of youth-led projects and international collaboration in the fight against climate change.

Universities increasingly realize how important it is to educate students to be global citizens and change agents in a society facing previously unheard-of environmental and social issues. A first-of-its-kind university collaboration program, the Sustainable Community Exchange program connects students from other nations and creates links between rural and urban areas. It enables a thorough exchange of knowledge, enabling participants to create community-driven projects focused on environmental preservation and social development, as well as learn from one another's sustainable practices.

The program's primary goals include knowledge sharing, sustainable practices, intercultural learning, and community-driven projects. Through this program, participating kids gain empathy, a sense of global citizenship, and cross-cultural comprehension. Their exposure to various sustainable techniques in community development, agriculture, resource management, and renewable energy broadens their understanding. It gives them the tools they need to tackle issues on a local and global scale. Facilitating the sharing of experiences among advocates for sustainability around the globe creates a worldwide network that allows communities to adopt best practices. The program aims to address local challenges through collaborative activities that prioritize social development and environmental conservation.

The activities of the Sustainable Community Exchange program aim to offer a holistic perspective on sustainability. Student exchanges offer a distinctive and intensive experience as participants briefly switch places and reside with host families in one other's areas. They can gain a thorough grasp of regional customs and challenges. In seminars led by regional specialists, students can talk about issues including community development, renewable energy, sustainable agriculture, and environmental preservation. Collaborative projects are a crucial part of the program that complements these courses. Students work together to pinpoint problems in their host communities and find solutions. This leads to various projects introducing renewable energy initiatives and sustainable farming practices. Students are also encouraged to participate in environmental efforts, cultural events, and customs that improve their comprehension of the host community and promote cross-cultural interactions.

The initiative for Sustainable Community Exchange has a wide-ranging and significant effect. Participants gain cross-cultural knowledge as global citizens, which promotes tolerance, cultural sensitivity, and a sense of interconnectedness throughout the world. Furthermore, due to the pupils' shared knowledge, environmentally beneficial behaviours are adopted, and quantifiable progress is made in environmental conservation within each community. Participants in the program design and carry out projects that improve social development by strengthening community cohesion and expanding access to education. The curriculum fosters a strong feeling of environmental and social responsibility and global citizenship. Many students are so moved by their experiences that they return to their hometowns as fervent supporters of sustainability, pushing for constructive change and inspiring others to adopt sustainable habits.

To sum up, the Sustainable Community Exchange program is an effective model for promoting sustainable behaviours, intercultural understanding, and student empowerment as engaged global citizens. This effort enhances the lives of those involved. It contributes to substantial and enduring environmental and social changes in various parts of the world by breaking down barriers between rural and urban populations and promoting information sharing. The initiative is a prime example of how global cooperation may change the world and lead to a more just and sustainable society.

International cooperation in sustainability education is crucial to advancing multicultural understanding, diversified problem-solving, and global consciousness. Through developing valuable relationships, they foster collaboration and give professionals and students a more comprehensive grasp of sustainability concerns, ultimately leading to the construction of a more sustainable world. These collaborations do, however, come with their own set of challenges. Language barriers can make it difficult to communicate effectively, which may call for translation and interpreting services. When coordinating operations across multiple time zones, careful planning and adaptability may be required. The availability of resources can be impacted by economic gaps among participating nations; therefore, fair and inclusive considerations are necessary. Cultural awareness is essential to avoiding miscommunication and cultural appropriation. It is necessary to resolve any regulatory or legal issues that may arise about funding, data sharing, and intellectual property using open contracts. Overcoming these challenges ensures that multinational collaborations achieve their goals and promote just and profitable global cooperation.

5. CROSS-DISCIPLINARY COLLABORATIONS

Cross-disciplinary collaborations in sustainability education involve using the knowledge and skills of different academic fields to address challenging environmental, social, and economic issues. This method recognizes the complexity and interconnectedness of sustainability concerns, which call for a comprehensive, interdisciplinary understanding. Cross-disciplinary partnerships combine knowledge from environmental science, engineering, social sciences, economics, and the humanities to offer a comprehensive approach to sustainability education. When working across disciplines, educators and students participate in research, discussion, and problem-solving that incorporates a variety of viewpoints, theories, and approaches. This method encourages creative problem-solving that considers the interactions of societal, economic, and environmental aspects, which eventually results in more lasting and efficient solutions (França et al., 2017; Carayannis & Morawska-Jancelewicz, 2022).

In sustainability education, diverse multidisciplinary collaborations are leading the way in addressing intricate global concerns. To create complete city plans, Initiatives for Sustainable Urban Planning brings together environmental scientists, social scientists, architects, and urban planners. Ecological infrastructure, public transportation, fair housing, and community involvement are prioritized in these initiatives to guarantee environmentally and socially responsible urban growth. Multidisciplinary research teams of scientists, economists, sociologists, and public health experts perform climate change impact studies. In order to mitigate these effects through data analysis, scenario modelling, and the creation of adaptive strategies, they look at how ecosystems, human populations, and economies are affected by climate change. Initiatives for a Circular Economy encourage engineers, designers, business students, and environmentalists to work together to create circular economy models for various businesses. These models place a strong emphasis on reducing waste, promoting resource efficiency, and establishing sustainable patterns of consumption and production. These cooperative projects perfectly capture the spirit of sustainability education, where a variety of knowledge bases are brought together to create creative answers for a future that is both resilient and sustainable.

In interdisciplinary collaboration for sustainability education, several best practices have been developed to support robust and productive alliances. Establishing clear objectives is the first step in ensuring that everyone involved has a common goal and vision. Effective communication, focusing on active listening, respecting different points of view, and freely expressing ideas is the cornerstone of these collaborations. Multidisciplinary workshops and seminars enable experts from many fields to discuss approaches, exchange ideas, and pinpoint locations where their distinct specialities might come together (Parry & Metzger, 2023; Crabbé, 2019). Last but not least, adaptability and flexibility are crucial elements that enable project plans to change and include fresh concepts and techniques that arise from cross-disciplinary contacts. These best practices foster creativity and problem-solving for a more sustainable future and raise the calibre and efficacy of sustainability education programs (Horn et al., 2022).

Collaborations across disciplines are essential to sustainability education because they offer a comprehensive strategy for tackling complex global issues. These partnerships encourage creativity by bringing together specialists from different fields, leading to more practical and long-lasting solutions. Clear objectives, good communication, flexibility, and respect for one another are necessary for these collaborations to succeed (Klein, 2020; Kim & Coonan, 2023). These programs do, however, face certain obstacles, such as difficulty assessing the overall impact of interdisciplinary work, power imbalances, time and budget limits, and impediments related to language and jargon. In order to effectively address urgent global issues and advance multidisciplinary sustainability education, these hurdles must be overcome. Collaborations across disciplines are essential to sustainability education because they offer a thorough and integrated method of tackling complex problems. By utilizing a range of viewpoints and areas of expertise, these partnerships promote creativity and result in more long-lasting and efficient solutions. The success of these kinds of partnerships depends on several factors, including mutual respect, flexibility, good communication, and defined goals. Navigating issues with language, resource limitations, power dynamics, and assessment methods is also critical.

6. ASSESSING COLLABORATIVE APPROACHES TO SUSTAINABILITY EDUCATION

Evaluating joint initiatives, including industry partnerships, community-based learning, cross-disciplinary collaborations, and international collaborations, is necessary to determine the effectiveness, impact, and outcomes of collaborative approaches to sustainability education (Lind & Lindström, 2023; Horn et al., 2023). These assessment techniques aim to gauge how well these strategies meet their learning goals, encourage sustainable behaviour, and foster favourable results for students, communities, and stakeholders. Assessment approaches in sustainability education often combine qualitative and quantitative data to provide a thorough evaluation. These could consist of surveys, interviews, focus groups, observations, performance reviews, and analysis of project results. These methods help teachers, organizations, and other interested parties determine the advantages and disadvantages of cooperative sustainability education programs (Stokols et al., 2008; Koehn & Uitto, 2014). Understanding the efficacy and significance of joint sustainability projects, for instance, requires an understanding of successful assessment procedures.

Diverse techniques are used in order to achieve a thorough comprehension. Initially, pre- and post-surveys are given to participants to gauge changes in their knowledge, attitudes, and behaviours linked to sustainability. These surveys offer numerical perceptions of the program's effectiveness. Second, participants can discuss their learning experiences and personal growth using qualitative methodologies like case studies and reflection journals. These accounts facilitate a more profound understanding of individual and collective growth throughout the partnership. Thirdly, specific metrics and evaluation criteria must be established for each project. This offers measurable information on the initiative's accomplishments, such as resource savings, environmental impact reduction, and community engagement levels. Finally, but just as importantly, gathering input from a wide range of stakeholders—including students, teachers, members of the community, and business partners—helps determine overall happiness and perceived efficacy. Their understanding of the advantages and disadvantages of the cooperative method is priceless. Combining various assessment methods results in a thorough understanding of the influence of cooperative sustainability projects, enabling ongoing future development and the development of more successful initiatives.

Evaluating cooperative sustainability education programs presents several difficulties that must be carefully considered. Given that changes in attitudes and behaviours may occur gradually over time, one of these problems is the need to quantify the complex and long-term benefits of these programs (Findler et al., 2018). As such, it is critical to develop evaluation strategies that can work with these more prolonged periods. Second, it is essential to carefully analyze and triangulate data from many sources to ensure accuracy and reliability because qualitative data, which are commonly used for evaluation, might be subjective and lack standards (Bryhn & Belgranò, 2023; Al-Kuwari et al., 2021). Thirdly, smaller educational institutions or community organizations with limited resources may find it prohibitive to collect and analyze data for collaborative approaches as they may require additional time and effort (Baumber, 2021; Herranen et al., 2020). Since the goals of many cooperative sustainability projects are complex, it is challenging to assess their efficacy fully. It is necessary to prioritize and manage several evaluation objectives to overcome this complexity. The outcomes of assessments may translate poorly to other contexts or educational initiatives since collaborative sustainability education is context-specific. For evaluations to be functional, they must consider the unique context and goals of each program. To fully understand collaborative approaches to sustainability education and their potential for good change, it is imperative to evaluate them. By integrating qualitative and quantitative assessment approaches, educators and stakeholders can obtain critical insights into the results and challenges of collaborative sustainability programs. When developing evaluation procedures, one must consider the complexity of long-term repercussions, the subjectivity of qualitative data, resource limitations, and numerous objectives. Thorough assessment techniques can support the growth and improvement of cooperative sustainability education programs.

7. FUTURE DIRECTIONS AND EMERGING TRENDS

Several new developments in the constantly evolving field of sustainability education are altering how professionals and students approach this critical subject. A significant trend in higher education is the proliferation of interdisciplinary degree programs provided by more and more schools. These programs combine environmental, social, and economic viewpoints to give students a thorough understanding of sustainability through a cross-disciplinary curriculum. Furthermore, the development of online and virtual collaborations by technical improvements has made it possible for people in different places to work together on sustainability projects and initiatives. Furthermore, to understand the interconnectedness of sustainability concerns and provide comprehensive solutions, interdisciplinary collaborations are increasingly using systems thinking techniques.

Furthermore, there is a rising focus on encouraging cultural diversity and inclusive problem-solving methods, as well as integrating indigenous knowledge and traditions into sustainability education. Last but not least, international alliances are emerging to tackle global sustainability concerns, encouraging cooperation among academic institutions, governments, non-governmental groups, and corporations worldwide. These patterns show that sustainability education takes a dynamic and ever-expanding approach that equips students to tackle challenging and urgent global concerns. Since the goals of many cooperative sustainability projects are complex, it is challenging to assess their efficacy fully. It is necessary to prioritize and manage several evaluation objectives to overcome this complexity. The outcomes of assessments may translate poorly to other contexts or educational initiatives since collaborative sustainability education is context-specific. For evaluations to be functional, they must consider the unique context and goals of each program.

Exciting and revolutionary paths are ahead for cross-disciplinary cooperation and sustainable education. A significant change will be the emphasis on action-oriented learning, in which students from many academic fields collaborate to solve actual environmental issues, promoting real-world problem-solving and hands-on learning. Furthermore, a greater focus on resolving social equity issues within interdisciplinary sustainability collaborations will result from realizing the close connection between environmental sustainability and social justice. Education on sustainability will become more widely available outside of traditional academic settings, providing chances for continuing education and lifelong learning. Professionals from various disciplines will be able to improve their sustainability competencies and adjust to the changing needs of a sustainable society. Integrating sustainability concepts into many academic subjects will be another significant change, promoting multidisciplinary problem-solving and thinking across many subject areas. Students will be more prepared to handle challenging global issues as sustainability becomes a core component of the school.

In addition, industry-driven curricula will become increasingly popular as partnerships with companies influence sustainability education initiatives. This guarantees that students learn the information and abilities needed by the labour market while successfully tackling sustainability issues. The future mentioned above approaches delineate a dynamic and comprehensive approach to sustainability education, equipping professionals and students to effect significant change in response to pressing global concerns. In sustainability education, cross-disciplinary collaborations present a wealth of opportunities that can significantly accelerate our efforts to create a more sustainable future. The opportunity for originality and imagination is one of the main benefits. When people with different academic backgrounds collaborate, their unique viewpoints and methods can produce creative answers and strategies for dealing with sustainability issues. When working on urban sustainability, for example, a team of environmental scientists, engineers, social scientists, and designers can combine their talents to create complete and successful designs that consider various environmental and social aspects.

Furthermore, by considering the interdependencies between environmental, social, and economic elements, these collaborations enable a comprehensive understanding of sustainability concerns. This strategy makes sure that all sustainability-related issues are addressed and that solutions are not divided into discrete areas. Another significant potential is the transfer of skills and knowledge between disciplines, which enables specialists in other domains to benefit from one another's knowledge and contribute to solving challenging issues. To investigate environmental patterns and forecast future trends, for instance, sustainability specialists working with data scientists can use advanced data analytics. This method makes sustainability decision-making more data-driven. Another beneficial result of cross-disciplinary interactions is interdisciplinary research, which inspires scholars to investigate novel fields of inquiry and use a range of approaches and theoretical frameworks to address challenging sustainability problems. Collaboratively, specialists from several disciplines, including climatologists, agronomists, economists, and sociologists, can conduct exhaustive investigations and devise tactics to tackle noteworthy obstacles, such as the consequences of climate change on food security. Together, these changes foster creativity, thorough comprehension, the transfer of skills and knowledge, and multidisciplinary research, which eventually helps to create more revolutionary and successful methods of teaching sustainability and addressing problems.

In order to promote a common understanding of sustainability concerns and approaches, cross-disciplinary collaborations provide an excellent forum for the sharing of knowledge, best practices, and experiences across academic disciplines, institutions, and geographies. Through these partnerships, participants can explore creative solutions that draw on the strengths of several disciplines and obtain a better understanding of sustainability concerns. They also stimulate the sharing of specialist knowledge, approaches, and insights from various domains. These partnerships also encourage the exchange of lessons learned and best practices, which helps projects learn from past mistakes and build on techniques that work for more effective and efficient sustainability programs. These partnerships frequently involve multidisciplinary workshops, seminars, and conferences, which give experts a forum to share and debate their research results, remain current on sustainability advancements, and spot possible joint venture prospects. Cross-disciplinary partnerships, particularly those involving multinational cooperation, also enrich global perspectives. The international knowledge exchange that occurs fosters intercultural comprehension and a more thorough approach to difficulties related to transnational sustainability.

Furthermore, regardless of a person's career stage, these partnerships support lifelong learning by enabling professionals, educators, and students to remain current on developments in sustainability research and practice. Knowledge sharing via interdisciplinary collaborations is an essential component of sustainability education. It promotes an inclusive, team-based approach to sustainability by utilizing information and experiences to tackle the world's complex problems.

One crucial potential provided by cross-disciplinary collaborations in sustainability education is to enhance real-world applicability. By integrating many disciplinary views and real-world experiences, these collaborations facilitate students' acquisition of a complete grasp of how sustainability concerns materialize in real-world contexts. This method gives students the tools to solve problems and create change in various professional contexts. A vital element of these relationships is experiential learning, which gives students chances to get their hands dirty in fieldwork, corporate partnerships, and community-based projects. Students can close the knowledge and skill gap between theory and practice by applying their knowledge and skills to real-world sustainability concerns. Because sustainability problems are intrinsically complex, they call for diversified approaches. Students collaborating on multidisciplinary projects that address real-world sustainability challenges are better equipped to confront these complexities head-on, thanks to cross-disciplinary collaborations. They are then more equipped to handle the complexities of their future employment.

A significant component of cooperative projects is frequent stakeholder participation, which involves community members, business experts, governmental bodies, and non-governmental groups. By interacting with stakeholders, students are exposed to a diverse array of viewpoints and concerns on sustainability-related decision-making processes, which cultivates a deeper understanding of inclusive and participatory methods. Understanding systemic effects is yet another significant result of interdisciplinary work. Sustainability problems often have broad systemic ramifications and are linked. These partnerships help students see these connections and comprehend how their actions in one area of sustainability might impact other aspects of sustainability. They can design and implement sustainable solutions with a broader perspective thanks to this systems thinking methodology. Students are exposed to successful and poor sustainability techniques when working on real sustainability concerns. Students can cultivate a resilient and flexible viewpoint by learning from these results, which is crucial for tackling the dynamic and ever-changing character of sustainability concerns.

Ultimately, students' professional preparedness is improved by the cross-disciplinary collaborations' emphasis on real-world application. Graduates with real-world experience, problem-solving abilities, and the capacity to function well in heterogeneous teams are highly valued by employers; these are attributes fostered by collaborative, sustainable education. These partnerships give students the tools they need to be proactive change agents. They provide them with the information, experiences, and skills they need to negotiate the intricacies of sustainability concerns and make meaningful contributions to society as responsible global citizens. A critical potential that interdisciplinary collaborations in sustainability education present is the development of a global perspective. International collaborations provide a forum for experts, educators, and students from various nations and cultures to come together, providing a rare chance to understand the transnational nature of sustainability concerns and the necessity of international cooperation. Exposure to cultural variety is a crucial result of these multinational interactions. Students who interact with people from different cultural backgrounds learn about different viewpoints and strategies for sustainability. This encounter fosters cultural awareness and comprehension, which is essential for tackling global issues in many cultural contexts. International collaborations frequently entail comparative analyses of sustainability policies, practices, and concerns across many locations. Students can facilitate knowledge exchange and mutual learning by adapting successful sustainability efforts from other nations to their situations through these analyses.

In order to effectively address global sustainability concerns, including resource depletion, biodiversity reduction, and climate change, coordinated worldwide initiatives are needed. Students can understand the interdependence of these global issues and the need for a coordinated response through international partnership. Additionally, by working together, students can create international networks of mentors, peers, and sustainability experts. These networks can create a sense of interconnection and shared purpose by being an excellent resource for information sharing, research collaborations, and future job prospects. Students participating in global sustainability initiatives are exposed to cutting-edge sustainability strategies and best practices implemented successfully in various settings. Their broad range of experiences improves their comprehension of practical sustainability tactics. International partnerships in sustainability education, above all, foster a sense of shared responsibility for the health of the world and its inhabitants, as well as global citizenship. Pupils are taught to think about the bigger picture of their actions and are made aware of their obligations as global citizens. Working across disciplines, students must get a global perspective to become influential sustainability leaders in a world of linked problems. Their perspectives are widened, empathy and intercultural understanding are encouraged, and cooperative efforts toward global sustainability goals are spurred.

For sustainable education to continue, interdisciplinary partnerships and collaborations must be expanded. Sustainability education needs to integrate social justice, embrace emerging trends, prioritize action-oriented learning, and offer chances for lifelong learning in order to address global concerns effectively. The many potentials that cross-disciplinary collaborations present—including encouraging innovation, facilitating knowledge exchange, and boosting real-world applicability—make them a promising path toward realizing a more resilient and sustainable future, notwithstanding the hurdles involved. It is imperative to acknowledge and tackle the obstacles that could impede the smooth execution of these joint ventures, such as organizational hurdles, limited resources, linguistic and cultural disparities, and the requirement for appropriate assessment models. Sustainability education may develop further and have a significant influence on international sustainability initiatives by overcoming these obstacles and seizing these chances.

8. DISCUSSION

The importance of interdisciplinary approaches in addressing complex environmental concerns is highlighted in this chapter's exploration of interdisciplinary partnerships and collaborations in sustainability education. Collaborative projects serve to bridge the gap between many academic fields, offering a thorough grasp of sustainability concerns and facilitating the creation of creative solutions that consider social, economic, and environmental factors. One of the key ideas covered is community-based learning, a powerful tool in sustainability education. Through engagement with local communities, students gain practical experience and direct exposure to real-world sustainability concerns. With the help of this tactic, which encourages civic engagement and cultural sensitivity, students can develop empathy and a greater understanding of the local environment in which sustainable solutions are applied. Emphasis is placed on the value of industry partnerships in sustainability education. Working together with businesses gives students practical experience and lets universities take advantage of the knowledge and resources that business offers. Industry alliances support environmentally friendly corporate practices, which helps create a more sustainable business environment.

The development of global collaborations is another essential facet of sustainability education. Educational institutions are exposed to a range of viewpoints and strategies regarding sustainability issues through the creation of international partnerships. It encourages intercultural understanding and equips students to work together to solve global sustainability challenges. It highlights how crucial it is to use efficient assessment methods when determining if collaborative initiatives are effective. By conducting thorough evaluations, educators and stakeholders may assess the effectiveness of sustainability efforts and make evidence-based decisions for ongoing development. Properly capturing the full potential of cross-disciplinary collaborations is ensured by developing assessment frameworks specifically designed for them. The analysis of the importance of interdisciplinary partnerships and collaborations brings to light the transformative potential of these strategies in constructing a sustainable future. Students are given the tools they need to become proactive change agents and contribute to creating a more resilient and sustainable world by embracing diverse information and viewpoints. Future directions and consequences for interdisciplinary cooperation and sustainability education emphasize the need for ongoing development and improvement. With the increasing popularity of new trends like virtual collaborations and multidisciplinary degree programs, sustainability education may embrace dynamic learning settings that efficiently tackle current and future issues.

The future of sustainability education will largely depend on how action-oriented learning and social justice are integrated. By stressing practical fixes and fair tactics, students can develop into champions for social and environmental change. Opportunities for lifelong learning support the ongoing growth and flexibility needed to meet changing sustainability problems for professionals across various disciplines. Cross-disciplinary collaborations offer tremendous benefits, but they also come with challenges in the form of institutional hurdles, cultural differences, and resource constraints. When working together, all participants must be dedicated to fairness, inclusivity, and respect for one another. The transforming power of interdisciplinary teams and educational alliances for sustainability is highlighted in this chapter. These cooperative methods promote integration, creativity, and international cooperation, laying the foundation for a resilient and sustainable future. By using these ideas, educators, institutions, and stakeholders may create a sustainable environment that benefits both present and future generations.

9. CONCLUSION

This chapter addresses complex global environmental concerns and emphasizes the critical role that cross-disciplinary relationships play in sustainability education. These relationships, which prioritize industrial ties, international collaborations, and community-based learning, serve as the foundation for sustainable education. Through interdisciplinary collaboration, environmental problems can be solved holistically, enabling students to participate in sustainable development. Interdisciplinary coalitions must prioritize social justice and action-oriented education while incorporating virtual collaboration, lifelong learning, and systems thinking to advance. While industry-driven courses and international collaborations are essential for building sustainable competencies, they are beset by obstacles such as resource constraints and institutional hurdles. To evaluate the impact and sustainability of these cooperation, efficient evaluation techniques will be necessary. In conclusion, transdisciplinary partnerships promote revolutionary change and a resilient, sustainable future.

REFERENCES

Al-Kuwari, M. M., Du, X., & Koç, M. (2021). Performance assessment in education for sustainable development: A case study of the Qatar education system. *Prospects*, *52*(3–4), 513–527. doi:10.1007/s11125-021-09570-w

Annan-Diab, F., & Molinari, C. (2017). Interdisciplinarity: Practical approach to advancing education for sustainability and the Sustainable Development Goals. *International Journal of Management Education*, *15*(2), 73–83. doi:10.1016/j.ijme.2017.03.006

Awasthy, R., Flint, S., Sankarnarayana, R., & Jones, R. (2020). A framework to improve university–industry collaboration. *Journal of Industry-university Collaboration*, *2*(1), 49–62. doi:10.1108/JIUC-09-2019-0016

Baumber, A. (2021). Transforming sustainability education through transdisciplinary practice. *Environment, Development and Sustainability*, *24*(6), 7622–7639. doi:10.1007/s10668-021-01731-3 PMID:34393621

Berglund, T., Gericke, N., Pauw, J. B., Olsson, D., & Chang, T. (2019). A cross-cultural comparative study of sustainability consciousness between students in Taiwan and Sweden. *Environment, Development and Sustainability*, *22*(7), 6287–6313. doi:10.1007/s10668-019-00478-2

Brodin, E., & Avery, H. (2020). Cross-disciplinary collaboration and scholarly independence in multidisciplinary learning environments at the doctoral level and beyond. *Minerva*, *58*(3), 409–433. doi:10.1007/s11024-020-09397-3

Bruce, B. C., & Bloch, N. (2013). Pragmatism and community inquiry: A case study of community-based learning. *Education and Culture*, *29*(1), 27–45. doi:10.1353/eac.2013.0004

Bryhn, A. C., & Belgrano, A. (2023). Teaching, learning and assessment methods for sustainability education on the land–sea interface. *Discover Sustainability*, *4*(1), 3. Advance online publication. doi:10.1007/s43621-023-00120-2 PMID:36694724

Carayannis, E. G., & Morawska-Jancelewicz, J. (2022). The futures of Europe: Society 5.0 and Industry 5.0 as driving forces of future universities. *Journal of the Knowledge Economy*, *13*(4), 3445–3471. doi:10.1007/s13132-021-00854-2

Cerreta, M., & La Rocca, L. (2021). Urban Regeneration Processes and Social Impact: A literature review to explore the role of evaluation. In Lecture Notes in Computer Science (pp. 167–182). doi:10.1007/978-3-030-86979-3_13

Chen, Y., Ren, X., Xu, X., Wang, R., & Xu, Y. (2023). Evaluation of Living Environment Quality in Urban Residential Areas under the Concept of Urban Renewal—A Case Study of Binjiang District, Hangzhou, China. *Sustainability (Basel)*, *15*(19), 14104. doi:10.3390/su151914104

Clewell, A. F., & Aronson, J. (2013). Ecological references. In Island Press/Center for Resource Economics eBooks (pp. 137–153). doi:10.5822/978-1-59726-323-8_7

Crabbé, A. (2019). Interdisciplinarity and sustainable development. In Springer eBooks (pp. 998–1006). doi:10.1007/978-3-030-11352-0_215

Daher, E., Maktabifard, M., Kubicki, S., Decorme, R., Pak, B., & Desmaris, R. (2021). Tools for citizen engagement in urban planning. In Springer eBooks (pp. 115–145). doi:10.1007/978-3-030-85566-6_7

Denham, D., Rozance, M. A., Malone, M., & Goodling, E. (2020). Sustaining future environmental educators: Building critical interdisciplinary teaching capacity among graduate students. *Journal of Environmental Studies and Sciences*, *11*(1), 101–114. doi:10.1007/s13412-020-00611-y

Findler, F., Schönherr, N., Lozano, R., & Stacherl, B. (2018). Assessing the Impacts of Higher Education Institutions on Sustainable Development—An Analysis of Tools and Indicators. *Sustainability (Basel)*, *11*(1), 59. doi:10.3390/su11010059

Finger, D. C., Drăghici, C., Perniu, D., Smederevac-Lalić, M., Hălbac-Cotoară-Zamfir, R., Sehic, A., & Solomun, M. K. (2021). The importance of international collaboration to enhance education for environmental citizenship. *Sustainability (Basel)*, *13*(18), 10326. doi:10.3390/su131810326

França, C. L., Broman, G., Robert, K. H., Basile, G., & Trygg, L. (2017). An approach to business model innovation and design for strategic sustainable development. *Journal of Cleaner Production*, *140*, 155–166. doi:10.1016/j.jclepro.2016.06.124

Fransen, J., Peralta, D. O., Vanelli, F., Edelenbos, J., & Olvera, B. C. (2021). The Emergence of Urban Community Resilience Initiatives During the COVID-19 Pandemic: An International Exploratory Study. *European Journal of Development Research*, *34*(1), 432–454. doi:10.1057/s41287-020-00348-y PMID:33456209

Giannakos, M. N., Mikalef, P., & Pappas, I. O. (2021). Systematic Literature Review of E-Learning Capabilities to Enhance Organizational Learning. *Information Systems Frontiers*, *24*(2), 619–635. doi:10.1007/s10796-020-10097-2 PMID:33551669

Glasson, J., & Wood, G. (2009). Urban regeneration and impact assessment for social sustainability. *Impact Assessment and Project Appraisal*, *27*(4), 283–290. doi:10.3152/146155109X480358

Goi, C., & Ng, P. Y. (2008). E-learning in Malaysia: Success factors in implementing e-learning program. *International Journal on Teaching and Learning in Higher Education*, *20*(2).

Graça, J., Roque, L., Guedes, D., Campos, L., Trüninger, M., Godinho, C., & Vinnari, M. (2022). Enabling sustainable food transitions in schools: A systemic approach. *British Food Journal*, *124*(13), 322–339. doi:10.1108/BFJ-11-2021-1188

Greening, H., Heck, K. L., McKinney, L., Diefenderfer, H. L., Boynton, W. R., Kleiss, B. A., Mishra, D. R., George, A. II, Kraft, B. C., Kling, C. A., & Windecker, L. A. (2022). Assessing the Effectiveness of Large-Scale Environmental Restoration: Challenges and Opportunities. *Estuaries and Coasts*, *46*(2), 293–301. doi:10.1007/s12237-022-01149-8

Hart, A., & Akhurst, J. (2017). Community-based learning and critical community psychology practice: Conducive and corrosive aspects. *Journal of Community & Applied Social Psychology*, *27*(1), 3–15. doi:10.1002/casp.2287

Herranen, J., Tolppanen, S., Vesterinen, V., & Aksela, M. (2020). Challenges and tensions in collaborative planning of a student-led course on sustainability education. *Nordic Studies in Science Education*, *16*(1), 18–36. doi:10.5617/nordina.6583

Holst, J. (2022). Towards coherence on sustainability in education: A systematic review of Whole Institution Approaches. *Sustainability Science*, *18*(2), 1015–1030. doi:10.1007/s11625-022-01226-8

Horn, A., Scheffelaar, A., Urias, E., & Zweekhorst, M. B. (2023). Training students for complex sustainability issues: A literature review on inter-and transdisciplinary higher education design. *International Journal of Sustainability in Higher Education*, *24*(1), 1–27. doi:10.1108/IJSHE-03-2021-0111

Huambachano, M., Arulingam, I., Bowness, E., Korzenszky, A., Mungai, C., Termine, P., & Wittman, H. (2022). Knowledge networks to support youth engagement in sustainable food systems. *Frontiers in Sustainable Food Systems*, *6*, 867344. Advance online publication. doi:10.3389/fsufs.2022.867344

Joseph, C., & Said, R. (2020). Community-Based education: a participatory approach to achieve the sustainable development goal. In Encyclopedia of the UN sustainable development goals (pp. 101–111). doi:10.1007/978-3-319-95870-5_67

Kim, E., & Coonan, T. (2023). Advancing Sustainability Education through a Cross-Disciplinary Online Course: Sustainability and Human Rights in the Business World. *Sustainability (Basel)*, *15*(6), 4759. doi:10.3390/su15064759

Klein, J. T. (2020). Sustainability and Collaboration: Cross-disciplinary and Cross-Sector horizons. *Sustainability (Basel)*, *12*(4), 1515. doi:10.3390/su12041515

Koehn, P. H., & Uitto, J. I. (2014). Evaluating sustainability education: Lessons from international development experience. *Higher Education*, *67*(5), 621–635. doi:10.1007/s10734-013-9669-x

Laurie, R., Nonoyama-Tarumi, Y., Mckeown, R., & Hopkins, C. (2016). Contributions of education for sustainable development (ESD) to quality education: A research synthesis. *Journal of Education for Sustainable Development*, *10*(2), 226–242. doi:10.1177/0973408216661442

Lawson, H. A. (2005). Empowering people, facilitating community development, and contributing to sustainable development: The social work of sport, exercise, and physical education programs. *Sport Education and Society*, *10*(1), 135–160. doi:10.1080/1357332052000308800

Lind, J., & Lindström, J. (2023). Towards a framework for exploring the indirect value of tourist attractions in place branding: The case of Tom Tits Experiment Science Center. *Scandinavian Journal of Hospitality and Tourism*, 1–23. doi:10.1080/15022250.2023.2233486

Lindenmayer, D. B. (2020). Improving restoration programs through more significant connection with ecological theory and better monitoring. *Frontiers in Ecology and Evolution*, *8*, 50. Advance online publication. doi:10.3389/fevo.2020.00050

Mariani, P. (2023). *Climate Change and International Cooperation — ALI Social Impact Review*. ALI Social Impact Review. https://www.sir.advancedleadership.harvard.edu/articles/climate-change-and-international-cooperation

Nikou, S., & Maslov, I. (2021). An analysis of students' perspectives on e-learning participation – the case of the COVID-19 pandemic. *Campus-Wide Information Systems*, *38*(3), 299–315. doi:10.1108/IJILT-12-2020-0220

Nyemba, W. R., Mbohwa, C., & Carter, K. F. (2021). Academia and Industry Collaborations: A research and professional perspective. In EAI/Springer Innovations in Communication and Computing. doi:10.1007/978-3-030-70493-3_4

Parry, S., & Metzger, E. P. (2023). Barriers to learning for sustainability: A teacher perspective. *Sustainable Earth Reviews*, *6*(1), 2. Advance online publication. doi:10.1186/s42055-022-00050-3

Payne, L., & O'Neil, J. K. (2019). Cultural sustainability in higher education. In Springer eBooks (pp. 1–7). doi:10.1007/978-3-319-63951-2_109-1

Rocha Beardall, T. (2023). Abolition as praxis and virtual community-based learning. *Teaching Sociology*, *51*(1), 79–91. doi:10.1177/0092055X221096663

Rodríguez, B. C., Durán-Zuazo, V. H., Soriano, M., García-Tejero, I. F., Ruíz, B. G., & Tavira, S. C. (2022). Conservation Agriculture as a Sustainable System for Soil Health: A review. *Soil Systems*, *6*(4), 87. doi:10.3390/soilsystems6040087

Stokols, D., Misra, S., Moser, R. P., Hall, K. L., & Taylor, B. K. (2008). The ecology of team science: Understanding contextual influences on transdisciplinary collaboration. *American Journal of Preventive Medicine*, *35*(2), S96–S115. doi:10.1016/j.amepre.2008.05.003 PMID:18619410

UN. (2023). *Global Partnerships - United Nations Sustainable development*. United Nations Sustainable Development. https://www.un.org/sustainabledevelopment/globalpartnerships/

Watson, R. T., Boudreau, M. C., & Chen, A. J. (2010). Information systems and environmentally sustainable development: Energy informatics and new directions for the IS community. *Management Information Systems Quarterly*, *34*(1), 23–38. doi:10.2307/20721413

Wei, X., Song, W., Shao, Y., & Cai, X. (2022). Progress of ecological restoration research based on bibliometric analysis. *International Journal of Environmental Research and Public Health, 20*(1), 520. doi:10.3390/ijerph20010520 PMID:36612842

Weißhuhn, P., Helming, K., & Ferretti, J. (2017). Research impact assessment in agriculture—A review of approaches and impact areas. *Research Evaluation, 27*(1), 36–42. doi:10.1093/reseval/rvx034

Wilhelmsson, M., Pelling, S., Ludvigsson, J., Hammar, M., Dahlgren, L. O., & Faresjö, T. (2009). Twenty years experience of interprofessional education in Linköping–ground-breaking and sustainable. *Journal of Interprofessional Care, 23*(2), 121–133. doi:10.1080/13561820902728984 PMID:19225972

Wisz, M. S., Satterthwaite, E. V., Fudge, M., Fischer, M., Polejack, A., St John, M., Fletcher, S., & Rudd, M. A. (2020). 100 Opportunities for More Inclusive Ocean Research: Cross-Disciplinary Research Questions for Sustainable Ocean Governance and Management. Frontiers in Marine Science, 7. doi:10.3389/fmars.2020.00576

Xie, F., Liu, G., & Zhuang, T. (2021). A comprehensive review of urban regeneration governance for developing appropriate governance arrangements. *Land (Basel), 10*(5), 545. doi:10.3390/land10050545

Yahiaoui, F., Aichouche, R., Chergui, K., Brika, S. K. M., Almezher, M., Musa, A. A., & Lamari, I. A. (2022). The Impact of e-Learning Systems on motivating students and enhancing their outcomes during COVID-19: A Mixed-Method Approach. *Frontiers in Psychology, 13*, 874181. Advance online publication. doi:10.3389/fpsyg.2022.874181 PMID:35967714

Yarime, M., Trencher, G., Mino, T., Scholz, R., Olsson, L., Van Ness, B., Frantzeskaki, N., & Rotmans, J. (2012). Establishing sustainability science in higher education institutions: Towards integrating academic development, institutionalization, and stakeholder collaborations. *Sustainability Science, 7*(S1), 101–113. doi:10.1007/s11625-012-0157-5

Yu, X., Paudel, K. P., Li, D., Xiong, X., & Gong, Y. (2020). Sustainable Collaborative Innovation between Research Institutions and Seed Enterprises in China. *Sustainability (Basel), 12*(2), 624. doi:10.3390/su12020624

Zalat, M. M., Hamed, M., & Bolbol, S. A. (2021). The experiences, challenges, and acceptance of e-learning as a tool for teaching during the COVID-19 pandemic among university medical staff. *PLoS One, 16*(3), e0248758. doi:10.1371/journal.pone.0248758 PMID:33770079

KEY TERMS AND DEFINITIONS

Community-Based Learning: Integration of community involvement with educational activities.

Cross-Disciplinary Collaborations: Working together across academic disciplines to tackle challenging problems.

Diverse Academic Disciplines: A variety of academic fields or subjects.

Effective Assessment Methods: Assessment methods that reliably measure learning objectives.

Industry Partnerships: Collaborations between academia and businesses.

International Collaborations: Educational partnerships with institutions abroad.

Sustainability Education: Teaching the ideas of social and environmental sustainability.

Chapter 3
Marketing as a Tool to Bridge the Gap Between Attitude and Sustainable Behavior

Jorge Lara-Guillén
https://orcid.org/0009-0008-9347-4948
Universitat Oberta de Catalunya, Spain

M. Dolores Méndez-Aparicio
https://orcid.org/0000-0003-1787-8031
Universitat Oberta de Catalunya, Spain

Ana Jiménez-Zarco
https://orcid.org/0000-0002-8980-6814
Universitat Oberta de Catalunya, Spain

ABSTRACT

The 2030 Agenda for Sustainable Development calls on all countries to adopt a series of measures to promote sustainability. The circular economy emerges as a new paradigm of production and consumption aligned with sustainable development represents a major challenge for all the agents involved. At the corporate level, it represents the inclusion of sustainable practices which put marketing in the spotlight of business strategy. In this context, education plays a decisive role in contributing to the sustainable development of societies, and this educational process can be promoted in different areas. The companies, for their potential to booster responsible demand from consumers, can be a powerful sustainability tool that contributes to reducing the gap between attitude and sustainable behavior. Based on a desk research methodology, this chapter pursues a twofold objective: to analyze whether there is a gap between consumers' stated attitude and sustainable behavior and to identify whether an appropriate marketing strategy can help overcome it.

DOI: 10.4018/978-1-6684-9859-0.ch003

OBJECTIVES, METHODOLOGY, AND FINDINGS

This chapter has a double objective: to corroborate the existence of a gap between expressed attitude and sustainable behavior of consumers and to identify whether an adequate marketing strategy, with its potential to educate in values, generate experiences and promote sustainable consumption, can help to overcome it.

In terms of the means used to obtain information, we have relied on documentary methodology and, taking into account the level of knowledge that is intended to be acquired in this phase, we have used secondary exploratory research. This methodology is often used for those topics that lack in-depth research, especially when it comes to issues related to products and markets, seeking a better understanding of them and identifying problems that may be the focus of future research. It is very suitable in any commercial strategy as it allows a focus on the market and can be used with a specific objective such as knowing areas of potential growth, alternative business models or prioritizing areas that require statistical research.

Faced with the challenge of mobilizing the human flow involved in consumer behavior through their daily habits and decisions, and as a result of this work, we can conclude that, according to the literature reviewed, the existence of a gap between attitude and sustainable behavior is confirmed and that marketing is configured as one of the most effective and efficient strategies that companies can develop to inform and educate consumers and achieve a multiplier effect of individual actions as an effective lever of change.

INTRODUCTION

For decades, companies tried to find a differentiated position in markets saturated with competitors (what we might call "red oceans"), or, alternatively, they sought out new markets deserted of competitors, launching totally innovative products or services (what has come to be called "blue ocean strategies"). But those strategies have become obsolete for a simple reason: there are no more free oceans, neither red nor blue. The oceans are depleted. The oceans are empty, with no raw materials to exploit and saturated with pollution and garbage (Zhexembayeva, 2014).

Since the 1970s, the world's population has doubled and gross domestic product has quadrupled. These trends have required large amounts of natural resources that have driven economic development and improved human well-being, yes, but at great cost to the environment that has become a threat to maintaining the well-being achieved. (Global Resources Outlook, 2019)

Exploitation of non-renewable resources that, after being introduced into production processes and used, accumulate as waste; basic resources for industrial activity that are being depleted; exploitation of renewable resources above the replenishment rate leading to deforestation, habitat destruction and biodiversity loss. As the population increases, the need for resources to sustain it grows, and if we add to this the climate change that is occurring as a result of global warming, it results in a growing tension that seriously threatens the health of the planet (Global Resources Outlook, 2019).

The Global Resources Outlook (GRO) warns that, if we continue with current consumption and production habits and levels, global material use would double compared to 2015 to reach 190,000 million tons by 2060. Per capita resource use would increase from 11.9 to 18.5 tonnes and this growth would cause stress on resource supply systems and unacceptable levels of environmental pressure that would exceed operating spaces for businesses and society.

"The second half of the 20th century is unique in the entire history of human existence on Earth. Many human activities reached take-off points sometime in the 20th century and accelerated sharply toward the end of the century. The last 50 years of the 20th century have undoubtedly seen the most rapid transformation of the human relationship with the natural world in all of human history. " It is the "Great Acceleration", led by developed countries and which poses two possible scenarios, "the great decoupling" or "the great collapse" (Steffen, et al., 2015).

The linear "produce, use, throw away" economic model that relies on the availability of large and easily accessible quantities of materials and energy, as well as cheap means of disposing of what is no longer of interest, which has been at the heart of industrial development and has generated an unprecedented level of growth, is reaching its physical limits. Such a model is not sustainable. (Steffen et al., 2015).

Another scenario is possible, yes, but a sustainable future will not happen spontaneously.

Appropriate policies, environmentally sound technologies, sustainable financing schemes, capacity building, awareness-raising and changes in social behaviour, and public-private partnerships can help chart a trajectory towards a sustainable future by decoupling economic activity and human well-being from the indiscriminate use of natural resources and destructive environmental impacts.

New rules of the game are imposed that governments, companies and society must face in order to successfully face the new economic order by adopting a series of principles, the first of which is to move from the linear vision to the circular vision of the value chain.

As a paradigm, sustainable development is a vision of humanity and an ideal of society. Humanity seeks to ensure that societies reach a condition in which the satisfaction of needs and economic prosperity, quality of life, equity and social well-being, and the protection of natural resources converge (Salas-Zapata & Ortiz-Muñoz, 2019). It is a goal that has not yet been achieved. The legacy left by a linear economy model confronts us with an abyss of environmental, economic and social disaster, and the solution to the global problems of unsustainability requires societies to introduce changes in the way people think and act, in their consciousness and values, as well as changes in the way they relate to others and to the environment (UNESCO, 2014a).

The latter implies cultural transformations, for which changes in people's educational processes are necessary to raise awareness and promote sustainable development.

For this reason, achieving changes in people's behaviours, particularly through education, has been one of the strategies promoted by UNESCO to achieve sustainable development (UNESCO, 2014).

Educating for sustainable development consists of guiding people's learning processes so that they acquire knowledge, values, attitudes and skills that allow them to make decisions and undertake actions aimed at protecting ecosystems and natural resources and maintaining a harmonious relationship with nature, thus contributing to sustainable development (UNESCO, 2014).

And we say, make decisions and take action. The former would be related to attitude and the latter to behavior. And, they are not always the same because a sustainable attitude may not materialize into a sustainable behavior. It is then that we speak of gap and that is why, in order to achieve this consistent coincidence between attitude and consumer behavior, it is necessary to train individuals as such and society as a whole.

In this context, sustainable marketing is confirmed as a tool to educate the consumer (Jiménez-Marín et al., 2019) and facilitate the transition from a linear economy model to a circular economy model.

SUSTAINABILITY: FROM A LINEAR ECONOMY MODEL TO A CIRCULAR ECONOMY MODEL

Sustainable development is defined as that which meets the needs of the present without compromising the ability of future generations to meet their own needs (WCED, 1987) and the circular economy incorporates principles for the use of resources that are implicit in the concept of sustainable development, so most researchers accept the close relationship between the two concepts (Geissdoerfer et al., 2017), sustainable development and circular economy.

Sustainable development sets goals to be achieved to solve environmental problems and their consequences and the circular economy is a tool to address some of the causes of these problems (Suarez Eiroa et al., 2019) that needs a bottom-up movement to achieve sustainable development (Genovese et al., 2017).

In this context, the circular economy emerges as a new paradigm of production and consumption aligned with sustainable development by decoupling the extraction of material resources and energy consumption from growth, promoting an efficient use of resources based on the recovery and recycling of used materials and the transformation of waste into new resources (Ghisellini et al., 2016) to achieve the goal of sustainability in its three dimensions: environmental, economic and social.

Conceptually, the circular economy is understood as an integrative framework comprising a variety of approaches and disciplines developed over the last thirty years around the idea of creating a model of production, distribution, and consumption that is regenerative and restorative by design (Hidalgo Carvajal et al., 2021) that can rely on digitalization and advances in information and communication technologies to decouple economic development from the use of natural resources, while boosting investment in infrastructure and job creation to make progress towards the Sustainable Development Goals (SDGs) and the 2030 Agenda's purpose of "leaving no one behind".

Faced with this radical change in the rules of the economic game, companies have two alternatives: do nothing, and then they will perish at the same time as the linear economy system; or they can face the problem and turn it into an opportunity (Zhexembayeva, 2014).

The shift from a linear economy model to a circular economy model is no longer an option, it is a necessity and an opportunity. In the era of Customer Centricity (Naim, 2022), attracting, retaining and promoting the recommendation of responsible consumers are essential business objectives (Ulaga, 2018; Fader, 2020) if they want to ensure sustainable growth of their businesses and only entrepreneurs who understand and lead the change that is taking place will be able to turn the new reality into disruptive innovation and sustainable competitive advantages. Only those who succeed will survive in the 21st century. Those who understand this will be able to launch new products, define new business models, or discover new markets... Those who do not understand it will perish (Nadya Zhexembayeva, 2014).

Companies must lead this transformation by designing intelligent strategies and systems and, ultimately, doing what they do best: creating cost-effective solutions that meet the needs and desires of an increasingly demanding, informed and aware consumer. Solutions that are tailored to both a global and local market, and that include the value and real cost of living consistently within the limits of the planet. It's business strategy in its purest form.

Innovation: New Circular Business Models

Putting the circular economy into practice requires creative innovation in system design and committed collaboration between and within value chains, as well as between different stakeholders. This disruptive behaviour is linked to technology, innovation and the implementation of new sustainable business models within the framework of a circular economy model. Technological advances enable entrepreneurial and collaborative approaches that were not possible before, and creativity and innovation give birth to new alternative and competitive business models with a lower impact on the environment. Less impact on the environment but greater impact on society and the companies that will succeed will be those that use business models to invite customers, employees, partners or the general public to collaborate in developing the path to the future.

The business strategy addresses reduction, reuse and recycling by innovating, both in its production and distribution processes, as well as in new business proposals, focusing on the use of fewer products, components, materials and energy during the phases of the product life cycle (from design to recovery), on regeneration, with the use of renewable and non-toxic materials, as well as the use of renewable energies; and information, which relies on the use of information technology to support the circular economy (Konietzko et al., 2020). The business strategy is therefore addressing sustainability through innovation.

A critical component of a company's innovative strategy is its business model. Not only is it a manifestation of the company's established logic for value creation, but it is also a key component of the roadmap when it comes to transforming a business.

Circular business models are seen as an innovative formula; But, for the circular economy system to become a reality, consumer support is needed. To achieve this, the company must activate marketing strategies to convey the message and purpose both internally (employees) and externally (users in general). It is about promoting the perception of the brand as a sustainable brand and with circular initiatives strengthening the brand and the reputation of the company and, in addition, to make pedagogy, to be a positive influence on the attitude and intention of users regarding sustainability because a pro-sustainable attitude is always a good omen for the acceptance and therefore the successful development of new business models.

If the sustainable commitment of a brand and the proposal of new business models in line with circularity were perceived by consumers as an added value, the transition to a circular economic model would be an achievable goal because, the circular economy is a viable and scalable growth model that can improve the productivity of resources. It reverses the destructive trend, fosters innovation and job creation and has the potential to transform the dynamics of competitiveness. In this transition, companies can lead the way by adopting new business models, seizing the opportunity, and taking the risk, but success will largely depend on consumer behavior.

CONSUMER BEHAVIOR

In an ecosystem where business success depends, to a large extent, on consumers, the "sustainable profile", but above all their behaviour, will mark the step towards moving from a linear economy (produce-use-throw away) to a circular economy based on the principles of repairing, reusing, recycling and maximising the advantages that technology provides us with in terms of saving resources. Not so much the profile but the sustainable behavior of the consumer is the factor capable of consolidating the sustainable initiatives of companies, favoring the development of the circular economy through the implementation of new business models that guarantee current development without putting the development of future generations at risk.

Pro-environmental behaviour has been defined as behaviour that harms the environment as little as possible or even contributes to benefiting it (Steg & Vlek, 2009, p. 309) or the set of deliberate and effective actions that respond to social requirements and/or individual convictions for the protection of the environment (Corral-Verdugo & Pinheiro, 2004) and is considered by many authors to be a decisive factor in curbing the impact caused on nature by the activity (production and consumption) and reverse a trend that has been described as suicidal (De Leeuw et al., 2015; Goh et al., 2017).

Pro-environmental behaviour is a phase that, going beyond awareness, attitude and intention, takes the form of responsible action because awareness of environmental problems does not always translate into sustainable consumption and the fact is that the behaviour of the organic consumer is the result of a complex process in which many and diverse factors influence it that enhances or inhibits it. Discovering the relationship between these factors and real sustainable consumer behavior is of vital importance to promote and extend responsible consumption practices (Cañones Castañón, 2018).

We said that sustainable consumer behaviour is a complex process that goes beyond awareness and attitude, but it is necessary for them to be aware of the consequences of their use and consumption habits, and to make them aware they need to be informed and marketing can play a key role in this aspect, offering consumers relevant information for their consumption decisions within the framework of a digital society Consumer awareness of the environmental impact of many products and services is still low. A high percentage of consumers do not know the environmental or social consequences of many of the products they buy, and this lack of knowledge about the environmental impact of the use/consumption of certain goods or services may be one of the causes of the observed gap between attitude and sustainable behavior.

To reverse the trend, correct the errors of the linear economy and try to redirect a situation that already threatens to be critical, the progress of the circular economy requires the involvement of multiple agents and, especially, companies and consumers because it is an economic system enabled by novel business models and responsible consumers and users" (Kirchherr et al, 2017).

That is why, from the point of view of companies, the current debate on the circular economy is mainly focused on designing future scenarios with business models that are desirable to the consumer because the lack of interest and awareness of consumers is a main impediment to a transition to the circular economy (Kirchherr et al., 2017) and the lack of support from demand networks can prevent or delay the implementation of innovations (Rizos et al., 2016). Hence the importance of involving consumers because the continuity over time and success of this economic model will only be possible if it is the market that demands innovative products, services and business models that incorporate circularity principles.

Innovation can be understood as the ability to force change through technical or economic means (Hard) but it is also associated with the ability to generate those changes by attracting others through institutional values and practices that shape their attitudes and preferences (Soft) (De Jesús & Mendoça, 2018). From the latter perspective, innovation in circular business models, such as social and collaborative innovation, can generate changes considering the values and environmental awareness of consumers and other stakeholders aware of the need to move from the commercial models of the linear economy to new business models that allow progress to be made in achieving the sustainable development goals.

Awareness: Temporal and Attitudinal Process

However, awareness, although necessary, may not be enough for the adoption and incorporation of sustainable use and consumption behaviors to take place, because since the process of adopting innovations is the one by which it is decided to use, or not, said innovation (Rogers, 1962), all innovation requires a temporal and attitudinal process as explained by Everett Rogers' Theory of Adoption of Innovations.

This theory graphically represents on a timeline the percentage of individuals who, according to its theory, adopt an innovation and form the typical Gaussian Bell depending on whether they are innovator-pioneers, early adopters, early majority, late majority or laggards (Rogers, 1995).

Regarding the attitudinal process, when individuals have to decide whether to accept or reject an innovation and, therefore, maintain or not a certain behavior of use or consumption, they face a psychological process that starts from obtaining and synthesizing information related to said innovation and, taking into account their own values and the social framework in which they find themselves, A series of beliefs are generated that will determine their attitude, their intention and, ultimately, their decision to adopt or not adopt that innovation and maintain or not to maintain that behavior.

The link between attitude and behavior has been studied by numerous authors, both in the field of psychology (Allport, 1935; Campbell & Stanley, 1963; Ajzen and Fishbein, 1980; Schifter and Ajzen, 1985; Fazio, 1986; Ajzen and Fishbein, 1991, among others) and marketing (Lehmann & Hulbert, 1971; Bass and Talarzyk, 1972; Bass and Wilkie, 1973; Bettman et al. 1975; Bagozzi and Warshaw, 1990; Lutz, 1991; Taylor and Todd, 1995), and has led to the development of different general models of consumer behaviour (Nicosia, 1966; Engel et al., 1968; Howard and Sheth, 1969; Engel et al. 1986) and various theories that seek to understand the process through which the behavior of individuals is forged.

A series of internal variables, marketing variables, and external variables are involved in the purchase decision process, as shown in Figure 1.

Figure 1. Variables involved in the consumer's purchase decision process
Source: Santesmases (2012), cited by Jiménez Parra (2014)

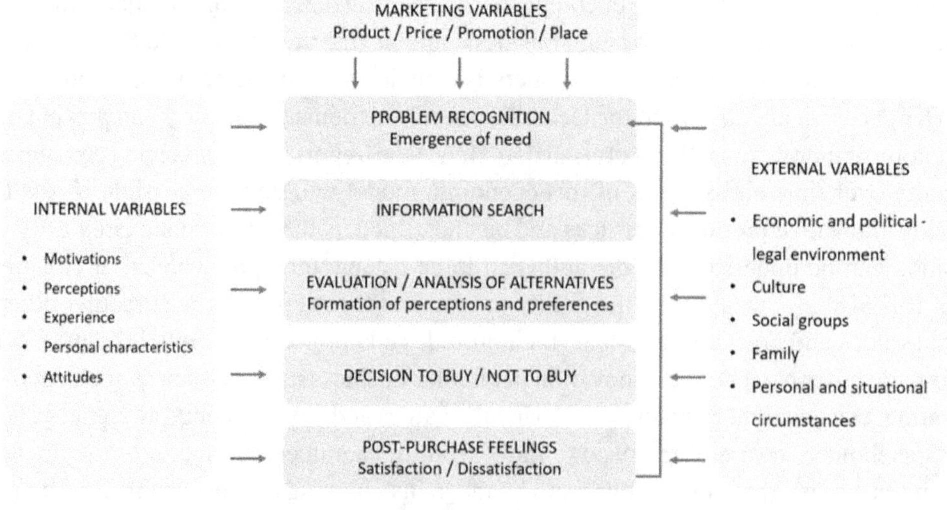

And as the European Economic and Social Committee (EESC) points out, the real challenge is to mobilise the human wealth of consumer behaviour through their daily habits and decisions, which will produce a multiplier effect on individual actions and be an effective lever for change (EESC, 2019).

GAP BETWEEN SUSTAINABLE CONSUMER ATTITUDE AND BEHAVIOR

We are talking about actions and not so much about attitudes and some authors have investigated whether attitude always materializes in a consistent behavior to conclude that this is not always the case and maintain the existence of a gap between attitude and sustainable behavior (Camacho et al., 2018)

In the 1950s, socially responsible personality began to be studied, mainly in the USA (Dueñas et al., 2014). As early as the 1960s, some authors spoke of social responsibility as a motivating aspect of certain consumer behaviors (Berkowitz & Daniels, 1964; Berkowitz and Lutterman, 1968, cited in Dueñas et al., (2014).

In the academic literature we find various definitions to refer to the conscious, responsible, ethical consumer. Some of them are shown in the table below:

From the 1970s onwards, measurement scales such as Webster's SIM model (1975) were also used to build profiles of environmentally conscious people (Moreno et al., 2012) as well as others proposed by different authors for various research works among target consumers in different countries (Webster, 1975; Tucker et al., 1981; Belch, 1982; Antil, 1984; Roberts, 1996; Mohr and Webb, 2008; Ismail & Akcali., 2009; Lee, 2009; Lee and Shin, 2010; Yan & She, 2011; Sudbury-Riley & Kohlbacher, 2015, cited in Palacios González, 2017). It is shown in the following table:

In the field of Spanish and Latin American academia, scales applied by Pérez-Barea, Montero-Simó and Araque-Padilla (2015) can be highlighted; Villa, Perdomo-Ortiz, Dueñas & Durán (2016) and Villa, Perdomo-Ortiz, Durán, Dueñas & Arredondo (2018). Some of these scales include items related exclusively to ecological behaviors and others also consider social and even labor aspects. Some are one-dimensional and others multi-dimensional, since socially responsible behaviour has a multidimensional character because consumers can express their social and environmental commitment in different ways, giving different importance to their purchasing and consumption actions (Palacios-González, Chamorro-Mera, 2019) according to their values.

It has been proven that values influence attitudes and a sustainable attitude would be a determinant of intention which, in turn, would translate into responsible buying/use behaviors (Kumar & Ghodeswar, 2015; López-Bonilla, 2018; Reyes-Rodriguez & López-Bonilla, 2018; Giefer, Peterson, & Chen, 2019;).

An example of the interest that this topic arouses among researchers is shown in the following tables:

It can therefore be said that there are numerous factors that can positively influence or inhibit sustainable consumption behaviour, they have been studied with diverse and sometimes contradictory results, but the authors agree that consumer behaviour will be decisive for the adoption of new proposals that represent progress in the change from a linear economy to a circular economy.

Therefore, it is of interest to distinguish between an informed attitude about consumer preferences and actual consumer behavior.

Table 1. Definitions of socially responsible consumers

Author	Expression	Given Definition
Webster (1975)	Socially conscious consumer	"The consumer who takes into account the public consequences of his or her private consumption habits, or who attempts to use his or her purchasing power to bring about social change.
Antil (1984)	Socially responsible consumption	"Those behaviours and purchasing decisions undertaken by consumers that are related to environmental problems and resource scarcity, whose motivation is not only the desire to satisfy a personal need, but also concern for the possible adverse consequences of their consequent effects".
Muncy and Vitell (1992)	Ethical consumer	"The consumer whose individual or group behaviour is guided by moral principles and standards in obtaining, using and disposing of goods and services".
Roberts (1993)	Socially conscious consumer	"A consumer who buys products and services that he or she perceives to have a positive (or less negative) effect on the environment or who favours companies that are trying to bring about positive social change".
Roberts (1995)	Socially conscious consumer	"A consumer who buys products and services that he or she perceives to have a positive (or less negative) effect on the environment or who favours companies that are trying to bring about positive social change."
Mohr et al. (2001)	Socially conscious consumer	"A person who bases their purchase, use and disposal of products on the desire to minimise or eliminate any detrimental effects and maximise the beneficial impact on society in the long term.
François - Lecompte (2005)	Socially responsible consumption	"One who purchases products and services that are perceived to have a positive (or less negative) impact on the individual's environment and/or uses their purchasing power to express their social or environmental concerns."
François - Lecompte and Roberts (2006)	Socially conscious consumer	"One who considers the welfare of stakeholders who may be affected by his or her purchasing decisions".
Hailes (2007)	Ecological consumption	"One who sees in his or her actions the opportunity to preserve the environment and the quality of life in society in a particular local context".
Lar (2009)	Socially responsible consumption	"Those voluntary acts that can be framed within the sphere of consumption that have been made on the basis of an awareness of the negative consequences for the outside world (excluding the functionality of the product or service and the interest shown in it)".

Source: Palacios González, (2017)

Informed Attitude

Environmental psychology differentiates between pro-environmental attitude and pro-environmental behavior (Amérigo et al., 2016) and empirical research shows that environmental attitudes are not necessarily related to environmental behaviors (Shrum, Lowrey, & McCarty 1994; Stern, 2000; Kollmuss et al., 2002).

The following chart, an example of an informed attitude, shows results from a survey conducted by Capgemini Research Institute in March 2020 among more than 7,500 consumers:

The same consultancy and as a result of a survey on circular economy, August-September 2021, published these data:

According to the same Capgemini Research Institute survey, consumers are often unaware of the environmental or social consequences of many of the products they buy, and adequate information about this would reinforce sustainable behavior. Information that can provide an adequate marketing strategy.

A study carried out among more than 3000 consumers by the Business, Economy and Sustainability Research Group (E-SOST) of the Comillas Pontifical University and Procter & Gamble (P&G) published

Table 2. Scales for measuring socially responsible consumption behavior

Research	Scala	Country
Berkowitz and Daniels (1964), Berkowitz and Lutterman (1968)	Social Responsibility Scale (SRS)	USA
Anderson and Cunningham (1972)	Social Responsibility Scale (SRS)	USA
Anderson et al. (1974)	Social and Ecologically Concerned Consumers Scale (SECC)	USA
Webster (1975)	The Social Involvement Model (SIM)	UK
Tucker etal. (1981)	Environmental Responsible Citizen Scale (ERC)	USA
Belch (1982)	Life Style Analysis cale (LSA)	USA
Antil (1984)	Sociality Responsible Consumption Behavior Scale (SRCB)	USA
Roberts (19964), Roberts (1996b), Straughan and Roberts (1999), Akehurst et al. (2012)	Ecologically Conscious Consumer Behavior Scale (ECCB)	USA & Portugal
Mohr and Webb (2005), Webb et al (2008).	Social Responsible Purchase and Disposal Scale (SRPD)	USA
Frangois-Lecompte and Valette-Florence (2006), François-Lecompte and Roberts	socially Responsible Consumption scale (SRC)	France
Ismail et al. (2006)	Environmental Consumerism and Purchasing Behavior Scale (ECPB)	Malaysia
Lee (2008)	Socially Responsible Consumption Scale (SRC)	Hong Kong
Lee and shin (2010)	Socially Responsible Consumption Scale (SRC)	South Korea
Yan and she (2011)	Socially Responsible Consumption Behaviour Scale China (src)	China
Sudbury-Riley and Kohibacher (2016)	Ethically Minded Consumer Behaviour (EMCB)	UK, Germany, Hungary and Japan

Source: Palacios González (2017)

Table 3. Variables that can act as drivers or inhibitors of sustainable consumption

Authors	Factors Studied (Drivers and Inhibitors)
(M. Laroche et al., 2001)	Demographic variables, knowledge, behaviours, attitudes, values, environmental attitudes, altruistic values, pro-environmental behaviour, household income levels and number of family members.
(Clark et al., 2003)	Environmental attitudes, altruistic values, pro-environmental behaviour, household income levels and family size.
Lee (2008)	Social influence, environmental concern and self-image.
(Pickett-Baker 8 Ozaki, 2008)	Pro-environmental values, product performance
Lee (2009)	Peer influence, environmental concern, environmental attitude, perceived severity of environmental problems, environmental concern
(Barbarossa et al, 2015)	Perceived barriers such as: lack of product availability, perceived higher prices and inadequate communication at time of purchase
P. Kumar 8 Ghodeswar, 2015	Consumers with strong support for environmental protection and responsibility, shopping experience and social influence
Paul, Modi, and Patel (2016)	Attitude, perceived behavioural control and environmental concern
(Onel 8 Mukherjee, 2016)	Knowledge
(Nguyen et al., 2017)	Subjective norms, attitudes and perceived barriers.

Source: Reyes-Rueda, 2020

Table 4. Academy findings on sustainable consumption and factors assessed

Classification	Assessed Factors/Constructs	Findings
Demographic factors	Age, gender, income, education level, employment status, home ownership, marital status, household size	Women have greater pro-environmental behaviours (Lee, 2009). However, there are other research shows no significant differences (Suren, Purbey, & Sadarangani, 2018).
		Generational studies have shown age to be a determining factor. Nielsen (2018b) concluded that millennials are more likely to be environmentally conscious than Generation X or baby boomers, which makes them more likely to engage in environmentally friendly behaviour.
		Among less members in the family and higher income, increases the probability of an environmental behavior (Clark et al., 2003).
Knowledge	Ecological literature	Have a greater knowledge about the environment in the intention to purchase green products (A. Shama 8 Foropon, 2019).
		People who have environmental knowledge usually have more friendly consumption/compra behavior with the environment (Maniatis, 2016).
Values	Individualism, collectivism.	Individuals who have collectivists values tend to make green shopping (Wang, 2014).
		People with collectivist values, usually have positive attitudes towards green products and these products once again generate purchase intention (Streen et al., 2018).
Behaviours	Considering environmental issues when making a purchase, recycling, buying environmentally friendly products.	Subjects who recycle have a greater predisposition to buy environmentally friendly products (Park 8.Ha, 2012).
Attitudes	Importance of being environmentally friendly, undesirability of being environmentally friendly, severity of environmental problems, level of corporate responsibility	Environmental attitudes generate intentions that in turn translate into pro-environmental purchasing behaviour (López - Bonilla, Reyes - Rodríguez and López - Bonilla, 2018).
		Environmental attitudes generate more pro-environmental behaviour (Giefer, Peterson and Chen, 2019).
		Supporting environmental protection positively influences consumers' green purchasing decisions (P. Kumar and Ghodeswar, 2015).
Other variables	Social influence	Social influence is the most important factor in purchasing behaviour among young people (Lee, 2008).
		Social attractiveness positively influences the purchase of green products (P. Kumar and Ghodeswar, 2015).

Source: Reyes-Rueda, 2020, based on the classification provided by M. Laroche et al., (2001)

in January 2023, concludes that environmental sustainability in Spain has the face of a woman over 55 years of age. Contrary to the belief that it is young people with university studies who have a higher/ better responsible consumption behavior, according to this study it would be women over 55 years of age who give more importance to sustainability in their purchasing decisions: 74% compared to 58% of those between 18 and 34 years old. The level of education has an influence, yes, but it would be conditioned by age and socio-economic situation. The higher the level of education, age and possibility of reconciling work and family life, the greater the commitment to sustainability.

Figure 2. Consumers' emotional connection to products and organizations they perceive as sustainable
Source: Capgemini Research Institute, March 2020

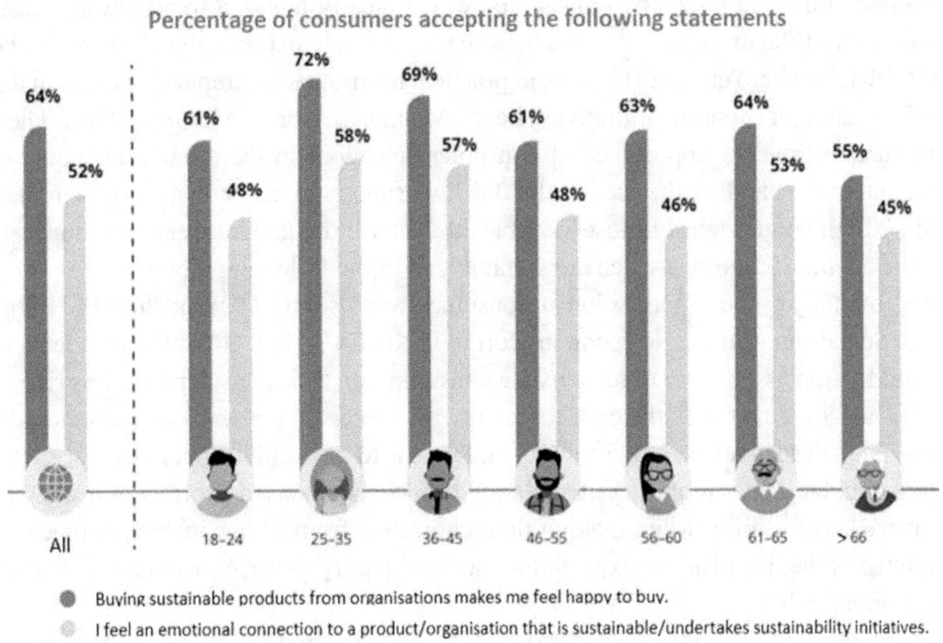

Figure 3. Consumer interest in buying exclusively from brands that practice circularity. Percentage by country.

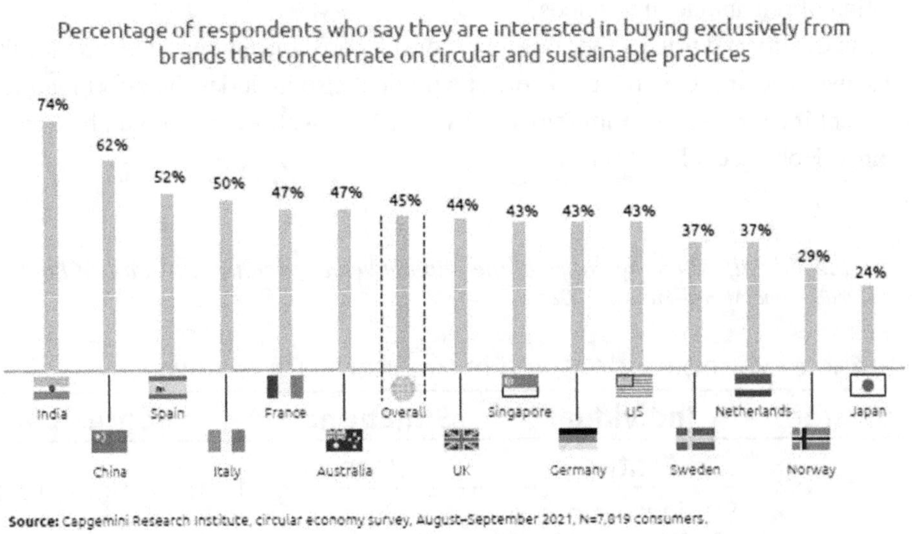

Real Consumer Behavior: Barriers to Sustainable Behavior

But, despite the attitude expressed by consumers, which clearly marks a trend towards sustainability, many authors warn of the existence of a gap between the attitude expressed and the actual behavior of consumers (Camacho-Otero et al., 2018) and, in practical terms, this discrepancy can mean the resounding failure of any circular economy initiative (Nieves-Mendoza, Alonso-Morales, 2022). These authors, Based on the unquestionable importance of consumer behaviour in the circular economy (Fogarassy et al., 2020), and understanding the factors that lead consumers to accept and adopt circular practices (Russo et al., 2019), try to identify those variables that cause the gap between consumer attitude and behavior in the circular economy, which are summarized in the following table:

According to a study by the Organization of Consumers and Users (OCU) and the NESI Forum for New Economy and Social Innovation (Neweconomyforum in Rueda, 2019), 60% of respondents understand the lack of information as a barrier to responsible consumption, 58% consider that the price of products and services is not consistent with the expectation of value received and, 54% express their difficulty in finding accessible alternatives due to proximity and access to the point of purchase.

Delivering circular value implies contemplating proposals to varied profiles of users who will not always prioritize sustainability when making their choice so, from a sustainable approach, companies will have to satisfy other priorities or expectations such as quality, price, convenience, identity, hygiene, time, skills... among others.

Value for money is the most important factor and, at the same time, the barrier to consumer participation in the circular economy (European Commission, 2018, p11) and therefore, organizations that want to successfully implement business models that advance circularity must ask themselves questions such as: What is the price that will encourage users to invest in a remanufactured or leased product? What information should be provided to Customers about the full lifecycle costs of a product? What economic incentives do they need to change their purchasing decisions? How can they ensure quality and change users' perception of remanufactured goods?

In addition to quality and price, other factors to consider are convenience and circular skills because, from the consumer's perspective, the total cost of a product also includes the effort required to acquire, repair, buy or rent it (European Commission, 2018, p 186) as well as what could be termed as a series of circular skills (Hobson et al., 2021).

Figure 4. Variables identified as the cause of the gap between consumer attitude and behavior
Source: Nieves-Mendoza and Alonso-Morales, 2022

Clustering of variables identified in the literature review

Dimension	Individual	Situational	Contextual
Variables	Attitude	Price	Demographic factors
	Subjective norm	Cost	
	Perceived behavioural	Knowledge	Culture
		Information	Incentives
	Awaraness		

Repairing and maintaining a product requires time, effort and skills, which could discourage some users, even if they initially show a sustainable attitude. Therefore, manufacturers and suppliers of circular products and services must induce responsible behaviours by educating, training and facilitating such behaviours (Hobson et al., 2021) so that they are perceived not only as sustainable, but also as convenient for the purposes of comfort and personal satisfaction in a way that encourages participation.

Meeting this need for user convenience should lead organizations to ask questions such as: How do we maintain ease of access and minimize customer effort to drive product repairs, second-hand product purchases, and product rentals or leasing? for the identification of the quality, maintenance and repair of the product? What factors and incentives can motivate users to take on changes, not as an individual resignation but as a collective advantage? How do you eliminate uncertainty and inspire confidence?

Because another essential factor in promoting circularity is trust. The importance consumers place on supplier reputation and trust is significantly related to perceived benefit and risk, which, in turn, affects the attitude and intention to purchase refurbished products (Agostini et al., 2021).

This should lead organizations to question how to use the knowledge provided by the information to provide the consumer with the assurance that circular products and services will work and the experience will be at the expected level.

An experience that includes overcoming reluctance due to hygiene issues, guaranteeing the customer that second-hand or refurbished appliances, for example, meet the same safety and hygiene standards as newly manufactured products.

On the other hand, in the Circular Economy, consumption becomes anonymous because people don't own products, they just use them... so that people can no longer define themselves by the goods they have (Camacho-Otero, Boks & Pettersen, 2018). In order for goods not to lose that value of social identity that sometimes turns them into simple "objects of desire" not because of their functionalities, but because of what their ownership represents socially, organizations must consider how to give social identity value to the circular model.

For the advancement of the circular economy, pro-environmental behaviors must be promoted, valued and considered key development indicators as they constitute a form of cooperation, a joint effort of individuals to achieve a shared benefit such as the preservation of natural resources (Klein et al., 2022, p. 182) and it is a matter of establish synergies between the social, environmental and economic dimensions, at the micro, meso and macro levels, since circularity implies systemic change (Stankevičienė et al. (2020). One of the biggest challenges that the circular transition must address is the acceptance of circular strategies by consumers (Calvo & Lévy, 2020) because, although academia highlights the need for more studies, it accepts that there is a discordance between concern-awareness-manifested pro-environmental attitude and sustainable behavior (Lakaktos et al., 2018; Grębosz & Siuda (2019).

In order to enhance the involvement of consumers and thus favour their decisive collaboration in the change of the economic model, the European Economic and Social Committee (EESC), in its 2019 report, proposed the adoption of measures such as promoting information campaigns related to new models of production and consumption, taking into account age and cultural factors, and, In this regard, marketing can play a critical role in educating, stimulating, and meeting sustainable consumer demand.

SUSTAINABLE MARKETING: EDUCATING THE CONSUMER

Sustainability is a current, relevant and cross-cutting issue in most sectors, companies, government agendas and institutions, as well as among consumers and society in general. It is considered a "megatrend" by various authors such as Mittelstaedt et al., (2014) and Tollin et al., (2015) and has not been alien to the interest of marketing strategies.

Historical Evolution of Marketing

Marketing's interest in the environment is not recent. The first publications date back to the 1970s (20th century) when some researchers, mainly from the United States, made the first contributions to the development of marketing with an ecological perspective. Most of these works focused on the study of the relationships between ecological concern and ecological behaviour and on the characterization of the ecological consumer (Kilbourne and Beckmann, 1998).

Market research has sought to integrate sustainability aspects from various perspectives. From green marketing, through green marketing, cradle-to-cradle design, and life cycle analysis (Kemper & Ballantine, 2019) to sustainable marketing that addresses the social consequences of marketing and aims to change consumer attitudes and behavior towards adopting sustainable lifestyles and offers more critical perspectives on consumption (Pereira Heath & Varey, 2010; Chatzidakis, 2012; Kemper & Ballantine, 2019).

After the publication of The Green Consumer Guide (Elkington and Hailes, 1988), which defined a green consumer as consumers who are aware and tend to avoid certain products with a negative impact on nature, there is a growing awareness that concern for the environment could influence consumer behaviour and the need for marketing to be oriented in this direction.

And this was the direction in which marketing was oriented, especially after the publication of the United Nations socio-economic report "Our Common Future" in 1987, known as the "Brundtland Report" which, with its idea of sustainability, contributed to the international awareness of the degradation of the environment, of the interrelationship between development and the environment as well as of the need to recognize that we live in a world in which that social inequalities are a reality.

Definition of Sustainable Marketing and Impact

Sustainable marketing is one of the most effective and efficient strategies that companies can develop. According to the WADA (n.d.), we can define sustainable marketing or Green Marketing as that which develops products considered environmentally friendly (for example, designed to minimize the damage caused to the physical environment or improve its quality) or value the effort involved in the production, promotion, packaging and demand of goods in a way that is respectful and responsible with ecological concerns. For Liao et al., (2020), sustainable marketing emerges as a market trend that allows consumers to attract the attention of competitors and has a direct impact on their behavior. Other authors highlight adequate communication as a fundamental element of sustainable marketing. That is, sustainable marketing as part of the company's communication strategy that allows it to positively influence consumers and achieve its goals (Nadanyiova et al., 2020).

From Responsible Consumers to Sustainable Consumers

From a marketing point of view, we have gone from talking about a responsible consumer to a sustainable consumer. Responsible consumers are those who consume consciously and critically with the environmental and resource situation (Nubia-Arias, 2016). Responsible consumer behaviour involves different behaviours taking into account the environmental impact at different stages of the purchasing process (Hoyer et al, 2013). A distinction can be made between two types of responsible behavior (Aldamiz-Echevarría & Vicente Molina, 2003): (a) Direct purchasing behavior. Consumers will only buy those products that they feel meet their sustainability requirements. (b) Post-purchase behavior. Responsible behaviour encompasses both the process of purchasing a product and the use, consumption and treatment of waste generated.

The sustainable consumer goes a step further, and not only bets on a change in individual consumption patterns, but also adopts a global vision, and seeks efficiency in consumption, both of resources and products. According to Kostadinova (2016), "the use of services and related products which respond to basic needs and bring a better quality of life while minimizing the use of natural resources and toxic materials as well as emissions of waste and pollutants over the life cycle of the service or product so as not to jeopardize the needs of future generations".

However, there are times when this relationship between attitude and behavior is not established directly, but is conditioned by some social or economic factors of the individual (Terlau and Hirsch, 2015; Bernardes et al., 2018) and faced with this situation, companies must implement marketing strategies that favour a change in consumer mentality by clearly and effectively highlighting the individual and social benefits of sustainable behaviour.

Benefits of Sustainable Marketing

In a society that is increasingly aware of environmental issues, companies committed to sustainability stand out and can attract a segment of consumers aware of the need to modify individual use and consumption habits to achieve collective environmental benefits (Jiménez-Marín et al., 2019).

Sustainable marketing has a number of benefits for both businesses and consumers. Differentiating themselves in the market, improving the sustainable reputation of the brand and reducing operating costs are some of the commercial benefits that it will bring to companies, which will also contribute to a more responsible and sustainable future.

Benefits that can be achieved by adopting a series of strategies. Like this:

To instill trust and make it easier for consumers to make informed decisions, a fundamental strategy in sustainable marketing is transparency by clearly and truthfully communicating the sustainable practices that the company implements in its production, supply and distribution chain.

Understanding the individual needs and preferences of consumers in order to offer specific solutions will require personalization and segmentation as strategies to optimize resources and reduce waste, adapting the message and products to each market segment and, always focusing the information on the formation of a sustainable conscience.

This goal can be achieved by creating relevant content that informs users about the benefits of a sustainable lifestyle and how the company is committed to contributing to sustainability with its products and services. With this strategy of engagement and educational content, it also seeks to empower the consumer so that they are the protagonists of change.

Commitment, not only to consumers, but to all stakeholders. Strategic partnerships with suppliers, distributors, and other companies and organizations that share the same sustainable efforts and values can amplify positive impact and help drive change across the industry.

Making a difference by educating and providing value to the consumer in terms of environmental responsibility does not necessarily require large and expensive actions, but rather knowledge of the market and creative proposals such as creating corporate identifiers that refer to all the small details in which the company has carried out eco-responsible actions or positively encouraging responsible consumer behavior and, naturally, communicate properly to make known what is being achieved. In today's era of social media, interaction and two-way communication, proper communication can produce a "pull effect", a multiplier effect of education and promotion of sustainable products and services.

Examples of Companies Pioneering Sustainable Marketing Strategies

In this line of action, companies must take initiatives on their marketing strategies that affect both products and services, regulatory pricing strategies, distribution logistics, and how to deal with new communication with their customers (Pomering, 2017; Pisitsankkhakarn, and Vassanadumrongdee, 2020). And to do this, it is essential to undertake a radical corporate change that affects both employees and the involvement of its customers.

As examples of pioneering companies in the implementation of sustainable marketing strategies, we can point out the following:

- Patagonia: This renowned outdoor clothing brand has built its identity around sustainability. Patagonia promotes responsible consumption, encouraging consumers to reflect on their purchases and opt for durability and repair rather than overconsumption.
- Tesla: Electric car manufacturer Tesla has revolutionized the automotive industry by offering high-performance electric vehicles. In addition, the company has created a network of electric charging stations that encourages the adoption of emission-free vehicles and contributes to the reduction of the carbon footprint in transport.
- Dove: Seeks to promote body acceptance and diversity through its advertising messages, challenging unrealistic beauty standards and fostering a positive self-image.
- Interface: This is a leading modular flooring manufacturing company, it is committed to being a completely sustainable company. They have implemented strategies such as recycling materials, using renewable energy, and reducing carbon emissions in their supply chain.
- Too Good To Go: A company that emerged in Barcelona in 2018 and has become a global movement in the fight against food waste with its slogan "food is not thrown away". Awarded in the first edition of the go! SDGs for its contribution to the achievement of SDG 2: Zero Hunger, won one of the sustainability awards of the CETT Alimara and, in 2023, it has been awarded the award for Best Sustainable Ecommerce at the Ecommerce Awards 2023 and one of the Infinito Communication Awards of the Spanish Metalgraphic Association for being an example of sustainable initiatives and demonstrating the potential of marketing campaigns to mobilise sustainable behaviour of the consumer.

These examples show how brands can incorporate sustainable marketing into their business approach and make a positive impact on the world while driving successful results. And they're not the only ones.

Companies such as Danone, a member of B Corp company since 2016 (B Corporation, 2022), are an excellent example of Green Marketing. Through the "One Planet. One Health", Danone presents a global and long-term approach that enables a more sustainable food system while moving towards the decarbonization of the planet (Danone, 2022d). This new strategy has guaranteed it a position as a company 100% aware of sustainability (IMF, 2022). As a starting point, Danone has promoted a change in its corporate culture, involving its employees. Through the creation of the "One Person, One Voice" program, since 2018, the company's more than 100,000 employees have been actively participating in the project in pursuit of the Danone 2030 goals (Danone, 2022a) in order to achieve the same vision of Circular Economy (WTWCo, 2021). In this way, employees actively participate in innovation and co-creation with customers, from an empowered and committed attitude, both in the definition of objectives and in the design of solutions.

In the case of BBVA, its social efforts have been aimed at objectives such as the eradication of poverty and hunger (SDGs 1 and 2), promoting and defending people's health through the "Together for Health" program (SDG 3), encouraging the training and involvement of young people through scholarships and donations to organizations for cultural and educational purposes (SDG 4) and the rehabilitation of the social fabric in the country. communities, mainly those affected by disasters (SDGs 8, 10 and 11), among others.

But the necessary social change is not possible without the example of the companies themselves in their commitment to the environment because, through an appropriate corporate social responsibility policy, they influence the economic system and also social and environmental aspects (García & Márquez, 2021) and this great effort cannot be undertaken alone but from the support provided to companies by various public and private institutions. leading "the way to a new stakeholder-driven model" (B Corporation, 2023a).

To achieve this interweaving of stakeholders, it is essential to combine the economic, environmental and social approaches under the same strategy and from a new corporate governance model, based on transparency and alliances (Bianchi, 2023). This is how private entities emerge that bring together companies that publish and share their best practices. It is worth mentioning associations such as B Lab (B Lab, 2023), a non-profit network, whose mission is to "transform the global economy to benefit all people, communities and the planet". Its objective is to certify companies that meet high standards of quality, performance, social responsibility, and transparency (see B Corporation directory, 2023b) and that it is supported by philanthropic donors, including foundations, government agencies, corporations, and individuals. Also, in the financial sector we find examples such as the United Nations (UN) partnership with the financial world (UNEP FINANCE, 2022) to contribute to the goal of creating a positive impact on people and the planet. A goal that will undoubtedly be easier to achieve by inspiring educators and companies to incorporate sustainable practices that equip students and consumers with the knowledge, skills and values necessary for a sustainable future. Practices that must be based on accountability and transparency.

Greenwashing

Because, not always, new marketing strategies imply more responsible, transparent and honest companies. Known as *Greenwashing (*de Freitas Netto et al., 2020)), a term that was first introduced in 1986 by activist Jay Westervel and related to the hotels' initiative on the reuse of towels in rooms as a propaganda action and the company's only ecological initiative (Gatti et al., 219). This "whitewashing" strategy is based on the theory that customers who perceive companies as socially responsible may be more willing to buy these companies' products at a higher price (Guo et al., 2014) or switch brands to a greener one (Nylasy et al., 2014). Faced with this situation, and aware of their business opportunity based on the expectations of these responsible customers, certain companies have promoted practices that, although they may minimize the carbon footprint in some aspects, have not contemplated the collateral damage caused to the environment or have remained their initiatives in mere headlines for the media. Ultimately, their initiatives have only happened to capture interest from their prospective customers. Instead of diversifying, researching and co-creating with the customer, they have reduced their 2030 agenda to offsetting carbon footprint or modifying small processes in their value chain (Xiong and Li, 2019). Aware of the spectacular growth that such practices have had in past decades (Ogilvy and Mather in Hsu, 2011), researchers have been interested in the impact that such actions have on customers' perception of this terrible performance from the lack of ethics and CSR of companies (Gatti et al., 2019), as well as the impact it has on consumers' skepticism about true green marketing (Hsu, 2011). 2011). Thus, TerraChoice (2010) has published a study that helps clients detect "the seven sins of greenwashing".

Limitations of Sustainable Marketing

These contributions of the research framework are essential for the establishment of a good business strategy that optimizes the advantages of sustainable marketing and manages to overcome the disadvantages. Because, more than criticisms, some limitations or disadvantages should be highlighted:

- High cost of implementation: Implementing a green marketing strategy can be costly for businesses, especially those that are already established.
- Unfair competition: Some companies use green marketing as an advertising tool and not as a form of true commitment to the environment.
- Market limitation: Some sustainable products and services may have a limited market due to higher prices.
- Difficulty measuring success: Sustainable marketing results can be difficult to measure, making it difficult to assess whether this marketing strategy is succeeding.

CONCLUSION

To contextualise the research, we exposed the environmental problem that demands urgent, courageous and committed responses from institutions, companies and society, such as the transition from a linear economy model to one based on circularity and new business models to promote sustainable change. Successful change requires the involvement of all stakeholders, but especially consumers. That's why we've raised the complexities of consumer behavior, including pro-environmental behavior and the attitude-behavior gap to connect with the central theme of the chapter, highlighting the potential of marketing to win and convince. Overcoming resistance to change and convincing people that individual disadvantages will result in social and environmental benefits.

The implementation of circular business models is a challenge for the business sector because it is not enough to offer, but it is necessary to educate and convince so that their proposals reach the critical mass necessary to demonstrate that they are operationally feasible, commercially viable and socially beneficial. This involves designing a strategy to deliver circular value. A strategy to keep in mind:

- Customer factors: designing initiatives that encourage and attract users to participate in the economy by adopting circular roles. The design of these initiatives should be based on customer knowledge and based on tests and pilot experiments that prove the attractiveness of product prototypes and new business models for users.
- business factors. The challenge for organizations is to secure the investment needed to scale and thus move beyond a mere add-on approach to the existing business model. To achieve this, it will be necessary to measure the success of the initiatives taken, for which the results of the pilot tests must be clear and easy to evaluate.
- System factors. Competitors with cheaper, more accessible business models with greater tradition and depth among consumers, lack of enabling infrastructure or legislation that still does not sufficiently encourage circular business models, are barriers that companies inform to advance in circularity. Barriers that could be overcome with policies that encourage more sustainable business practices and consumption behaviors.

Nowadays, generating experience is one of the main objectives set by marketing managers. The aim is to instill trust and emotional connection in the customer with the products/services and with the brand and to induce future behaviours and interactions (Torrent-Sellens, 2009) with a relationship of quality and warmth in communication that turns it into an opportunity.

In order to generate experiences that allow overcoming the sustainable attitude-behavior gap, organizations are proposed:

- Rethink their business models and define alternative revenue channels that aren't solely driven by product sales.
- Adopt circular design principles in your production processes, evaluating the product portfolio and identifying interventions.
- Rethinking value chains to "close the loop".
- Invest in infrastructure to ensure a stable supply of recycled materials.
- Reorganize supply chains to manage product recovery.
- Strengthen data sharing across the value chain.
- Leverage emerging technologies such as the Internet of Things (IoT), blockchain, and Artificial Intelligence (AI) that can significantly help track product and material flows within your supply chains.
- Collaboration between organizations in the business ecosystem, as well as externally with academics and think tanks.
- Foster a circular mindset within the organization by changing employee culture, increased leadership accountability, skills development, and adopting metrics and KPIs to track progress.
- Encourage the adoption of circularity by consumers by providing information and generating trust and environmental awareness for which valuable tools will be used, customer experience (CX) design focused on facilitating circular behaviors, roles and relationships and a marketing strategy to promote the values of circularity and incentivize suppliers and users to accept the additional cost associated with some circular business models.

In just a few years, sustainability has gone from being considered an alternative theory, an option, to being the corporate model to be integrated and promoted within organizations and disseminated among potential customers, which places sustainable marketing in the sights of companies.

In a global, digital, technological, and productive environment that respects the environment, sustainable marketing is presented as an effective tool capable of informing, but also educating consumers and is oriented to the creation of value for the customer starting with understanding the needs and desires of the consumer, deciding which target markets the organization can best serve and developing a convincing value proposition through which the company can attract, maintain and meet the expectations of its customers and achieve market share and profits (Kotler & Armstrong, 2017).

Supported by neuroscience and data science, it must be able to support the design of strategies to change or create habits for a responsible consumer society (Werner et al, 2022), which would involve communicating to educate and train competent leaders to lead the transition to a sustainable society (Maturén & Pizzolante, 2020) and influencing markets to evolve, contributing to the creation of a new social architecture where industry, stimulated by governments and public policy incentives, adopts sustainable production practices and consumers abandon legacy schemes and adopt smart consumption.

As long as there is not a majority of consumers committed and aware of sustainability who, in addition, trust in the environmental contribution of products and in the willingness and ability of companies to advance in that goal, it will be all the more necessary to create marketing strategies that help reduce the gap between attitude and sustainable behavior.

LIMITATIONS

Primary research among the academic community on pedagogical strategies to educate in sustainable values and to train professionals to materialize them from the business field would have been desirable.

Research among marketing professionals to learn about real current strategies and future trends would have been of great interest.

FUTURE LINES OF RESEARCH

Overcome the aforementioned limitations and address other areas of interest such as: Delving into the psychology of sustainable consumption by exploring concepts such as cognitive dissonance, social identity theory and the role of emotions in decision-making by looking for global trends in sustainable consumption, commonalities or differences in the attitude and behavior of consumers in different regions of the world; explore the sharing economy, fair trade, ethical sourcing, and ethical dilemmas that consumers face in their choices, as well as study the impact on consumer perceptions and choices of corporate initiatives and corporate social responsibility programs.

The challenges and barriers to implementing sustainable marketing strategies in a more comprehensive way that include educational contributions for consumers, transparency in communication by companies and their impact measurement methodologies, as well as the future challenges of sustainable marketing are aspects that also require attention and research and, in this sense, to delve into the Nudge Theory (Thaler & Sunstein, 2017) which, in the field of behavioral economics, argues that nudging pushes people to make beneficial decisions in the long term by looking for the most appropriate solution rather than the most immediate one.

REFERENCES

Agostini, L., Bigliardi, B., Filippelli, S., & Galati, F. (2021). Seller reputation, distribution and intention to purchase refurbished products. *Journal of Cleaner Production*, *316*, 128296. doi:10.1016/j.jclepro.2021.128296

Aguirre, M. S., Aldamiz-Echevarría González de Durana, C., Charterina, J., & Vicente-Molina, M. (2003). El consumidor ecológico. *Distribución y consumo, 13*(67), 41-54.

Ajzen, I., & Fishbein, M. (1980). *Understanding attitudes and predicting social behavior*. Prentice-Hall.

Ajzen, I., & Fishbein, M. (1991). The Influence of Attitudes on Behavior. The Handbook of Attitudes, 32.

Allport, G. (1935). Attitudes. In Handbook of social psychology. Clark University Press.

Antil, J. (1984). Socially Responsible Consumers: Profile and Implications for Public Policy. *Journal of Macromarketing*, *4*(2), 18–39. doi:10.1177/027614678400400203

Arias, B. (2016). El consumo responsable: educar para la sostenibilidad ambiental. *Aibi Revista de Investigación, Administración e Ingeniería, 4*, 29-34. doi:10.15649/2346030X.385

Bagozzi, R. P., & Warshaw, P. R. (1990). Trying to consume. *The Journal of Consumer Research*, *17*(2), 127–140. doi:10.1086/208543

Bakal, I., & Akcali, E. (2009). Effects of Random Yield in Remanufacturing with Price-Sensitive Supply and Demand. *Production and Operations Management*, *15*(3), 407–420. doi:10.1111/j.1937-5956.2006.tb00254.x

Barbarossa, C., Beckmann, S. C., De Pelsmacker, P., Moons, I., & Gwozdz, W. (2015). The impact of different social media channels on consumers' buying decision. *Journal of Marketing Management*, *31*(1-2), 136–154. doi:10.1080/0267257X.2014.965148

Bass, F., & Wilkie, W. (1973). A Comparative Analysis of Attitudinal Predictions of Brand Preference. *JMR, Journal of Marketing Research*, *10*(3), 262–269. doi:10.1177/002224377301000306

Bass, F. M., & Wayne Talarzyk, W. (1972). An Attitude Model for the Study of Brand Preference. *JMR, Journal of Marketing Research*, *9*(February), 93–96. doi:10.1177/002224377200900121

Berkowitz, L., & Daniels, L. R. (1964). Affecting the salience of the social responsibility norm: Effects of past help on the response to dependency relationships. *Journal of Abnormal and Social Psychology*, *68*(3), 275–281. doi:10.1037/h0040164 PMID:14126841

Berkowitz, L., & Lutterman, K. G. (1968). The Traditional Socially Responsible Personality. *Public Opinion Quarterly*, *32*(2), 169–185. doi:10.1086/267597

Bernardes, J., Ferreira, F., Marques, A. D., & Nogueira, M. (2018). "Do as I say, not as I do" - a systematic literature review on the attitude-behaviour gap towards sustainable consumption of Generation Y. *IOP Conference Series. Materials Science and Engineering*, *459*, 012089. doi:10.1088/1757-899X/459/1/012089

Bettman, J. R., Capon, N., & Lutz, R. J. (1975). Multiattribute measurement models and multiattribute attitude theory: A test of construct validity. *The Journal of Consumer Research*, *1*(4), 1–15. doi:10.1086/208602

Bianchi, C., & Bianchi, C. (2023). YPF Sustainability: Renovation of Service Stations and YPF Full Stores in Responsible Consumption and Sustainability. In *Case Studies from Corporate Social Responsibility, Social Marketing, and Behavioral Economics*. Springer Business Cases.

Bielawska, K., & Grębosz-Krawczyk, M. (2021). Consumers' Choice Behaviour Toward Green Clothing. *European Research Studies*, *24*(2), 238–256. doi:10.35808/ersj/2124

Blackwell, R. D., Miniard, P. W., & Engel, J. F. (1986). *Consumer Behavior. Better World Books*. Estados Unidos de America.

Boudon, R., & Bourricaud, F. (1989). *Dictionnaire critique de la sociologie*. Presses universitaires de France.

Calvo, P., & Lévy, J. P. (2020). *Foundations of augmented cognition: Directing the future of adaptive systems*. Springer.

Camacho-Otero, J., Boks, C., & Pettersen, I. (2018). Consumption in the Circular Economy: A Literature Review. *Sustainability (Basel)*, *10*(8), 2758. doi:10.3390/su10082758

Camacho-Otero, J., Boks, C., & Pettersen, I. N. (2018). Consumption in the Circular Economy: *A Literature Review. Sustainability (Basel)*, *10*(8), 2758. doi:10.3390/su10082758

Campbell, D. T., & Stanley, J. (1963). *Experimental and quasi-experimental designs for research*. Rand McNally.

Canavari, M., Dimech, M., & Caputo, V. (2011). Attitudes of Maltese Consumers Towards Quality in Fruit and Vegetables in Relation to Their Food-Related Lifestyles. *The International Food and Agribusiness Management Review*, 14.

Capgemini Research Institute. (2020). *Consumer products and retail. How sustainability is fundamentally changing consumer preferences*. Author.

Castaño, L., Perdomo-Ortiz, J., Dueñas-Ocampo, S., & Durán, W. (2016). Socially responsible consumption: An application in Colombia. *Business Ethics (Oxford, England)*, *25*(4), 460–481. doi:10.1111/beer.12128

Castañón, J. (2018). *El consumo sostenible en España. Influencia de la información sobre la acción ecológica entre los consumidores millennial*. Universidad Pontificia de Comillas.

Chamorro, A., & Palacios-González, M. (2020). Analysis of Socially Responsible Consumption: A Segmentation of Spanish Consumers. *Sustainability (Basel)*, *12*(20), 8418. Advance online publication. doi:10.3390/su12208418

Citysem. (2023). *Marketing sostenible, estrategias y ejemplos para un impacto positivo.* https://citysem.es/marketing-sostenible-estrategias-y-ejemplos/

Clark, W., & Dickson, N. (2003). Sustainability science: The emerging research program. *Proceedings of the National Academy of Sciences of the United States of America, 100*(14), 8059–8061. doi:10.1073/pnas.1231333100 PMID:12794187

Comisión Mundial sobre el Medio Ambiente y el Desarrollo. (1987). *Nuestro futuro común.* Fondo de Cultura Económica.

CorporationB. (2023a). *About B Lab.* https://www.bcorporation.net/en-us/movement/about-b-lab/

Corporation, B. (2023b). *Find a B Corp: Search here to buy from, work with, or learn more about Certified B Corporations around the world.* https://www.bcorporation.net/en-us/find-a-b-corp/?refinement%5BbftwCategories%5D%5B0%5D=Community

Corral-Verdugo, V., & Pinheiro, J. (2004). Aproximaciones al estudio de la conducta sustentable. *Medio Ambiente y Comportamiento Humano, 5,* 1–26.

Damgaard, M., & Nielsen, H. (2018). Nudging in education. *Economics of Education Review, 64,* 313–342. Advance online publication. doi:10.1016/j.econedurev.2018.03.008

Danone. (2023a). *Danone Goals.* https://www.danone.es

Danone. (2023b). *Raison D'Être.* https://www.danone.com/about-danone/sustainable-valuecreation/danone-entreprise-a-mission.html

de Freitas Netto, S. V., Sobral, M. F. F., Ribeiro, A. R. B., & Soares, G. R. D. L. (2020). Concepts and forms of greenwashing: A systematic review. *Environmental Sciences Europe, 32*(1), 1–12. doi:10.1186/s12302-020-0300-3

De Jesus, A., & Mendonna, S. (2018). Lost in Transition? Drivers and Barriers in the Eco-Innovation Road to the Circular Economy. *Ecological Economics, 145,* 75–89. doi:10.1016/j.ecolecon.2017.08.001

Díaz-Chao, Á., & Torrent-Sellens, J. (2010). ¿ Pueden el uso de las TIC y los activos intangibles mejorar la competitividad? Un análisis empírico para la empresa catalana. *Estudios de Economía Aplicada, 28*(2), 1–22.

Elkington, J., & Hailes, J. (1988). *The green consumer guide.* Grafton Books.

Engel, J. F., Kollat, D. T., & Blackwell, R. D. (1968). *Consumer Behavior.* Holt, Rinehart & Winston.

Europea, C. (2015). *Comunicación de la comisión al parlamento europeo, al consejo, al comité económico y social europeo y al comité de las regiones. Cerrar el círculo: un plan de acción de la UE para la economía circular.* Bruselas.

European Commission. (2018). *Behavioural study on consumers' engagement in the circular economy: executive summary.* https://data.europa.eu/doi/10.2818/921596

Fader, P. (2020). *Customer centricity: Focus on the right customers for strategic advantage.* University of Pennsylvania Press.

Fazio, R. H., Sanbonmatsu, D. M., Powell, M. C., & Kardes, F. R. (1986). On the automatic activation of attitudes. *Journal of Personality and Social Psychology*, *50*(2), 229–238. doi:10.1037/0022-3514.50.2.229 PMID:3701576

Fogarassy, C., & Finger, D. (2020). Theoretical and Practical Approaches of Circular Economy for Business Models and Technological Solutions. *Resources*, *9*(6), 76. doi:10.3390/resources9060076

Gallardo-Vázquez, D., Barroso-Méndez, M., Pajuelo-Moreno, M., & Sánchez-Meca, J. (2019). Corporate Social Responsibility Disclosure and Performance: A Meta-Analytic Approach. *Sustainability (Basel)*, *11*(4), 1115. doi:10.3390/su11041115

Gatti, L., Seele, P., & Rademacher, L. (2019). Grey zone in–greenwash out. A review of greenwashing research and implications for the voluntary-mandatory transition of CSR. *International Journal of Corporate Social Responsibility*, *4*(1), 1–15. doi:10.1186/s40991-019-0044-9

Geissdoerfer, M., Vladimirova, D., & Evans, S. (2018). Sustainable business model innovation: A review. *Journal of Cleaner Production*, *198*, 401–416. Advance online publication. doi:10.1016/j.jclepro.2018.06.240

Genovese, A., Acquaye, A., Figueroa, A., & Koh, S. C. L. (2015). Sustainable Supply Chain Management and the transition towards a Circular Economy: Evidence and some Applications. *Omega*, *66*, 344–357. Advance online publication. doi:10.1016/j.omega.2015.05.015

Ghisellini, P., Cialani, C., & Ulgiati, S. (2016). A review on circular economy: The expected transition to a balanced interplay of environmental and economic systems. *Journal of Cleaner Production*, *114*, 11–32. doi:10.1016/j.jclepro.2015.09.007

Giefer, J., Peterson, M., & Chen, Y. (2019). Understanding perceived corruption in business: Views from the United States and China. *Journal of International Business Ethics*, *2*(1), 24–34. doi:10.18352/joibe.37

Global Resources Outlook. (2019). *Natural Resources for the Future We Want*. United Nations Environment Programme.

Goh, D., Muskat, B., & Tan, A. H. T. (2017). The nexus between sustainable practices in hotels and future Gen Y hospitality students' career path decisions. *Journal of Teaching in Travel & Tourism*, *17*(4), 237–253. Advance online publication. doi:10.1080/15313220.2017.1362971

González, J. S., & Sellens, J. T. (2009). Knowledge, networks and economic activity. Revisiting the network effects in the knowledge economy. *UOC Papers, e-Journal on the Knowledge Society,* (8).

Guo, R., Tao, L., & Gao, P. (2014). The research on greenwashing brands' rebuilding strategies and mechanism of brand trust after biochemical and other pollutions. *Biotechnology*, *10*(9), 3270–3279.

Hidalgo-Carvajal, D., Carrasco-Gallego, R., & Morales, G. (2021). From Goods to Services and from Linear to Circular: The Role of Servitization's Challenges and Drivers in the Shifting Process. *Sustainability (Basel)*, *13*(8), 4539. doi:10.3390/su13084539

Hobson, K., Holmes, H., Welch, D., Wheeler, K., & Wieser, H. (2021). Consumption Work in the circular economy: A research agenda. *Journal of Cleaner Production*, *321*, 128969. doi:10.1016/j.jclepro.2021.128969

Howard, J., & Sheth, J. (1969). *The Theory of Buyer Behavior.* . doi:10.2307/2284311

Hoyer, W., Macinnis, D., & Pieters, R. (2016). *Consumer Behavior*. Cengage Learning.

Hoyer, W., Macinnis, D., & Pieters, R. (2016). *Consumer Behavior*. Academic Press.

Hsu, T. (2011). *Skepticism grows over products touted as eco-friendly*. https://www.latimes.com/business/la-xpm-2011-may-21-la-fi-greenwash-20110521-story.html

IMF. (2023). *Danone, una empresa 100% concienciada con la sostenibilidad*. https://blogs.imf-formacion.com/blog/mba/danone-empresa-sostenible/

Inglehart, R. (2011). *Modernization, cultural change, and democracy: The human development sequence*. Cambridge University Press.

Jiménez-Marín, G., Bellido-Pérez, E., & López-Cortés, Á. (2019). Marketing sensorial: El concepto, sus técnicas y su aplicación en el punto de venta. *Vivat Academia*, (148), 121–147. doi:10.15178/va.2019.148.121-147

Johansen, K., & Werner, S. (2022). Something is sustainable in the state of Denmark: A review of the Danish district heating sector. *Renewable & Sustainable Energy Reviews*, *158*, 112117. doi:10.1016/j.rser.2022.112117

Kemper, J. A., & Ballantine, P. W. (2019). Moral identity and green consumption. *Journal of Business Research*, *95*, 236–247.

Kilbourne, W. E., & Beckmann, S. C. (1998). Review and critical assessment of research on marketing and the environment. *Journal of Marketing Management*, *14*(6), 513–532. doi:10.1362/026725798784867716

Kirchherr, J., Reike, D., & Hekkert, M.P. (2017). Conceptualizing the Circular Economy: An Analysis of 114 Definitions. SSRN *Electronic Journal, 127*. doi:10.2139/ssrn.3037579

Klein, N., Ramos, T., & Deutz, P. (2020). Circular Economy Practices and Strategies in Public Sector Organizations: An Integrative Review. *Sustainability (Basel)*, *12*(10), 4181. doi:10.3390/su12104181

Klein, N., Ramos, T., & Deutz, P. (2022). Factors and strategies for circularity implementation in the public sector: An organisational change management approach for sustainability. *Corporate Social Responsibility and Environmental Management*, *29*(3), 509–523. Advance online publication. doi:10.1002/csr.2215

Kollmuss, A., & Agyeman, J. (2002). Mind the gap: Why do people act environmentally and what are the barriers to pro-environmental behavior? *Environmental Education Research*, *8*(3), 239–260. doi:10.1080/13504620220145401

Konietzko, J., Bocken, N., & Hultink, E. J. (2020, January 5). ATool to Analyze, Ideate and Develop Circular Innovation Ecosystems. *Sustainability (Basel)*, *12*(1), 417. doi:10.3390/su12010417

Kostadinova, E. (2016). *Sustainable consumer behaviour: A literature review*. Economic Alternatives.

Kotler, P., & Armstrong, G. (2017). *Principles of marketing*. Pearson.

Kumar, V., & Ghodeswar, B. M. (2015). Impact of corruption on firm's competitive performance: An analysis of Indian firms. *Paradigm*, *19*(1), 20–33.

Kumar, V., & Ghodeswar, B. M. (2015). Importance of corporate social responsibility to societies. *Journal of Human Values, 21*(1), 63–75. doi:10.1177/0971685814558677

Laroche, M., Bergeron, J., & Barbaro-Forleo, G. (2001). Targeting consumers who are willing to pay more for environmentally friendly products. *Journal of Consumer Marketing, 18*(6), 503–520. doi:10.1108/EUM0000000006155

Laroche, M., Bergeron, J., & Barbaro-Forleo, G. (2001). Targeting Consumers Who Are Willing to Pay More for Environmentally Friendly Products. *Journal of Consumer Marketing, 18*(6), 503–520. doi:10.1108/EUM0000000006155

Lee, J. W. Y. (2008). A review of the theories of corporate social responsibility: Its evolutionary path and the road ahead. *International Journal of Management Reviews, 10*(1), 53–73. doi:10.1111/j.1468-2370.2007.00226.x

Lee, K.-H. (2009). Why and How to Adopt Green Management into Business Organizations? The Case Study of Korean SMEs in Manufacturing Industry. *Management Decision, 47*(7), 1101–1121. doi:10.1108/00251740910978322

Lee, K.-H., & Shin, D. (2010). Consumers' responses to CSR activities: The linkage between increased awareness and purchase intention. *Public Relations Review, 36*(2), 193–195. Advance online publication. doi:10.1016/j.pubrev.2009.10.014

Leeuw, A., Valois, P., Ajzen, I., & Schmidt, P. (2015). Using the theory of planned behavior to identify key beliefs underlying pro-environmental behavior in high-school students: Implications for educational interventions. *Journal of Environmental Psychology, 42*, 128–138. Advance online publication. doi:10.1016/j.jenvp.2015.03.005

Lehmann Donald, R. (1972). *Hulbert James*. Are Three-Point Scales Always Good Enough. doi:10.1177/002224377200900416

López-Bonilla, J. M. (2018). The effect of corruption on economic growth in Mexico. International *Journal of Economics, Commerce and Management, 6*(10), 45–55.

Lutz. (1991). *The role of attitude theory in marketing*. Academic Press.

Maniatis, P. (2015). Investigating factors influencing consumer decision-making while choosing green products. *Journal of Cleaner Production, 132*, 215–228. Advance online publication. doi:10.1016/j.jclepro.2015.02.067

Maturén, R., & Pizzolante, F. A. (2020). Entrepreneurial intentions of university students in a developing country: The case of Venezuela. *Journal of Entrepreneurship in Emerging Economies, 12*(1), 1–23. doi:10.1108/JEEE-04-2019-0038

Mittelstaedt, J. D., Moore, T., & West, J. E. (2014). Sustainability and strategic management decision-making: The role of TBL accounting in translating sustainability concepts into business practices. *Journal of Business and Management, 20*(1), 26–45.

Nadanyiova, M., Gajanova, L., & Majerova, J. (2020). Green Marketing as a Part of the Socially Responsible Brand's Communication from the Aspect of Generational Stratification. *Sustainability (Basel), 12*(17), 7118. doi:10.3390/su12177118

Naim, A. (2022). Understanding the customer centric approach to add value to social ecrm (SECRM). American. *Journal of Technology and Applied Sciences, 3*, 17–31.

Nguyen, H., Lobo, A., & Greenland, S. (2017). An empirical analysis of corporate social responsibility (CSR) disclosure in an emerging economy: The case of Vietnam. *Journal of Business Ethics, 144*(2), 399–411. doi:10.1007/s10551-015-2826-1

Nielsen Media Research. (2015). https://www.nielsen.com/us/en/insights/reports/2015/the-sustainability-imperative.html

Nieves, L., & Morales, W. (2022). Una mirada a la brecha entre actitud y comportamiento del consumidor en la economía circular. *Tendencias, 23*(1), 372–394. doi:10.22267/rtend.222301.194

Nyilasy, G., Gangadharbatla, H., & Paladino, A. (2014). *Perceived greenwashing: the interactive*. Academic Press.

Onel, N., & Mukherjee, A. (2016). Corporate social responsibility, consumer evaluations, and purchase intentions in a developing country. *Journal of Business Research, 69*(12), 5721–5729. doi:10.1016/j.jbusres.2016.05.016

Palacios-González, M., & Chamorro, A. (2021). Analysis of the predictive variables of socially responsible consumption. *Business Strategy & Development, 5*(3), 187–196. Advance online publication. doi:10.1002/bsd2.189

Palavecinos, M., Amérigo, M., Ulloa, J. B., & Muñoz, J. (2016). Preocupación y conducta ecológica responsable en estudiantes universitarios: Estudio comparativo entre estudiantes chilenos y españoles. *Psychosocial Intervention, 25*(3), 143–148. doi:10.1016/j.psi.2016.01.001

Paul, J., Modi, A., & Patel, J. (2016). Predicting green product consumption using theory of planned behavior and reasoned action. *Journal of Cleaner Production, 135*, 732–739. doi:10.1016/j.jclepro.2016.06.139

Pereira Heath, T., & Chatzidakis, A. (2012). A dialectical perspective on the social life of sustainable consumption: Exploring the dyadic tensions between individual and political expressions of sustainable consumption. *Journal of Business Ethics, 111*(2), 203–215.

Pisitsankkhakarn, R., & Vassanadumrongdee, S. (2020). Enhancing purchase intention in circular economy: An empirical evidence of remanufactured automotive product in Thailand. *Resources, Conservation and Recycling, 156*, 104702. doi:10.1016/j.resconrec.2020.104702

Pomering, A. (2017). Marketing for sustainability: Extending the conceptualisation of the marketing mix to drive value for individuals and society at large. *Australasian Marketing Journal, 25*(2), 157–165. doi:10.1016/j.ausmj.2017.04.011

Rangel-Lyne, L., Alberto Sánchez-Jaime, R., Cuellar-Hernández, L., & Morales-Ramírez, D. (2023). Socially Responsible Actions for a Sustainable Life Model: The Case of the BBVA Financial Group in Mexico in Responsible Consumption and Sustainability. In *Case Studies from Corporate Social Responsibility*. Social Marketing, and Behavioral Economics. Springer Business Cases.

Reyes-Rodriguez, J. F., & López-Bonilla, J. M. (2018). The relationship between corruption and economic growth: An empirical analysis. *Journal of Economics and Sustainable Development*, *9*(15), 15–22.

Rizos, V., Behrens, A., van der Gaast, W., Hofman, E., Ioannou, A., Kafyeke, T., Flamos, A., Rinaldi, R., Papadelis, S., Hirschnitz-Garbers, M., & Topi, C. (2016). Implementation of Circular Economy Business Models by Small and Medium-Sized Enterprises (SMEs): Barriers and Enablers. *Sustainability (Basel)*, *8*(11), 1212. doi:10.3390/su8111212

Roberts, J. (1996). Green Consumers in the 1990s: Profile and Implications for Advertising. *Journal of Business Research*, *36*(3), 217–231. doi:10.1016/0148-2963(95)00150-6

Rogers, E. M. (1962). *Diffusion of Innovations*. Free Press.

Rogers, E. M. (1995) *Diffusion of Innovations (4th ed.)*. The Free Press.

Rojas-Berrio, S., & Rueda, F. (2020). Sustainable consumption: Associated factors and scale validation to measure appliance purchase intentions. *International Journal of Green Economics*, *14*(3), 191. doi:10.1504/IJGE.2020.111469

Russo, R., Cacciatore, F., Lagazio, C., & Renga, F. (2019). Enhancing green consumption: A dynamic framework based on cognitive dissonance theory. *Journal of Cleaner Production*, *225*, 234–243. doi:10.1016/j.jclepro.2019.03.042

Salas Zapata, W., & Ortiz Muñoz, S. (2019). Analysis of meanings of the concept of sustainability. *Sustainable Development (Bradford)*, *27*(1), 153–161. doi:10.1002/sd.1885

Schifter, D., & Ajzen, I. (1985). Intention, Perceived Control, and Weight Loss: An Application of the Theory of Planned Behavior. *Journal of Personality and Social Psychology*, *49*(3), 843–851. doi:10.1037/0022-3514.49.3.843 PMID:4045706

Sharma, R., & Foropon, C. (2019). Blockchain technology in supply chain management: A review. *International Journal of Production Research*, *57*(7), 2117–2139. doi:10.1080/00207543.2018.1533261

Shrum, L. J., Lowrey, T. M., & McCarty, J. A. (1994). Buyer characteristics of the green consumer and their implications for advertising strategy. *Journal of Advertising*, *23*(2), 21–31. doi:10.1080/00913367.1994.10673464

Sreen, N., Purbey, S., & Sadarangani, P. (2017). Impact of culture, behavior and gender on green purchase intention. *Journal of Retailing and Consumer Services*, *41*, 177–189. Advance online publication. doi:10.1016/j.jretconser.2017.12.002

Stankevičienė, K., Mockevičienė, D., & Giedraitis, V. (2020). The role of digital marketing in the contemporary business environment. *The Engineering Economist*, *31*(1), 103–114. doi:10.5755/j01.ee.31.1.21578

Steffen, W., Broadgate, W., Deutsch, L., Gaffney, O., & Ludwig, C. (2015). The trajectory of the Anthropocene: The Great Acceleration. *The Anthropocene Review*, *2*(1), 81–98. doi:10.1177/2053019614564785

Steg, L., & Vlek, C. (2009). Encouraging pro-environmental behaviour: An integrative review and research agenda. *Journal of Environmental Psychology*, *29*(3), 309–317. doi:10.1016/j.jenvp.2008.10.004

Stern, P. C. (2000). Toward a coherent theory of environmentally significant behavior. *The Journal of Social Issues*, *56*(3), 407–424. doi:10.1111/0022-4537.00175

Suárez-Eiroa, B., Fernández, E., Martínez, G., & Soto-Oñate, D. (2019). Operational principles of Circular Economy for Sustainable Development: Linking theory and practice. *Journal of Cleaner Production*, *214*, 952–961. doi:10.1016/j.jclepro.2018.12.271

Sudbury-Riley, L., & Kohlbacher, F. (2015). Ethically minded consumer behavior: Scale review, development, and validation. *Journal of Business Research*, *69*(8), 2697–2710. Advance online publication. doi:10.1016/j.jbusres.2015.11.005

Sustainability Report. (2021). *YPF*. https://sustentabilidad.ypf.com/index_en.html

Taylor, S., & Todd, P. (1995). Decomposition and Crossover Effects in the Theory of Planned Behavior: A Study of Consumer Adoption Intentions. *International Journal of Research in Marketing*, *12*(2), 137–155. doi:10.1016/0167-8116(94)00019-K

Terlau, W., & Hirsch, S. (2015). Managing complexity in stakeholder networks: A study on the evolution of sustainability partnerships. *Journal of Cleaner Production*, *108*, 1069–1078. doi:10.1016/j.jclepro.2015.07.002

TerraChoice. (2010). *The sins of greenwashing: home and family edition*. Retrieved from http://sinsofgreenwashing.org/findings/the-seven-sins/

Thaler, R. H., & Sunstein, C. R. (2017). *Un pequeño empujón: El impulso que necesitas para tomar mejores decisiones sobre salud, dinero y felicidad*. Taurus.

Tollin, N., Küller, R., Hartig, T., & Sarlöv-Herlin, I. (2015). The restorative effects of sensory experiences in nature. *Environment and Behavior*, *47*(2), 119–147.

Too Good To Go. (n.d.). https://www.toogoodtogo.com/es

Ulaga, W. (2018). The journey towards customer centricity and service growth in B2B: A commentary and research directions. *AMS Review*, *8*(1-2), 80–83. doi:10.1007/s13162-018-0119-x

UNEP FI. (2022). *Acerca de*. Retrieved from https://www.unepfi.org/about/

UNESCO. Director-General, 2009-2017 (Bokova, I.G.) (2014). *Estrategia de educación de la Unesco, 2014 - 2021*. https://unesdoc.unesco.org/ark:/48223/pf0000231288_spa

Varey, R. J. (2010). Sustainable marketing: Myth or reality? *Journal of Business Research*, *63*(4), 111–126.

Villa Castaño, L. E., Perdomo-Ortiz, J., Durán León, W. F., Dueñas Ocampo, S., & Arredondo Trapero, F. G. (2018). Measuring socially responsible consumption: A study of Colombia–Mexico. *Academia (Caracas)*, *31*(3), 553–568. Advance online publication. doi:10.1108/ARLA-12-2016-0356

Webb, D., Mohr, L., & Harris, K. (2008). A re-examination of socially responsible consumption and its measurement. *Journal of Business Research*, *61*(2), 91–98. doi:10.1016/j.jbusres.2007.05.007

Webster, F. E. Jr. (1975). Determining the Characteristics of the Socially Conscious Consumer. *The Journal of Consumer Research*, *2*(3), 188–196. doi:10.1086/208631

Wolsko, C., Marino, E., Doherty, T. J., Fisher, S., Green, A. S., Goodwin, B., Reese, R., & Wirth, A. (2016). Systems of access: A multidisciplinary strategy for assessing the social dimensions of sustainability. Sustainability: Science. *Sustainability*, *12*(1), 88–100. doi:10.1080/15487733.2016.11908156

WTWCo. (2021). *How employee insights power Danone's sustainable growth agenda*. Retrieved from https://www.wtwco.com/en-CM/Insights/2021/09/how-employee-insights-power-danones-sustainable-growth-agenda

Wu, W.-Y., & Pham, T. (2020). Examining the Moderating Effects of Green Marketing and Green Psychological Benefits on Customers' Green Attitude, Value and Purchase Intention. *Sustainability (Basel)*, *12*(18), 7461. doi:10.3390/su12187461

Xiong, Z., & Li, H. (2019). Ecological deficit tax: A tax design and simulation of compensation for ecosystem service value based on ecological footprint in China. *Journal of Cleaner Production*, *230*, 1128–1137. doi:10.1016/j.jclepro.2019.05.172

Yan, J., & She, Q. (2011). Developing a trichotomy model to measure socially responsible behavior in China. *International Journal of Market Research*, *53*(2), 253–274. doi:10.2501/IJMR-53-2-253-274

Zhexembayeva, N. (2014). *Overfished Ocean Strategy: Powering Up Innovation for a Resource-Deprived World*. Berrett-Koehler Publishers.

Chapter 4
Academic Design Thinking Pedagogy as an Approach to Foster Sustainability Skills in Higher Education

Minna-Maarit Jaskari
https://orcid.org/0000-0002-5412-7176
University of Vaasa, Finland

ABSTRACT

The United Nations' Sustainable Development Goals (SDGs) provide a collaborative framework for sectors worldwide to pursue a just, prosperous, and environmentally sustainable future. This chapter investigates the application of design thinking as a pedagogical strategy to equip students with the skills necessary to meet sustainability challenges. Rooted in a human-centered approach, design thinking seeks to intimately comprehend complex challenges to create solutions that bolster the welfare of both individuals and communities, thus enabling students to act as agents of societal change. Introducing academic design thinking pedagogy (ADTP), this chapter is tailored to the context of higher education and articulates its phases to provide educators with actionable insights on its implementation.

"The mission of design thinking is to translate observations into insights, and insights into solutions that will improve lives." -Tim Brown

In today's world, humanity grapples with numerous sustainability challenges, each with far-reaching and global implications. Climate change stands as a salient example of such challenges. Recognizing the urgency, the United Nations (UN, 2023) published a set of 17 Sustainable Development Goals (SDGs) in 2015 to foster global human well-being. These SDGs encompass a wide range of interconnected issues that are crucial for achieving sustainable development worldwide. By outlining these goals, the United

DOI: 10.4018/978-1-6684-9859-0.ch004

Nations provides a framework, urging governments, organizations, and individuals to collaboratively stride towards a just, prosperous, and sustainable future.

Given the scale and complexity of sustainability challenges, a collective effort encompassing individuals, communities, corporations, and governments is crucial. Higher business education stands at the forefront of this initiative. Leading business schools globally are fostering the development of leaders poised to effect transformative changes across societal sectors (AACSB, 2023). Subsequently, the integration of sustainability into business education curricula has intensified, echoing its escalating importance. As educators, we are tasked with weaving sustainability into our pedagogical practices to ensure graduates are primed to address these challenges and emerge as future change agents.

The subsequent query for educators then becomes: what are the precise 21st-century competencies that these graduates must possess to navigate the uncertainties of the future? The Center for Curriculum Redesign (CCR) provides a detailed blueprint, outlining a multi-dimensional framework (CCR 2015; 2019). This emphasizes the synergies between knowledge, which spans traditional disciplines and modern interdisciplinary fields; skills, particularly the 4C's: creativity, critical thinking, communication, and collaboration; character, emphasizing ethical behaviors and qualities such as mindfulness and resilience; and meta-learning, centered on self-awareness and adaptability. Cumulatively, this framework equips graduates with a diverse skill set, positioning them to confront the intricate sustainability challenges of today.

With this context in mind, the purpose of chapter is to delve into the transformative potential of design thinking as a means of addressing these sustainability issues. Rooted in a human-centric philosophy, design thinking accentuates the significance of individual experiences and emotions throughout the problem-solving process (Brown, 2008). By championing this approach, we cultivate a profound understanding of human needs, thereby ushering in solutions that are both ground-breaking and contextually relevant. Moreover, immersion in design thinking endows learners with pivotal abilities that supersede conventional problem-solving paradigms, fostering a confluence of creativity, collaboration, and proactive innovation, which are essential for confronting complex sustainability challenges (CMI, 2021; Enell-Nilsson et al., 2023; OECD, 2019).

Design thinking has been recognized as a pedagogical approach, empowering students to become proactive agents of change. They are not only equipped with knowledge but also poised to instigate transformative actions across their organizations, communities, and even larger societal structures (Razzouk & Shute, 2012; von Thienen et al., 2017). Existing literature comprehensively identifies the myriad benefits of embedding design thinking in education, such as the cultivation of empathy, the promotion of critical thinking, immersion in flow experiences, and fostering a profound sense of achievement and purpose (Jaskari, 2013b; von Thienen et al., 2017). However, like any approach, its implementation is not without challenges. Notably, while design thinking is widely celebrated for its unparalleled ability in fostering ideation and creativity, its occasionally abstract orientation does attract criticism. It's essential to understand that the mere generation of ideas, in isolation, is not adequate. For genuine impact, these ideas demand rigorous feasibility evaluation, and they must be anchored in robust, pragmatic business rationales before any business initiatives can genuinely materialize and succeed (Panke & Harth, 2018).

In light of this discourse, this chapter introduces the Academic Design Thinking Pedagogy (ADPT) model. Developed over 15 years of consistent action research, the ADPT model is presented in the form of a case study. It aims to bolster human-centered innovation while simultaneously maintaining a critical lens. This model not only draws from the rich history of design thinking but also refines it to better suit the academic context. It posits design thinking as an invaluable pedagogical instrument, priming students with the tools and mindset to effectively navigate sustainability challenges, both in the present context and as they emerge in the future. However, while it harnesses the strengths of design thinking, the chapter also forthrightly acknowledges its inherent limitations.

The chapter is structured as follows: It begins with an introduction to design thinking, exploring its origins, foundational principles, and established frameworks. Subsequently, the discussion shifts to the merits and drawbacks of employing design thinking in educational settings, placing particular emphasis on the pivotal dual role of educators in guiding students through the design process. The chapter then presents the Academic Design Thinking Pedagogy (ADPT) model, accompanied by a case study underscoring its application in higher education. The chapter culminates with recommendations specifically crafted for educators in higher education.

DESIGN THINKING IN SOLVING SUSTAINABILITY CHALLENGES

Design thinking is an interdisciplinary approach that employs the designer's sensibility and methods to effectively address the alignment of people's needs with technological feasibility and viable business strategies (Brown, 2009). It points out the importance of empathy, creativity, and iterative problem-solving in order to develop innovative solutions that meet the needs of both users and businesses.

Design thinking has been around for a long time, and it has been extensively used to tackle a variety of both mundane and complex problems, not least social innovations and sustainability (Brown & Wyatt, 2010; Earle & Leya-de la Hiz, 2021). By integrating the perspectives of various stakeholders, design thinking seeks to bridge the gap between desirability, feasibility, and viability, ultimately driving the creation of successful products, services, and experiences (e.g., Brown, 2009). Over the years, design thinking has moved from product design to process and service design, to vital part of strategy and business design.

Design thinking has gained significant popularity in the business literature primarily through case studies, empirical illustrations, and narratives showcasing the work undertaken by design organizations such as IDEO (Matthews & Wrigley, 2017). It can be claimed that design thinking has undergone commercialization and has been conceptualized into various frameworks, methods, canvases, and design tools. However, the principles behind the process are the key to understanding and appreciating design thinking.

Principles Behind Design Thinking

The design thinking process is characterized by several key principles. These principles serve as guiding tenets that shape the approach and mindset of designers when tackling complex problems and creating innovative solutions.

Firstly, human-centrism and empathy is a fundamental principle of design thinking and emphasized whenever design thinking is discussed (Brown & Katz, 2011). Designers strive to deeply understand the needs, desires, and challenges of the people they are designing for. Brown (2008) describes the empathetic characteristic of design thinkers – having an ability to imagine world through the eyes of different stakeholders such as consumers, clients, end-users, citizens and so on. He further elaborates how the "people first" approach is used to envision solutions that are both desirable and meet people's explicit or implicit needs. In this way design thinking prioritizes the well-being of the individuals. Moreover, Bertão et al. (2023) note that due to the human-centered approach, design thinking enables companies to enhance the capabilities of their employees as agents of innovation, thereby facilitating the delivery of customer-centric solutions to their businesses.

The second principle behind design thinking is its iterative, non-linear nature. According to Brown (2008), the design process is best characterized metaphorically as "a system of spaces", rather than a rigid sequence of predetermined and linear steps. This metaphorical framework acknowledges that design is a complex and iterative process, where designers move between different spaces of ideation, research, prototyping, testing, and refinement, rather than adhering to a fixed and linear progression. Brown (2008) describes how good designers do not only rely on analytical processes but they also can see all of the salient - and sometimes contradictory - aspects of complex or even wicked problems and create novel solutions that go beyond and significantly improve on existing alternatives. Moreover, successful designers are quick to experiment through which they can test their solutions and develop continuously (Micheli et al., 2019) The process of design thinking may appear disorderly and even overwhelming to individuals who are encountering it for the first time. However, throughout the duration of a project, individuals involved gradually develop an understanding and appreciation for the process, despite its deviation from the conventional linear and milestone-oriented approaches commonly observed in other business activities (Brown, 2008).

Third, despite its iterative and nonlinear nature, the design thinking process has some sequential aspects. For example, discovering the problem to be worth solving and understanding the users are prior to coming up with solutions. The different design thinking frameworks emphasize this aspect by framing different phases of the process and emphasizing the need for problem recognition and user understanding. However, it can be challenging for analytics-oriented business managers and business students to spend enough time in the first phases of the process, to "love the problem" and not jump into conclusions too quickly (Glen et al., 2014; Jaskari, 2013b; Titus, 2007).

Fourth, design thinking processes typically involve collaboration where people from different units or organizations are brought together in order to gain more perspective and viewpoints to advancing innovation (Micheli et al., 2019) According to Glen et al. (2014, p. 660) cross-functional and multidisciplinary teams can "address project complexity, ensuring that technical, business, and human dimensions of a problem are all represented".

Fifth, design thinking processes are at their best holistic in nature, referring to systemic approach and embracing complexity (e.g. Micheli et al. 2019). The holistic principle has grown in importance due to the need for sustainability. This makes designers and other innovators have a wider look of the consequences or their actions (Dunne & Martin, 2006).

Models of Design Thinking Process

Various models of the design thinking process have been developed and documented by scholars and practitioners (Liedtka, 2021; Micheli et al., 2019). Among these models, the IDEO model, the British Design Council model, and the Stanford Design School *d.school* model have emerged as particularly influential. All of these various models adhere to the design thinking principles outlined earlier, but they have implemented the process in distinct phases. Subsequently, this section delves into the various stages and unique characteristics of these models. The summary of the models is in Table 1.

IDEO Model

IDEO is a global design consultancy that is widely recognized for its pioneering efforts in effectively applying human-centered design principles beyond the traditional scope of design. According to Brown (2008, 2009, 2011), design projects must ultimately go through three spaces that are labeled: "inspiration", "ideation", and "implementation". The process is highly iterative, as there are revisits to these spaces, especially the first two, multiple times as ideas are further developed and new paths are explored.

Inspiration: The focus of this space is on comprehending the problem at hand. Through the process of observation and active interaction, designers acquire valuable insights about the needs and wants, desires, requirements, actions, and affective states of users. The objective is to cultivate empathy towards users in order to identify underlying needs.

Ideation: Using user insights from the inspiration stage as a springboard, designers come up with a wide range of creative solutions. This space is dedicated to the exploration of innovative ideas, unconventional thinking, and the questioning of established beliefs. During this phase, a variety of ideas are generated and subsequently refined.

Implementation: This is the space at which the concepts developed during ideation are put into action. Prototypes undergo development and testing, and subsequent to receiving feedback, they are subject to either iterative improvements or complete reworking. The objective is to reach a resolution that aligns with the preferences of users and effectively caters to their requirements.

Brown (2009) suggests that in order for an idea to be successful, it should meet three essential criteria: desirability, feasibility, and viability. Desirability refers to the level of demand or interest that individuals may have for the proposed concept. Feasibility examines the technical and practical aspects of transforming the idea into a tangible product or service. Viability encompasses the long-term sustainability and profitability of the idea as a business venture. Furthermore, a successful concept achieves a harmonious equilibrium among the three aforementioned criteria. The convergence of desirability, feasibility, and viability signifies the optimal point where innovative solutions, which place emphasis on user requirements and long-term sustainability, are frequently discovered (Brown, 2008, 2009).

The Double Diamond Design Process

The British Design Council first presented the Double Diamond design process model in 2005 as a visual representation of the design process. The framework aims at helping designers and non-designers to tackle complex problems such as sustainability in a large sense: social, economic and environmental challenges. This model is known as the "design process model." It divides the design process into four distinct phases, each of which is represented by a separate diamond: the first diamond in called "Diverge" and it covers "Discover and Define," while the second diamond is called "Converge" and it covers "Develop and Deliver" (Design Council, 2023).

Discover: involves identifying the issue by observing and thinking about a wide variety of issues, looking at the world from a number of different points of view, and gathering a wide variety of insights. The key is to clear your mind and look at things from a broader perspective.

Define: involves narrowing in on a specific issue that needs to be resolved and concentrating on that issue. The process of defining a particular design challenge by synthesizing all of the information that has been gathered is known as design synthesis.

Develop: Once the nature of the issue has been elucidated, it is time to investigate the various options that are available. During this phase, you will be coming up with ideas and iterating on various design concepts as well as prototyping and testing.

Deliver: This is the final phase, and it's the time to finalize and produce the best solutions that were identified in the develop phase. It includes refining and testing the solution, in addition to getting it ready for the market or for implementation.

The visual representation of these phases as two diamonds brings attention to the concept that design begins by broadening the thinking (diverging) before focusing in on specific problems and solutions (converging) (de Bono, 1970). The design thinking process is characterized by this pattern of branching off in different directions and then coming back together again.

D.School: The Stanford Design School Model

The "*d.school model*" was developed at the Hasso Plattner Institute of Design at Stanford University. This model breaks down the design thinking process into five stages, often presented in a non-linear fashion to underscore the iterative nature of the process: "empathize," "define," "ideate," "prototype," and "test" (d.school, 2023).

Empathize: This foundational stage involves designers immersing themselves in the problem space. They engage directly with users to grasp their emotional experiences, needs, motivations, and challenges. The main goal is to understand the users. This may include for example collecting empirical data in a form interviews and observations.

Define: After gathering rich insights during the empathize stage, this phase is about synthesizing that data. The objective is to pinpoint and articulate a clear, actionable problem statement viewed from the user's perspective.

Ideate: With a clear problem defined, the ideate stage centers on brainstorming a multitude of potential solutions. It encourages unconventional thinking and values the diversity and quantity of ideas. This phase may include the use of several different kinds of divergent ideation methods as well as more convergent evaluation methods (de Bono, 1970).

Prototype: In this phase, the most promising ideas from the ideate stage are transformed into tangible, experimental representations. These might be rudimentary or quickly sketched versions, but they serve the purpose of illustrating potential solutions and conceptualizing ideas. In the marketing classroom visual mock-ups of different marketing tools are typically used, such as event flyers, mobile application mock-ups and posters.

Test: Here, prototypes undergo testing with real users. Based on feedback, solutions undergo refinement. Insights might also circle back to previous stages, sparking further iterations.

In addition to the design thinking principles, the aforementioned design thinking models share common characteristics. Cochrane and Munn (2016) assert that all models place significant importance on user comprehension, achieved through the utilization of observational research, visual sense-making, and rapid prototyping. Furthermore, although various models may pertain to distinct phases, all models consist of three key stages in some form, namely: data collection pertaining to user needs, generation of ideas, and testing (Liedtka, 2021).

Table 1. The classical design thinking models

Name of the Model	Developer	Phases
Ideo design thinking process	IDEO	Inspiration Ideation Implementation
Double Diamond Design Process	British Design Council	Discover Define Develop Deliver
D-School model	Standford Design School	Empathy Define Ideate Prototype Test

DESIGN THINKING AS A PEDAGOGICAL APPROACH IN HIGHER EDUCATION

Design thinking has garnered consistent recognition for its valuable contributions to the field of business and management based on its ability to generate ideas and innovations (Brown, 2008). Consequently, there has been a rise in the proliferation of higher education initiatives that offer instructional courses on design thinking specifically tailored for business students, managers, and executives (Glen, 2014; Matthews & Wrigley, 2017). Indeed, the ability to generate novel concepts and adapt to environmental demands is closely associated with the endeavors of future marketing professionals (Greenacre et al., 2017; Jaskari, 2013b; Jaskari, 2015). Hence, it is of utmost significance to acquire the necessary competencies associated with innovation during the educational process.

Design thinking is established as a pedagogical approach (Razzouk & Shute, 2012; Wrigley & Straker, 2017) and it has been implemented in all levels of education (Panke, 2019) and in wide variety of academic subjects such as geography (Carroll et al., 2010), business and management (Matthews & Wrigley, 2017), marketing (Jaskari, 2013b), and engineering (Enell-Nilsson, et al. 2019) and in several informal learning settings (Panke, 2019). To illustrate, Guaman-Quintanilla et al. (2023) used design thinking methodology in higher education in order improve students problem-solving and creativitys skills in real-life problem-solving.

Earlier research has pointed out several benefits of design thinking practices in education (Panke, 2019) that are highlighted in Table 2. Jacobs (2016) and Groth (2017) pinpointed that these practices encourage tacit experiences by promoting the construction of physical analogues or prototypes, effectively enhancing communication and interaction. Furthermore, the inherent human-centric approach of design thinking not only augments empathy among participants but also integrates user understanding into the innovation process (Panke, 2019). Liedtka (2015) argued that design thinking practices can mitigate cognitive biases, from projection bias to the empathy gap, providing a more balanced evaluation of ideas. A unique feature of design thinking, emphasized by Ferguson et al. (2019) and Panke (2019), is its ability to instill a sense of playfulness in learning environments, subsequently fostering innovation. This is complemented by findings from Primus and Sonnenburg (2018), von Thienen et al. (2017), and Jaskari (2013b), who collectively highlighted the association of design thinking with flow experiences, both at individual and group levels. Collaboration, being central to design thinking, is underscored as a potent tool to amplify multidisciplinary teamwork and interdisciplinary cooperation (Enell-Nilsson

Table 2. Benefits of design thinking practices in education

Benefits of Design Thinking Practices in Education	Illustration From Earlier Research
Enhancement of Tacit Experiences	Design thinking fosters tacit experiences through the construction of physical analogues or prototypes, enhancing communication and interaction (Jacobs, 2016; Groth, 2017).
Augmentation of Empathy	The human-centric focus of design thinking amplifies empathy among participants, and integrates user understanding into the innovation process (Panke, 2019).
Reduction of Cognitive Biases	Design thinking practices can minimize cognitive biases like projection bias and empathy gap (Liedtka, 2015).
Promotion of Playfulness	Design thinking encourages a playful learning environment, thereby fostering innovation (Ferguson et al., 2019; Panke, 2019).
Fostering Flow Experiences	Design thinking activities are associated with enhancing flow experiences at both individual and group levels (Primus and Sonnenburg, 2018; von Thienen et al., 2017; Jaskari, 2013b).
Encouragement of Multidisciplinary Collaboration	The collaborative nature of design thinking nurtures multidisciplinary teamwork and interdisciplinary cooperation (Enell-Nilsson et al., 2019).
Appreciation of Failures as Opportunities	Design thinking allows for the appreciation of failures or constraints as prospects for future development, enhancing students' resilience (Micheli et al., 2019).
Induction of Positive Emotional Responses	Design thinking can induce emotions such as surprise, delight, pride, and happiness (Elsbach and Stigliani, 2018; Jaskari, 2013b).
Growth in Creative Confidence	It offers a platform for growing creative confidence, allowing for imagination beyond traditional boundaries and constraints (Carroll et al., 2010).

et al., 2019). A paradigm shift noted by Micheli et al. (2019) is the refashioning of failures and constraints into potential opportunities within the design thinking framework, thereby bolstering students' resilience. Adding to the richness of emotional responses, studies by Elsbach and Stigliani (2018) and Jaskari (2013b) affirm the ability of design thinking to evoke feelings of surprise, delight, and pride. Lastly, Carroll et al. (2010) contended that the practices inherent to design thinking serve as catalysts, augmenting creative confidence and unbridling the imagination from conventional constraints.

However, also several limitations, drawbacks or potential negative outcomes of design thinking in educational settings are identified (Panke, 2019). These are showcased in Table 3. Firstly, the design thinking process can sometimes result in superficial solutions. This could be due to students' lack of creative confidence or because the design thinking approach has inadequately bolstered their belief in their creative abilities (Ohly et al., 2017; Valentim et al., 2017). On the flip side, certain design thinking workshops might inflate creative confidence, causing students to overly rely on workshop outcomes, especially if they lack the necessary user research skills (Taheri et al., 2016).

Moreover, the inherent nature of the design thinking process can sometimes be chaotic, leading students to feel overwhelmed, anxious, frustrated and bewildered (Glen et al., 2015; Jaskari, 2013b). These feelings are not unique to educational contexts; Brown (2008) documented similar sentiments among workshop participants. However, in a business context, participants usually have a vested interest in resolving the problem. In academic settings, there can be a challenge to ignite the same passion in students, especially if some are merely aiming to pass the course rather than deeply engage with the design process.

Another pitfall is the potential for team conflicts, especially in multidisciplinary and culturally diverse teams (Panke, 2019). While such diversity is advocated to spur innovation, it can also introduce

Table 3. Potential limitations and drawbacks of design thinking practices in educational setting

Drawbacks of Design Thinking Practices in Education	Illustration From Earlier Research
Superficial Solutions	The process can sometimes result in mundane or shallow solutions due to students' lack of creative confidence. (Ohly et al., 2017; Valentim et al., 2017)
Over-reliance on Workshop Outcomes	There's a potential for inflated creative confidence, leading students to rely too heavily on workshop outcomes, especially if they lack user research skills. (Taheri et al., 2016)
Chaos and Overwhelm	The design thinking process can be chaotic, causing feelings of frustration and confusion among students. (Glen et al., 2015; Jaskari, 2013b; Brown, 2008)
Motivational Issues in Educational Settings	In academic contexts, there might be challenges motivating students, especially if they're primarily aiming to pass a course. (Brown, 2008)
Team Conflicts	Especially in multidisciplinary and culturally diverse teams, potential conflicts can arise from cultural differences, varying disciplinary knowledge, and disparities in teamwork skills or motivations. (Henje & Jaskari, 2013; Enell-Nilsson et al., 2019)
Time Constraints	Limited durations, like in a bootcamp, might not provide students ample time to conduct comprehensive research, which is vital for idea validation. (Panke & Harth, 2018)
Alignment Challenges	There can be difficulties aligning classroom projects with academic standards, curricular content, and design thinking principles. (Carroll et al., 2019; Panke, 2019)
Varied Perceptions of Tools and Methods	Some participants may perceive the tools and methods used in design thinking as juvenile or trivial, potentially leading to a lack of commitment to the process.
Misconceptions	The seeming simplicity of the design thinking process might cause misconceptions, with some viewing the tools as mere play and not serious approaches to business problems.

challenges stemming from cultural variances, differences in disciplinary knowledge, and discrepancies in teamwork or motivational skills (Henje & Jaskari, 2013; Enell-Nilsson et al., 2019).

In certain scenarios, the limited duration of the process, such as during a design thinking bootcamp, may not provide students with ample time for comprehensive background research, which is crucial for validating their ideas (Panke & Harth, 2018). Broadly speaking, design thinking in academia is more beneficial for user understanding, option exploration, and priority setting rather than converting preliminary ideas into approved solutions (Grots & Creuznacher, 2016). Moreover, aligning classroom projects with academic standards, curricular content, and design thinking principles can sometimes be a daunting task (Carroll et al., 2019; Panke, 2019). However, it is utmost important for learning.

Lastly, participants may have varied perceptions of the tools and methods used in the design thinking process. While some might find them engaging and invigorating, others might view them as juvenile or trivial. The simplicity of the design thinking process may give an illusion of ease, potentially leading some students to perceive creativity-enhancing tools as mere play, thus sidelining 'real business issues.' Such misconceptions might result in a lack of commitment to the process, or it might inhibit students from fully embracing their creative identities.

In summary, design thinking as a pedagogical approach offers numerous advantages, but it can also present challenges and feel daunting or intimidating to some students. The uncertainty inherent in creating something new can deter students, with the fear of failure sometimes preventing them from fully committing to a project. Furthermore, students across different disciplines often have varying comfort levels with ambiguity and risk-taking. For instance, while art and design students are generally more open to diving into creative processes, business students tend to be more cautious, reserved, and goal-driven. Consequently, educators have a pivotal role in setting the context of the challenge, fostering a supportive learning atmosphere, and equipping students with tools for creative and critical problem-solving (Jaskari, 2012; Henje & Jaskari, 2013; Enell-Nilsson et al., 2019). Next the role of educators is discussed in more detail.

THE ROLE OF EDUCATORS IN DESIGN THINKING COURSES

The design thinking process is typically a facilitated process, even more so in educational settings. The facilitator plays a crucial role in guiding learners to perceive constraints as sources of inspiration (Brown & Wyatt, 2010). By reframing constraints as opportunities rather than limitations, learners are encouraged to think creatively and explore innovative solutions. This approach fosters a mindset that embraces challenges and leverages them to drive the design process forward. Ultimately, this mindset shift enables learners to unlock their creative potential and develop novel solutions that address complex problems effectively.

The role of the facilitator is pivotal to ensure that the outcomes of the design processes extend beyond mere "quick-fixes" (Buchanan, 1992) or hasty conclusions, often termed "quick-closures" (Biggs & Collins, 1982). The focus should be on deriving solutions anchored in human-centered empathy and iterative methodologies. Ideally, educators facilitate integrative thought processes encompassing empathy, idea generation, and concept creation. This involves considering multiple options simultaneously, refraining from premature conclusions, and iterating on the concept until it best aligns with the overarching goal. Occasionally, students manage to hold two opposing ideas within their cognitive framework. Rather than settling for either option A or B, they strive for a creative third option that melds the best elements from both options. This integrated approach, not only incorporates the positive aspects of both options but also enhances them, resulting in a more effective and superior solution (Martin, 2007).

Jaskari (2013b) posits that especially in business education context, educators play a dual role in design thinking processes, requiring them to strike a balance between two support mechanisms that are somewhat contradictory, yet complementary. She labels these mechanisms as the "facilitation process" and the "coaching process," each reflecting a distinct aspect of the educator's role. These two processes resemble lateral and vertical thinking as discussed by de Bono (1970).

The facilitation process encourages the students to use their lateral thinking skills in a generative and iterative learning process that involves making side steps, taking alternative approaches, and considering perspectives that may not have been initially anticipated or even imagined. It promotes a safe creative environment, fosters cognitive flexibility and support working in uncertainty. The process helps the students to think out of the box, to let go and to bear uncertainty (Titus 2000, 2007). The task at hand may be difficult for the students and needs to be facilitated by means such as creating a trusting atmosphere, bringing in creative tools and pushing the students to think from different angles (Jaskari, 2013b). Indeed, one of the primary challenges encountered by educators is to motivate students to engage in complex and demanding challenges that require substantial investments of time, effort, and energy (Titus, 2007).

The coaching process encourages students to employ their vertical thinking skills. It refers to a problem-solving process with a business focus, aimed at assisting students in achieving their objectives and generating market-valued and profitable solutions (Jaskari, 2013b). This process places emphasis on various managerial factors such as the orientation towards achieving goals, the market feasibility of the solutions, adherence to timetables, and the documentation of completed tasks – all of which are important skills for business graduates (Jaskari, 2013b). These elements are also approached in a sequential manner, following a predefined logic (De Bono 1970, 8).

ACADEMIC DESIGN THINKING PEDAGOGY: A CASE STUDY

The following case study is anchored in the tradition of action research. Action research is characterized by its close collaboration with the subject of study and emphasizes practical problem-solving with an intent to bring about positive changes and enhancements in practice (Eriksson & Kovalainen, 2008: p. 193). It is inherently iterative, marking an ongoing, generative, and transformative process (McNiff & Whitehead, 2006). While this type of research demands profound collaboration, bordering on an ethnographic method, it consistently upholds academic principles that foster a continuous dialogue between theory and practice.

In line with this, the insights presented in this case are shaped by comprehensive review of theories in design thinking, creative problem-solving, innovation, and experiential pedagogies and my substantial professional journey of over 15 years in teaching marketing, creative problem-solving and design thinking in a business school context. Through this period, I have both experienced the transformative potential of design thinking and navigated the complexities of mastering the method.

Tracing the Model's Development Journey

The history and evolution of academic design thinking process case traces back fifteen years. In its nascent stages, the emphasis was predominantly on elucidating the 'fuzzy front end' of innovations. The foundational teaching framework was conceived by integrating business development processes with creative processes (Jaskari, 2013b, p.80). Also, from the pedagogical perspective, an integrative approach was adopted, incorporating tenets of experiential learning, client-centric teaching, innovation, multidisciplinary teamwork, and creative problem-solving. Collaborations were fostered with design schools, harnessing design thinking tools to navigate client-specific business dilemmas (Henje & Jaskari, 2013).

Over time, this method underwent continuous refinement and learning design adaptations. It was tested and refined gradually during several client-based project courses. Also, some of its overreaching principles were used in other courses, while the design thinking was not in their core (Enell-Nilsson et al., 2019; Enell-Nilsson et al., 2023). At the moment the course is titled "Design Thinking in Marketing-Project". It caters to Master's level students and is annually conducted at the University of Vaasa, Finland. Moreover, there has been a paradigm shift from a rigid client-focused approach to a more flexible framework, which now encompasses collaborative initiatives with university and community development projects.

The Academic Design Thinking Pedagogy (ADTP)

The Academic Design Thinking Pedagogy (ADTP) is design thinking model that aims to combine the creative, human centred approach from classical design thinking theories to business thinking. The model is described using a design thinking course in higher marketing education as a case study. The aim of this process is to equip students with design thinking skills so that they can not only come up with innovative solutions for their coursework but can later act as change agents in their own organizations.

The approach encompasses two different levels of student learning outcomes. Initially students are tackling a sustainability challenge and learning the skills for creative problem-solving and decision-making. Concurrently, on a meta-level, they familiarize themselves with the design thinking process. The combination of these two learning processes yields a comprehensive and profound comprehension of the design thinking process and its application in problem-solving.

The Academic Design Thinking Pedagogy (ADTP) different from other design thinking applications as it combines theoretical academic literature with practical design thinking tools in the chosen context, as suggested by Carroll et al. (2020). Design thinking is not just a step-by-step process; it is rather a mindset that encourages students to adopt a designer's perspective and pushes them to venture beyond their comfort zone throughout the process.

The case study course "Design Thinking for Marketing" is a 5 ECTS[1] course for marketing students. The course runs for 10 weeks, and it includes 6 face-to-face workshops as well as extensive teamwork outside lecture times. The extensive time span allows students to do background research and validate their ideas in real life, which have been a critique of short-term design thinking spurts or boot camps (Panke & Harth, 2018). There are typically 20-25 students in the course and the students work in teams of 4-5 people. The evaluation of the course is based on project work and learning diary (Jaskari, 2013b).

The initial rendition of the model was crafted by Jaskari (2013a), with a particular focus on experiential, client-centered teaching. Subsequent iterations were deeply influenced by the Stanford Design School's five-phase framework and the Double Diamond model, which emphasizes both lateral and vertical thinking processes. To ensure alignment with the academic requirements of higher education curricula, ADPT model has been expanded to encompass eight distinct phases: kick-off, challenge, academic understanding, empathy, define, ideation, concept development, testing, and telling. The

Figure 1. Academic design thinking pedagogy

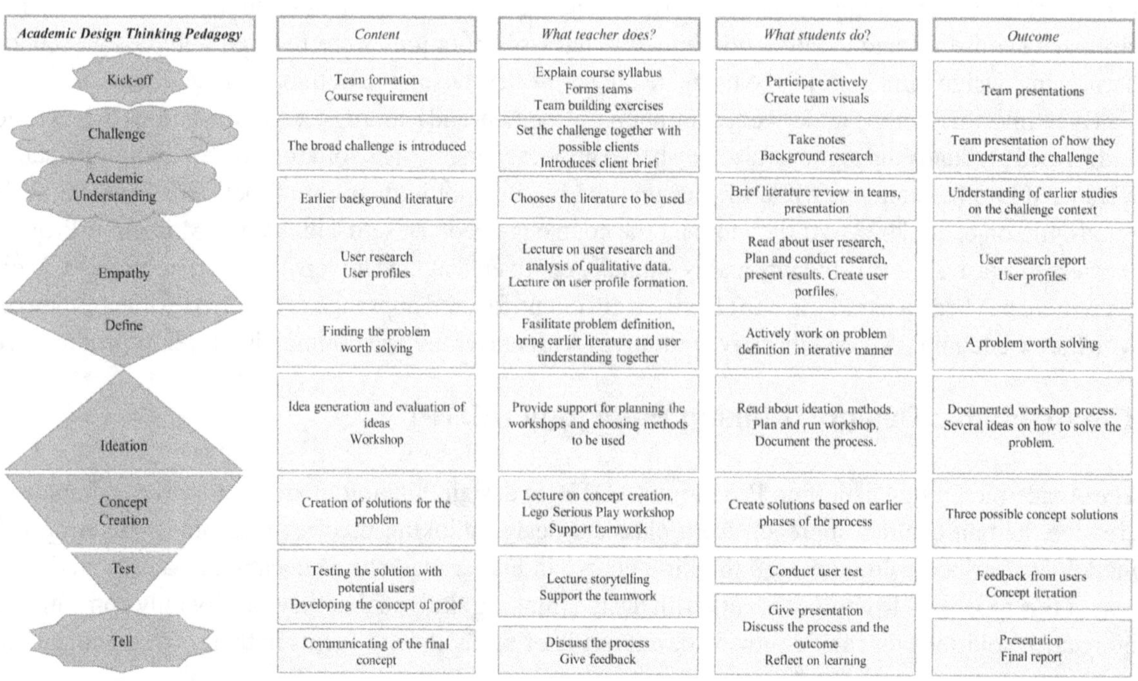

model is presented in Figure 1 explicating the phase of the design thinking process, the content, actions of teachers and students as well as the outcome of each phase. Next, each phase is presented in detail.

Kick-Off

The kick-off session lays the groundwork for the upcoming creative process. This phase is used in our framework to inspire students and set their expectations for the unexpected. At this point, music or drama can be used to set the expectations to the unexpected.

During the first face-to-face session, teams are formed, and students are given the task of imagining themselves as a marketing agency that develops concepts such as new product or service concepts, developing customer value in concepts, or developing branding concepts, for example. This fictitious marketing agency task is used to foster students' sense of community as well as their self-efficacy in marketing. At its best, this phase shifts their mindset from student to marketing professional. They prepare a PowerPoint presentation of their team as a concrete output - they are asked to name their company and create a quick visual brand identity for their agency.

Challenge

Students are then introduced with the designated design challenge. The complexity of the challenges can vary having a direct impact on how easy or difficult it is to grasp the challenge (Jaskari, 2013b). To gain a thorough understanding of their challenge, students might need to undertake an initial context analysis. For instance, in a particular year, a student teams were working with identifying potential applications and target markets for the biocarbon process. Given that these students had marketing studies as a background, they initially had to source relevant information to fully grasp the project's scope. This phase is often a mix of inspiration and frustration for students. While they are typically enthusiastic about venturing into novel territories, the unfamiliarity of such projects can also evoke apprehension.

Over time, I've noticed that when dealing with a complex challenge, it's often best to assign the same challenge to the entire class. This approach allows each team to develop its own unique solution. By the end, students see first-hand how a single challenge can yield a variety of solutions, highlighting the value of diverse perspectives and approaches. Also, this is very much what happens in the real life.

The student group is asked to present one PowerPoint slide (using their marketing agency brand visuals) on how they understand the challenge at this stage and how they plan to use the design thinking process as a guide.

Academic Understanding

In the realm of design thinking, academic knowledge provides a robust foundation that encompasses an understanding of the existing body of work on a given topic. However, it's crucial to recognize that this theoretical framework is not confining; rather, it serves as a tool that empowers students to cultivate a critical perspective. For instance, when students are tasked with enhancing retail brand experiences, established theories can offer valuable insights into the mechanics of brand experience formation and the role of multisensory marketing within servicescapes.

Beyond this, academic theory also imparts methodological clarity, guiding students on best practices for conducting rigorous user research. Thus, academic knowledge not only equips students with a rich backdrop against which they can juxtapose their innovative ideas but also fosters an environment where informed critical thinking thrives.

Empathy

Empathy is all about understanding potential users, customers, or other stakeholders. The potential user groups are first identified. Several user groups can usually be identified, and the students must decide which one to target. Second, the student groups "put themselves in other people's shoes" in order to gain a thorough understanding of the users' challenges, wishes, needs, and aspirations. This stage is typically very experiential, as students interact with actual users. They familiarize themselves with academic qualitative research examples and other field study materials to give them ideas on how to conduct the research. The actual research can vary greatly depending on the topic, selected user groups, and students' motivation and effort.

During this phase, students engage in a teacher-led workshop on "analyzing user data." Then the insights from the gathered data are shared with the entire class. Over time, this sharing process has emerged as a pivotal element of the course. Given that every team tackles the same challenge, they each gather empirical data and then share their findings with the class. This collaborative approach allows the class to amass a substantial amount of data, far more than any single group could gather on its own. This collective knowledge ensures that all student teams are well-informed with customer insights, which in turn enhances the quality of their subsequent design processes.

After concluding their research, students craft user profiles for chosen user groups. These profiles are qualitative, merging research findings with narrative techniques to keep users consistently engaged in the design process. Depending on the project's nature, students might also develop customer journey maps, drawing from their observations and interviews.

This phase typically includes two assignments for the students: an individual task focusing on user research methodologies and a group project centered on their collective user research.

Define

Define refers to reducing the wider problem context to a more manageable form so that it can be solved at a later stage in the process. The process of empathizing with others and articulating a problem that is important enough to solve is iterative by nature. At this phase the role of the educators is to challenge students' thinking that they do not choose a problem that is shallow or easy to solve.

Up to this juncture, student teams have met together in a shared classroom. However, following this, teams transition to operate more like independent marketing agencies. This shift subtly introduces an element of competition, as teams strive to outperform one another in crafting superior solutions. Nonetheless, no formal competition is declared, recognizing that not all students thrive in a competitive environment (Jaskari & Syrjälä, 2023). Instead, the emphasis remains on fostering a learning atmosphere that is conducive to experimentation, mutual learning, and fostering creativity in a supportive environment (Jaskari, 2012).

Ideation

In ideation phase students are encouraged to come up with creative and even crazy ideas by using various ideation methods. Students study different ideation methods and choose the best ones for their project. They also learn how to run ideation workshops, after which they organize anywhere from one to two workshops. Participants in the workshops might include users, clients, or even other stakeholders. During this stage, students are encouraged to come up with tens of ideas and use at least as many methods as there are students in the group. The more ideas and methods they come up with, the better. In this way, it is necessary for each student to become familiar with the method, as well as how to implement the method in a workshop setting. After the workshop, the students hand in a report presenting the process and the outcomes of their workshop.

Concept Creation

Concept creation is a prototyping phase, where students begin to form their concept solution for the problem. It begins with a three-concept phase where the students are encouraged to make three different solutions to the same problem. The idea behind this is that when the students are able to show their first solutions, they can be more courageous, braver, and visionary with others. The three concepts are discussed with the teacher, and the students are typically encouraged to integrate the concepts into one. During this phase, the solution becomes a concept-of-proof that can be tested with the potential users in the next phase.

The actual prototype can be a for example a 3D -printed product, a service map canvas, branding concept, a communication storyline – the main idea is to think, what is the best way to communicate the solution to other people. While working on this phase the students take part into a @Lego Serios Play workshop where their common idea is crystallized in a metaphorical level. Our finding is aligned with Priimus and Sonnenburg (2018) as this has proven to be a very useful tool to make sure that the teams understand their own idea in a similar way.

Testing

The prototype is taken to potential users or other stakeholders to be evaluated and discussed. The collected feedback is used to modify the solution. This phase is highly iterative in nature. Also, during this phase some of the students and teams are getting tired of the lengthy project.

Telling

The final session is all about presenting the solutions that have been created. The students are advised on storytelling techniques and pitching beforehand. If there have been clients or other outside members, they are invited to this session. The aim is to give a professional presentation of how their marketing agency has solved the given problem. The student teams do not usually know about the solutions of other teams. Thus, there is another interesting aspect of all presentations – how did the other group solve the same challenge? This is one of the main learning points in this session. During the session the solutions are discussed with each team. Feedback is given by peers, teachers and potential clients or other members.

As part of the final phase, each team hands in a written report of their project. Later, students write an individual assignment where they reflect their learning in a form of a learning diary.

The assessment of the course is based on both individual and project work conducted in teams. Individual assignments form 40% of the course grade and the team project 60%. Design thinking project is evaluated using the solo taxonomy approach for creative problem solving (Jaskari, 2013a), that emphasizes the role of process in assessment besides merely the outcome.

EVALUATING STUDENT PERSPECTIVES ON ACADEMIC DESIGN THINKING PEDAGOGY

Curriculum development in higher education strives to prepare graduates with the skills necessary for the future (OECD, 2019; Greenacre et al., 2017). Given the intricate and unpredictable nature of the future, pinpointing these requirements can be challenging. Numerous reports addressing the future of education highlight the importance of skills such as creativity, critical thinking, collaboration, and communication (CCR, 2015). The Academic Design Thinking Pedagogy is a suggested framework designed to offer students the opportunity to cultivate and hone these essential skills.

Although curriculum development encompasses more than just the views and desires of students, their feedback remains a pivotal component. For instance, formal practises such as anonymous feedback surveys or course evaluations can provide a structured way for students to voice their experiences, suggestions, and concerns. On the other hand, more informal practices, like incorporating feedback elements within learning assignments or encouraging students to maintain learning diaries, can shed light on their thoughts and feelings during the course. By integrating both official and unofficial feedback practices, educators can develop a more holistic understanding of student perspective.

The learning objectives of the case study course were twofold: firstly, to provide students with design thinking skills to generate innovative solutions for their coursework, and secondly, to prepare them to serve as change agents for sustainable future within their respective organizations in the coming years. Student feedback is gathered in three structured ways, all aligned with the learning objectives. First, throughout the course, there is continuous reflection on the relationship between the theoretical and practical aspects of the design thinking process, as well as on team dynamics. These reflections are discussed openly in coaching sessions. In this context, the educator's role is paramount. Establishing an open and critical line of communication demands a safe learning environment (Jaskari, 2012). Second, students submit learning diaries containing reflective essays. In these diaries, they ponder the course's influence on their perceptions of the design thinking process. Typically, these diaries also address practical aspects, with students discussing factors that either facilitated or hindered their learning. Finally, to ensure that all students feel secure in providing their perspectives, formal, anonymous feedback is collected.

Based on the collected feedback in 2021 and 2022, students found considerable value in the integration of real-world working methods and business projects into the academic curriculum. Many appreciated the diversification of course offerings, with one student pointing out, "It's nice that even in universities there are such diverse courses and real business projects," illustrating the desire for more practical, hands-on experiences in the academic setting.

While the innovative teaching methodology was largely applauded, it wasn't without its critiques. Students occasionally struggled with ambiguous instructions, highlighting the need for clearer guidelines. One feedback particularly mentioned, "At times, it was difficult to delve into different tasks because the instructions were occasionally unclear." Nevertheless, it's important to note that this ambiguity was recognized by some students as an intentional pedagogical approach meant to mimic the unpredictability and ambiguity of real-world problem-solving situations. This feedback is aligned with earlier research that illustrates how design thinking process can be confusing and chaotic, especially in the beginning (Glen et al., 2015; Jaskari, 2013b)

Personalized feedback and direct engagement with instructor were frequently noted as powerful learning enablers. As one student insightfully shared, "The teacher's group-specific coaching sessions, or the personalized guidance, enhanced learning," emphasizing the positive impact of tailored guidance. This underlines the importance of reflective and responsive teaching approach, where educators not only deliver content but also mentor and guide students through their learning journeys helping the students to reflect on their learning.

DISCUSSION AND CONCLUSIONS

This chapter explores the adoption of design thinking as a pedagogical approach to develop skills for solving a variety of sustainability challenges. While previous research has affirmed the efficacy of design thinking in this realm, it has not been immune to criticism, particularly concerning its tendency to generate excess of ideas at the expense of actionable solutions.

Design thinking has proven beneficial also educational purposes, with research spotlighting its multiple advantages in educational settings. Nonetheless, it presents certain obstacles that educators should consider. Familiarity with both the strengths and limitations of design thinking is crucial for instructors assessing its suitability for their specific curricular needs and teaching contexts. One of its biggest challenges has been its detachment from academic content.

This chapter introduces the Academic Design Thinking Pedagogy as a tailored approach for higher education, aimed at addressing the noted challenges. This pedagogical model integrates design thinking with academic content in innovative ways, offering a structured example for educators to weave it into their classroom instruction. Characterized by its adaptability, this methodology fosters the reconciliation of varied ideas, acknowledges individual and collective aspirations, promotes the incorporation of multiple perspectives, cultivates empathy, and helps in crafting a shared vision.

While both earlier research and the case study presented in this chapter acknowledge that students sometimes perceive the design thinking process as time-consuming, chaotic, and frustrating, educators too can feel overwhelmed during its implementation. Facilitating a creative process requires educators to dive in and exhibit courage themselves. Educators must encourage students, patiently enduring their moments of uncertainty and frustration. They need to convince students that it is worth investing time and effort into the process. Facilitating the creative process can bring unexpected challenges, such as carefully planned workshops not going as intended. However, overseeing a creative process can also be deeply rewarding when educators witness students fully engage, venture beyond their comfort zones, and embrace new experiences. Many times, these moments last for years, both for students and educators, showing the meaningfulness of the learning experience.

Design thinking flourishes in multidisciplinary environments, capitalizing on its interdisciplinary foundation to promote cohesive learning experiences. While it has found its way into various multidisciplinary curriculums, university structures and inflexible curricula often impede this progression, as observed by Henje & Jaskari (2013). Therefore, to facilitate this integration, university strategies and resources must actively support such endeavors. Yet, there's still a significant opportunity to weave design thinking not only within specific academic disciplines but also at the intersections of these fields.

There is also ethical consideration that educators need to think about, especially with the rising technologies. To illustrate, the potential consequences of design decisions can span areas from personal privacy to job markets and the students need to reflect on how their designs may produce unintended outcomes. Also, while ethically inattentive solutions may offer short-term benefits, they risk long-term setbacks, including societal backlash or legal repercussions, reinforcing the imperative of ethical considerations for sustainable outcomes.

The Future of Design Thinking in Education

Design thinking in education has experienced significant growth and adaptation since its introduction and the adaptation is likely to continue.

Centering on its empathetic approach, design thinking stands as a pivotal tool in addressing pressing sustainability issues, such as social inequities, climate change, or public health crises. As these challenges magnify, the role of design thinking in education becomes increasingly crucial, equipping students with the essential skills they'll need as future leaders to confront and navigate these complex, real-world issues. Moreover, design thinking has a possibility to establish itself as a tool for lifelong learning, through which it may empower individuals of all ages.

While technological tools, including the Artificial Intelligence (AI), immersive realms of Virtual Reality (VR) and Augmented Reality (AR), are expanding the horizons of design thinking methodologies, it's pivotal to strike a balance. One inherent strength of design thinking lies in its hands-on approach, whether it's crafting physical prototypes, engaging with tangible tools like Legos, or making mock-up flyers. Thus, as we use the opportunities offered by digitalization, it's essential to ensure that we don't loose these tactile, experiential elements that truly capture the essence of design thinking.

REFERENCES

AACSB. (2023). *Let's Lead Boldly: Elevating the Value of Business Schools.* Available online: https://www.aacsb.edu/insights/briefings/lets-lead-boldly-elevating-the-value-of-business-schools

Bertão, R. A., Jung, C. H., Chung, J., & Joo, J. (2023). Design thinking: A customized blueprint to train R & D personnel in creative problem-solving. *Thinking Skills and Creativity*, *48*, 101253. doi:10.1016/j.tsc.2023.101253

Biggs, J. B., & Collins, K. F. (1982). *Evaluating the Quality of Learning: the SOLO Taxonomy.* Academic Press.

Brown, T. (2008). Design thinking. *Harvard Business Review*, *86*(6), 84. PMID:18605031

Brown, T. (2009). *Change by Design: How Design Thinking Transforms Organizations and Inspires Innovation*. HarperCollins Publishers.

Brown, T., & Katz, B. (2011). Change by design. *Journal of Product Innovation Management*, *28*(3), 381–383. doi:10.1111/j.1540-5885.2011.00806.x

Brown, T., & Wyatt, J. (2010). Design thinking for social innovation. *Development Outreach, 12*(1), 29–43. doi:10.1596/1020-797X_12_1_29

Buchanan, R. (1992). Wicked Problems in Design Thinking. *Design Issues, 8*(2), 5–21. doi:10.2307/1511637

Carroll, M., Goldman, S., Britos, L., Koh, J., Royalty, A., & Hornstein, M. (2010). Destination, Imagination and the Fires within: Design Thinking in a Middle School Classroom. *International Journal of Art & Design Education, 29*(1), 37–53. doi:10.1111/j.1476-8070.2010.01632.x

CCR. (2015). *Skills for the 21st Century: What should students learn?* Center for Curriculum Redesign. Available at https://4dedu.org/wp-content/uploads/2022/10/CCR-Skills-for-the-21st-Century-5-2015.pdf

CCR. (2019). *Competencies / Subcompetencies Framwork, Rev 1.0.* Available at https://curriculumredesign.org/framework/

CMI. (2021). *Work ready graduates: Building employability skills for a hybrid world.* The Chartered Management Institute. Available online: https://www.managers.org.uk/wp-content/uploads/2021/09/employability-skills-research_work-ready-graduates.pdf

Cochrane, T., & Munn, J. (2016). EDR and Design Thinking: Enabling Creative Pedagogies. In *Proceedings of EdMedia 2016—World Conference on Educational Media and Technology* (pp. 315–324). Vancouver, BC, Canada: Association for the Advancement of Computing in Education (AACE). Retrieved August 1, 2023 from https://www.learntechlib.org/p/172969/

d.school. (2023). *Design thinking bootleg*. Retrieved August 1, 2023, from https://dschool.stanford.edu/resources/design-thinking-bootleg

De Bono, E. (1970). *Lateral Thinking: Creativity step by step*. Perennial Library.

Design Council. (2023). *The double diamond*. Retrieved August 1, 2023 from https://www.designcouncil.org.uk/our-resources/the-double-diamond/

Dunne, D., & Martin, R. (2006). Design thinking and how it will change management education: An interview and discussion. *Academy of Management Learning & Education, 5*(4), 512–523. doi:10.5465/amle.2006.23473212

Earle, A. G., & Leyva-de la Hiz, D. I. (2021). The wicked problem of teaching about wicked problems: Design thinking and emerging technologies in sustainability education. *Management Learning, 52*(5), 581–603. doi:10.1177/1350507620974857

Elsbach, K. D., & Stigliani, I. (2018). Design thinking and organizational culture: A review and framework for future research. *Journal of Management, 44*(6), 2274–2306. doi:10.1177/0149206317744252

Enell-Nilsson, M., Jaskari, M. M., & Kantola, J. (2019). How to use interdisciplinary team innovation to foster crucial engineering competencies? In *Global Advances in Engineering Education* (pp. 101–115). CRC Press. doi:10.1201/b22184-5

Enell-Nilsson, M., Jaskari, M.-M. & Borisov, P. (2023). Students' Perception of the Development of Generic Competencies in a User Innovation Course Context – A Longitudinal Study Comparing Different Teaching Modes in Higher Education. Nordic Journal of Business, 72(4), 218-236.

Eriksson, P., & Kovalainen, A. (2008). *Qualitative Methods in Business Research.* Sage. doi:10.4135/9780857028044

Ferguson, R., Coughlan, T., Egelandsdal, K., Gaved, M., Herodotou, C., Hillaire, G., ... Misiejuk, K. (2019). *Innovating Pedagogy 2019.* Open University Innovation Report 7.

Glen, R., Suciu, C., & Baughn, C. (2014). The need for design thinking in business schools. *Academy of Management Learning & Education, 13*(4), 653–667. doi:10.5465/amle.2012.0308

Greenacre, L., Freeman, L., Jaskari, M. M., & Cadwallader, S. (2017). Editors' Corner: The "work-ready" marketing graduate. *Journal of Marketing Education, 39*(2), 67–68. doi:10.1177/0273475317712411

Groth, C. (2017). *Making sense through hands: Design and craft practice analysed as embodied cognition.* Aalto University publication series. Doctoral Dissertations 1/2017. Available online: https://aaltodoc.aalto.fi/handle/123456789/24839

Grots, A., & Creuznacher, I. (2016). Design Thinking: Process or Culture? In Design Thinking for Innovation (pp. 183–191). Springer.

Guaman-Quintanilla, S., Everaert, P., Chiluiza, K., & Valcke, M. (2023). Impact of design thinking in higher education: A multi-actor perspective on problem solving and creativity. *International Journal of Technology and Design Education, 33*(1), 217–240. doi:10.1007/s10798-021-09724-z

Henje, C., & Jaskari, M.-M. (2013). Student and teacher experiences of multidisciplinary cross-border co-operation. *Reformation, Revolution, Evolution : Universitetslärandet Ur Ett Tidsperspektiv : Konferensrapport*, 153–154.

Jacobs, C. D. (2016). Making Is Thinking: The Design Practice of Crafting Strategy. In Design Thinking for Innovation (pp. 131–140). Springer.

Jaskari, M.-M. (2012). Using pictorial collages in user-centred construction of a creative and effective learning environment. In Veer, Ballantine, P. W. & Ozanne, L. K. (Eds). Asia-Pacific Advances in Consumer Research Vol. 10, (pp. 284–292). Association for Consumer Research.

Jaskari, M. M. (2013a). The Challenge of Assessing Creative Problem Solving in Client-based Marketing Development Projects: A SOLO Taxonomy Approach. *Journal of Marketing Education, 35*(3), 231–244. doi:10.1177/0273475313485586

Jaskari, M.-M. (2013b). *Teaching the future marketers through experiential client-based projects: marketing knowledge and skills in the context of early phases of service and product development.* Acta Wasaensia, 290. https://osuva.uwasa.fi/handle/10024/7377

Jaskari, M.-M. (2015). Teaching the Fuzzy Front End of Innovation: Real-Life Application With Cross-Functional And International Teams. In *Marketing Dynamism & Sustainability: Things Change, Things Stay the Same.... Developments in Marketing Science: Proceedings of the Academy of Marketing Science.* Springer. 10.1007/978-3-319-10912-1_106

Jaskari, M.-M., & Syrjälä, H. (2023). A Mixed-Methods Study of Marketing Students' Game-Playing Motivations and Gamification Elements. *Journal of Marketing Education, 45*(1), 38–54. doi:10.1177/02734753221083220

Liedtka, J. (2015). Design Thinking. *Journal of Product Innovation Management, 32*(6), 925–938. doi:10.1111/jpim.12163

Martin, R. (2007). How successful leaders think. *Harvard Business Review, 85*(6), 60. PMID:17580648

Matthews, J., & Wrigley, C. (2017). Design and Design Thinking in Business and Management Higher Education. *Journal of Learning Design, 10*(1), 41–54. doi:10.5204/jld.v9i3.294

McNiff, J., & Whitehead, J. (2006). *All you need to know about Action Research.* Sage.

Micheli, P., Wilner, S. J., Bhatti, S. H., Mura, M., & Beverland, M. B. (2019). Doing design thinking: Conceptual review, synthesis, and research agenda. *Journal of Product Innovation Management, 36*(2), 124–148. doi:10.1111/jpim.12466

OECD. (2019). *OECD Future of Education and Skills 2030.* OECD Learning Compass, 2030.

Ohly, S., Plückthun, L., & Kissel, D. (2017). Developing Students' Creative Self-Efficacy Based on Design-Thinking: Evaluation of an Elective University Course. *Psychology Learning & Teaching, 16*(1), 125–132. doi:10.1177/1475725716681714

Panke, S. (2019). Design thinking in education: Perspectives, opportunities and challenges. *Open Education Studies, 1*(1), 281–306. doi:10.1515/edu-2019-0022

Panke, S., & Harth, T. (2018). Design Thinking for Inclusive Community Design:(How) Does it Work? In EdMedia+ Innovate Learning (pp. 284–296). Association for the Advancement of Computing in Education (AACE).

Primus, D. J., & Sonnenburg, S. (2018). Flow Experience in Design Thinking and Practical Synergies with Lego Serious Play. *Creativity Research Journal, 30*(1), 104–112. doi:10.1080/10400419.2018.1411574

Razzouk, R., & Shute, V. (2012). What is design thinking and why is it important? *Review of Educational Research, 82*(3), 330–348. doi:10.3102/0034654312457429

Taheri, M., Unterholzer, T., Hölzle, K., & Meinel, C. (2016). An educational perspective on design thinking learning outcomes. In *ISPIM Innovation Symposium* (p. 1). The International Society for Professional Innovation Management (ISPIM).

Titus, P. A. (2000). Marketing and creative problem-solving process. *Journal of Marketing Education, 22*(3), 225–235. doi:10.1177/0273475300223006

Titus, P. A. (2007). Applied creativity: The creative marketing breakthrough model. *Journal of Marketing Education, 29*(3), 262–272. doi:10.1177/0273475307307600

UN. (2023). *United Nations: The 17 goals.* Retrieved in August, 1, 2023 from https://sdgs.un.org/goals

Valentim, N. M. C., Silva, W., & Conte, T. (2017). The students' perspectives on applying design thinking for the design of mobile applications. In *Proceedings of the 39th International Conference on Software Engineering: Software Engineering and Education Track* (pp. 77–86). IEEE Press. 10.1109/ICSE-SEET.2017.10

von Thienen, J., Royalty, A., & Meinel, C. (2017). Design thinking in higher education: How students become dedicated creative problem solvers. In *Handbook of research on creative problem-solving skill development in higher education* (pp. 306–328). IGI Global. doi:10.4018/978-1-5225-0643-0.ch014

Wrigley, C., & Straker, K. (2017). Design thinking pedagogy: The educational design ladder. *Innovations in Education and Teaching International*, *54*(4), 374–385. doi:10.1080/14703297.2015.1108214

ENDNOTE

1 The European Credit Transfer and Accumulation System (ECTS) is used to represent student workload based on defined learning outcomes. 60 ECTS credits are the equivalent of a full year of study or work.

Chapter 5
The Contribution of Experiential Learning to Self-Directed Learning Approaches in Higher Education

Celestin Mayombe
North-West University, South Africa

ABSTRACT

There have been trends in experiential learning (EL) to implement self-directed learning (SDL) in higher education institutions in the 21st century. Due to the necessity of a real-life, interactive environment in the lecturing approaches in higher education, the problem is to ensure that the learning becomes interactive and meaningful to the students. Students commit to learning when their goals and objectives are considered realistic and important, though they may face socio-economic challenges. Therefore, applying knowledge in the real world is important and relevant to the student's needs. This chapter explores the implications of experiential learning theory to gain work-related skills and knowledge for the work environment. The implication of the experiential learning theory (ELT) to higher education is that lecturers should design course materials that empower students to acquire knowledge and skills to solve real-life problems. Thus, ELT points to teaching practices that aim at enabling students to perform activities that are close to those found in the workplace.

1. INTRODUCTION TO THE STUDY

Higher education institutions (HEIs) play an important role in improving human capital to meet the demands of society by empowering graduates to solve socio-economic problems. To this end, enhancing the ability of students to apply skills, knowledge and solve real-life problems is one of the principal aims of higher education. The acquisition of necessary skills and knowledge is not enough for a graduate to solve their everyday problems; it is also important to provide them with opportunities to practice skills in real contexts reflective of the possibilities in future employment (Huang & Jiang, 2020; Kalyuga et al.,

DOI: 10.4018/978-1-6684-9859-0.ch005

2010). In other words, graduates should solve problems within the context of their field specialisation. For example, they must apply skills and knowledge in problem-solving and demonstrate professional competencies that help them function in the future. In this connection, the ability of students to develop practical skills and competencies depends on the teaching and learning approaches that lecturers use to create and enable students' active participation in their learning process. This chapter focuses on describing ways to design and deliver experiential learning, with specific interest in the learning approaches in higher education.

2. BACKGROUND TO THE STUDY

There have been growing trends in experiential learning to implement self-directed learning (SDL) in higher education in the 21st century. The effects of the Covid-19 global pandemic in 2020 compelled educational institutions to redirect teaching and learning approaches that enhanced students' acquisition of practical skills that they could practically apply in immediate employment after graduating (Lantu, et al., 2022). Parents and students have become mindful of the importance of experiential as part of education and training at HEIs. The rise of tuition fees leads to students becoming more agitated consumers of education who insist on about the levels of experiential learning opportunities obtainable from institutions they consider applying. Studies show that students who participate in practical activities during their learning process and take responsibility to reach their learning goals are more likely to achieve higher levels of skills and knowledge that they could utilise in the labour market (Wang, *et al*, 2021; Xie and Derakhshan, 2021). This argument suggests that HEIs are expected to motivate students to use their learning opportunities to actively participate in generating and honing employability skills.

To ensure that the students gain the required knowledge and skills for their future workplace, it is vital to create conducive environments that reflect and mirror real-world situations. Creating opportunities for practicing knowledge and skills to resolve problems in a real-work environment assists students to become acquainted with the challenges of their future careers (Huang and Jiang, 2021). So, in the 21st-century HEIs should focus on hands-on training and skills development. Experiential education contributes to empowering students with real-world skills for future employment (Lantu, et al., 2022). This argument is developed in detail in the section below, dealing with the importance of experiential learning approaches in higher education.

3. STATEMENT OF THE PROBLEM

Graduates from higher education institutions face challenges when trying to find entry-level jobs in the industries of their specialisation (Helyer and Lee, 2014; Lantu, *et al*., 2022; Sam & Van der Sijde, 2014). The fact is that industries and organisations perceive it as risky to employ graduates who were not exposed to work settings. On one hand, industries call for higher education institutions (HEIs) to shift from instructional teaching practices and promote experiential learning approaches (Girvan et al., 2016). In other words, HEIs should review the teaching and learning approaches to enhance student employability and use different modalities that increase and strengthen this practice, with an emphasis on delivering both theoretical and experiential learning approaches in curriculum delivery (Ritter *et al*., 2018; Pardo-Garcia & Barac, 2020). The implication is that the instructionist approach is no longer suitable for 21st-century colleges and universities because it does not train students in practice-based epistemologies.

On the other hand, many HEIs have realised the problem of a high number of graduates who remain unemployed after graduation (Helyer and Lee, 2014), clearly demonstrating the disjuncture in the competencies and skills of the graduates which constitute a mismatch between the labour market requirements and university pedagogy (Lantu, *et al.*, 2022). The is often a mismatch between students´ expectations and their training programmes at higher education institutions (Helyer and Lee, 2014). In fact, students desire to learn any type of skills through practical activities - whether in the classroom or workplace environments outside their higher education institution, instead of theoretical and scientific knowledge to which they are exposed in the academy (Green *et al.,* 2017; Villarroel, *et al*, 2020). The instrumentalist SDL approach would build their confidence in their abilities to perform tasks when offered the job after graduation (Sam and Van der Sijde, 2014). All the latter entail that the teaching and learning environment requires a real-life, interactive environment within the higher institutions. The problem lies in ensuring that the learning becomes interactive, effective and meaningful to the students. Students commit to learning when their goals and objectives are considered realistic and important, though they may face socio-economic challenges. This quickens the argument for SDL to meet the students' needs and aspirations.

To solve those problems, HEI requires a learning approach that enhances the relevant skills and competencies for the real workplace. The chapter addresses the problem of graduates who remain unemployed after graduation by exploring the possibilities of implementing core principles of the experiential learning theory (ELT) into the teaching and learning in higher education. The theory suggests that active learning is effective in skills acquisition when it occurs in the context in which it is applied in the work environments. In addition, experiential learning emphasises the role of experience in the learning process, such as "learning-by-doing" or "hands-on learning" with practical application in the real-world of work. In other words, experiential learning allows students to learn skills through "doing, learning while experiencing, and learning through 'hands-on practice' and 'reflection'" (Sampietro, 2019, p.721). The objectives of this chapter are designed to show:

- How experiential learning theory makes higher education learning an active process that assists students to apply the information in a practical situation.
- How experiential learning makes curricula meaningful to promote a real-life, interactive environment in higher education.
- What experiential learning contributes to the design of course materials.

4. METHODOLOGY

The chapter uses a pragmatic research paradigm for an in-depth exploration of the phenomenon (Creswell (2014; Leedy and Ormrod, 2014) on the possibilities of implementing the core principles of experiential learning theory into the learning processes in higher education. To achieve the research objectives, the study adopted the document analysis research method. The major focus of document analysis was on comparative content analysis (Nieuwenhuis, 2020; Sarantakos, 2005). Comparative content analysis entails comparing data and texts of different authors and analysing the validity, reliability and credibility of the information (Leedy & Ormrod, 2014; Sarantakos, 2005; Bowen, 2009). The researcher analysed primary and secondary data available from books, journal articles and electronic sources on the roles of EL in higher education. Other key issues in the literature review were the types and contributions of ELT to higher education in the design of course materials linked to the acquisition of specific skills and competencies. The critical review of available literature on the topic (Cohen et al., 2018; Nieuwenhuis, 2020) helped generate themes such as the importance of using experiential learning approaches in higher education, types of experiential learning, using ELT to engage with the real-life environment, and ultimately the implications of experiential learning for higher education.

5. DEFINING EXPERIENTIAL LEARNING

Experiential learning is learning by and through concrete experience. Due to the necessity of an interactive environment and practical activities in the lecturing approach, higher education institutions (HEIs) have embraced experiential learning to engage active participation of students in their learning process. In the context of higher education, Villarroel, *et al* (2020) argue that by participating in practical activities, students learn to apply the theoretical knowledge learnt from the classroom and textbooks to their real-life situations. Similarly, to ensure that students acquire the relevant skills and knowledge, it is important to allocate enough time for the training sessions to allow students to develop such competencies and acquired skills and knowledge in the real-world environments that are related to their future workplace (Huang and Jiang, 2020). Therefore, experiential learning (EL) facilitates the development of students' abilities for work readiness.

Experiential learning is defined as a learning approach that is based on "learning by doing, experiencing and reflecting" (Kolb, 1984, p.41). The concept describes a learning process that occurs when students see, listen, observe, touch, reflect, remember and interpret information. In Kolb's view, learning is more effective when adult learners are actively involved in the learning process rather than passively receiving knowledge transmitted by teachers. According to Kolb (1984, p.41), experiential learning theory (ELT) perceives learning as "the process in which knowledge is created through the transformation of experience. Knowledge results from the combination of grasping and transforming experience." Experiential learning focuses on the important role that experience plays in the learning process of students (Kolb, Boyatzis and Mainemelis, 1999; Herod, 2012). The core argument is that, by participating in practical activities during the teaching and learning, whether in class or outside of respective institutions, students competently transform the knowledge learnt into their understanding.

Furthermore, Simon Fraser University (quoted in Bates, 2019) has defined EL as:

The strategic, active engagement of students in opportunities to learn through doing, and reflection on those activities, which empowers them to apply their theoretical knowledge to practical endeavours in a multitude of settings inside and outside of the classroom.

Emphasising the important role of experience in the learning process, Lewis and Williams (1994, p.5) define EL in the following terms:

In its simplest form, experiential learning means learning from experience or learning by doing. Experiential education first immerses learners in an experience and then encourages reflection about the experience to develop new skills, new attitudes, or new ways of thinking.

It is significant that EL is an amalgam of strategies that enable students to actively engage in practical activities through doing. The strategies consist of helping students to link the learning to the real-world situation or scenario by applying concepts, ideas and theories to interactive settings (Greene, 2011). It should not be understood as a set of tools used to provide students with a variety of experiences, but rather an arsenal of instruction methods t in an educational context that values the student's abilities to undertake learning in various ways whether campus-based, project-based, industry-based or community contexts. The learning process achieves its goals when lecturers or facilitators encourage the students to reflect on their actions and experiences to develop new skills or new ways of doing certain tasks (Harvey *et al.*, 2016). For this very reason, experiential learning is often connected to practical activities of learning such as work-based learning, cooperative education, collaborative learning, service learning, work-integrated learning, internships and practicums.

6. STAGES OF THE EXPERIENTIAL LEARNING CYCLE

Experiential learning is a continuous process in which the students bring individual learning needs, experiences and repertoires to acquire skills and knowledge. The new knowledge acquired through the transformation of real experiences originates from abstractions and applications. Kolb (1984) created the "experiential learning cycle" which sheds more light on how a programme could be delivered in higher education. He developed four stages of EL in which students are required to participate actively to gain a deep understanding of the concepts being learnt. Figure 1 presents four stages of active learning (Kolb, 1984, p.41).

As Figure 1 shows, active learning is a continuous process and does not end after achieving objectives. To achieve meaningful learning in higher education, Dewey (in Kolb and Kolb, 2011, p.43) advises that the teaching-learning process should be directed at facilitating students' engagement in real-life activities that should develop their learning capabilities. The four stages of ELT are explained in the following paragraphs.

Concrete experience (CE): During this stage, students actively participate in the learning experience through concrete matters instead of passively receiving knowledge from a lecturer (Herod, 2012; Kolb, 1984). They participate in real-life activities by being involved in the actions as the lecturers or industrial mentors provide "hands-on" training and work-based learning to gain concrete experience. For example, the concrete experience for student-teachers may consist of visiting a classroom where a teacher interacts with learners, teaching learners under the supervision of a mentor, and assisting learners with learning challenges. While participating in concrete activities, the student may ask questions such as:

- What did I do to obtain this result? How did I begin and proceed to the end?
- What techniques, strategies and procedures did I use?
- What are the suitable tools in conducting this activity?

Reflective observation (RO): This stage entails thinking critically about the concrete experience. The student deeply reflects on that experience (Kolb, 1984; Kolb, *et al.*, 1999). Furthermore, the act of reflecting on experience provides a connection to praxis (Harvey *et al.*, 2016). Some of the questions that the student may ask are:

Figure 1. Experiential learning cycle
Source: Kolb (1984)

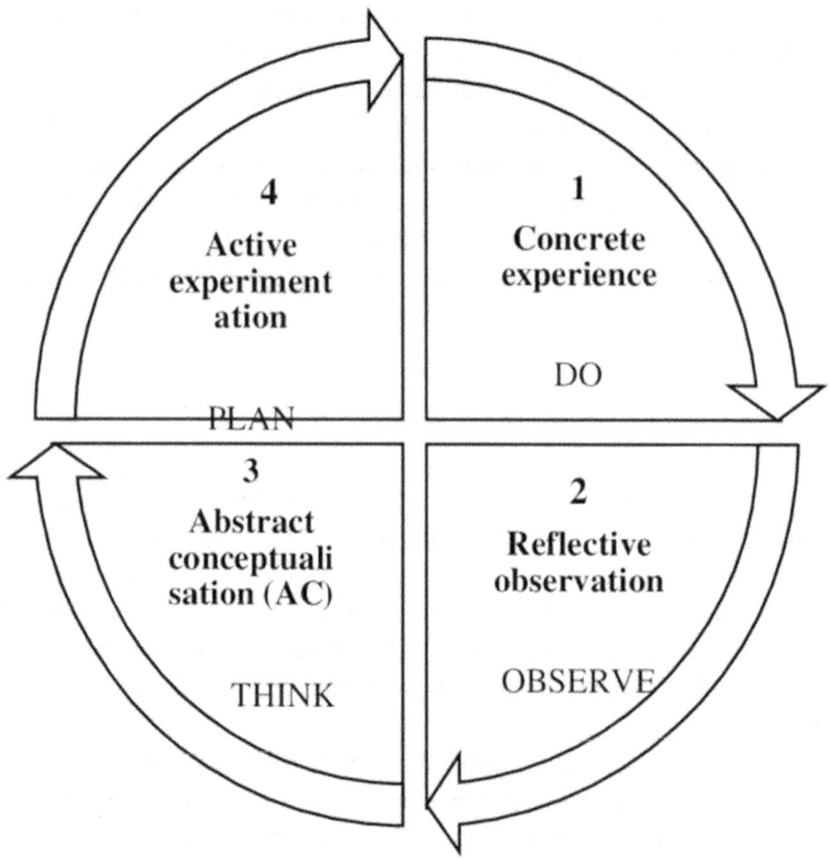

- What were the significant components of the activities?
- What actions worked for me and what did not?
- What did I learn from the experience as new skills and knowledge to be transferred in a new work environment?

Abstract conceptualisation (AC): This stage entails connecting the experience to the concept and theory learnt in the classroom (Kolb, *et al.,* 1999). The student conceptualises a theory from what was observed during the experience. Based on the observations and reflections about the concrete experience, the student strives to generalise and create abstract concepts about it.

Active experimentation (AE): This stage consists of applying the new knowledge in real-life situations and new workplace conditions (Kolb, 1984; Kolb, *et al.*, 1999). The students plan to test a theory on an upcoming experience (Villarroel, *et al.*, 2020). In this stage, the focus is on practical application and implication of the experience that students return to the "ground."

In summary, the stages of the cycle indicate that experiential learning is a teaching method that enhances students learning while Doing, Reflecting, Thinking and Applying. In EL, students participate in "a tangible experience (Do), reproduce that experience and other evidence (Reflect), cultivate theories in line with experiences and information (Think), and articulate an assumption or clarify a problem (Apply)" (Butler *et al.*, 2019, p. 36). The analysis of the information in this section confirms the view of Guo et al. (2016) that EL is a useful tool for positive transformations in HEIs and allows students to apply the knowledge gained in class to real-world situations (Guo *et al.*, 2016). This approach to learning consists of giving students more authority and responsibility to achieve the needs and objectives of learning and enables them to utilise the acquired skills in the labour market.

Having explained the stages of the experiential learning cycle to promote a real-life environment in higher education, the next section elaborates the importance of using an experiential learning approach in higher education. It outlines some contributions of experiential learning to teaching and learning processes at universities and technical and vocational education and training (TVET) colleges to achieve their goals and objectives.

7. THE IMPORTANCE OF USING AN EXPERIENTIAL LEARNING APPROACH IN HIGHER EDUCATION

As mentioned earlier, experiential learning (EL) facilitates active learning by providing real-world experiences to the students. The approach allows students to develop their capability to apply acquired skills and knowledge in real-world conditions to resolve problems related to their future occupations (Huang and Jiang, 2021). However, Green *et al.* (2017) contend that if EL occurs in the classroom, students may not be exposed to practical training for skills development; in reality, they obtain theoretical and academic education. EL consists of the creation of a variety of out-of-classroom activities including project-based learning, work-based training, participation in research and other activities which aim at providing students with work experiences.

In the context of higher education, the main argument is that 'experience is important' whether acquired prior to graduating or post-graduating and students can learn from it by reflecting on it. Using an experiential learning approach in HEI has a positive impact on student recruitment in the study programme, retention and completion rate. In addition, it increases the likelihood of employment outcomes for the graduates. Purposed towards producing competent graduates for the workforce, the experiential learning approach has gained traction in the higher education sector. Bartle (2015) notes that HEIs need to prepare students to become active members and leaders of local, national and international communities, who can apply their skills and knowledge to solve real-world problems. From a HE perspective, incorporating experiential learning into degree programmes affords students an opportunity to apply theory into practice in real-world work settings, hence expanding students' knowledge and understanding of the course (Cooper, Orrell and Bowden, 2010).

Students benefit from EL as it helps them learn in a real-life situation with hands-on experience. De Giancomo (2002, p.246), and Slavich and Zimbardo (2012, p.594.) note the following benefits of experiential learning to students:

- *As an alternative mode of learning*: the experiential learning approach provides meaning and relevance to the learning process by enhancing the integration of theoretical learning in the classroom with actual application in the workplace.

- *Linking theoretical concepts to practice*: experiential learning provides students with opportunities to experience concepts in a concrete manner hence giving students a richer understanding of key concepts of the course and operating in the real world of work.
- *Enhancement of the quality of the course programme*: quality enhancement happens when students are actively engaged in solving problems that are related to the field of study.
- *A practical understanding of the profession*: practical tasks and reports help the student to understand the profession and its characteristics, thereby developing realistic expectations about this future career.
- *Providing graduate employment opportunities*: job-orientated learning helps students to gain practical skills and experience that empower them to compete with other job seekers in the labour market upon graduation (Kong, 2021).
- *Facilitating a connection between higher education and professional experience*: EL has the capacity to create links between higher education and professional experience (Earnest *et al.,* 2016). In other words, it connects students to the real world of work (Friedman and Goldbaum, 2016).

Experiential learning helps students develop multiple competencies for future workplaces through concrete experience. If HEIs require a learning approach that enhances the real-works skills and abilities for the future employment of their students, experiential learning achieves this objective. EL garners student empowerment, self-confidence and motivation. To emphasise the contribution of experiential learning to HEIs, Kendall, Duley, Little, Permaul and Rubin (1986, p.19) note that "the experiential approach is a powerful motivator for learning because it is positive, meaningful, and real." Students' motivation subsequently increases the value of higher education because the more they are motivated, the more they succeed in their activities (Derakhshan et al., 2020). Thus, experiential education contributes to the preparation of a well-educated and productive workforce.

8. INTEGRATING EXPERIENTIAL LEARNING IN THE TEACHING AND LEARNING PROCESS

Through experiential learning, lecturers help students achieve their learning needs by participating in real-life activities. Students should effectively transform the theoretical knowledge learnt from the textbooks into practice, hence facilitating a deep understanding. Therefore, experiential learning involves mastering all four learning modes - experience, reflection, theorising and application (Kolb & Kolb, 2005). The capability to integrate all four learning stages is an indicator of the growth of students and effective teaching and learning at HEIs. This section presents selected types of EL approaches.

8.1 Types of Experiential Learning

Experiential learning may take the form of field-based learning (structured work experience) and classroom-based learning. During field-based learning, students are familiarised with the world of work within a higher education programme (O'Shea, 2014). Examples of field-based learning are internships, cooperative education, fieldwork, practicums, service-learning, project-based learning community-of-practice learning, team-based learning, apprenticeship, on-the-job training, situated learning, place-based learning, and workplace learning. Below are explanations of some selected field-based learning approaches that can be easily used at higher education institutions.

8.1.1 Field-Based Experiential Learning

The following types of field-based experiential learning have been used by several HEIs to ensure there is a match between training programmes and industries. The field-based learning approaches build students' confidence in their abilities to perform tasks when offered a job after graduation (Sam and Van der Sijde, 2014). These appear to be more effective than classroom-based learning because they occur in the context where the skills and knowledge are used in the students' work environments and emphasise the role that experience contributes to the acquisition of skills and knowledge, such as learning by doing.

Internships: This entails work experience conducted in a work environment under the guidance of a mentor or an experienced professional. It is generally conducted over an extended period to allow for "deep learning and development as a professional" (O'Shea, 2014, p. 8). Internship programmes provide a reality of the specific job requirements in terms of knowledge, skills and competencies.

Cooperative education: This variety of EL involves the work experience of students under the supervision of a skilled mentor for the purpose of developing the professional and relevant skills of a student. O'Shea (2014, p. 8) notes that it normally occurs as part of a professional cooperative education programme that "provides alternating full-time study with full-time employment." In cooperative education, students are exposed to "diverse work situations and development in work experience at multiple points across the length of the academic programme" (O'Shea, 2014, p.8). In simple terms, cooperative education consists of a programme which "alternates periods of academic study with periods of work experience in appropriate fields of business, industry, government and social services" (O'Shea, 2014, p.18). Students alternate work and study during their academic programmes, usually spending several weeks in campus-based learning (typically full-time) and a number of weeks in employment away from campus (typically full-time).

Fieldwork: Fieldwork comprises experiences in which the facilitator exposes the students to the work environment through direct involvement in work tasks, helping them participate in projects, field trips and site visits (O'Shea, 2014). Students gain experience as fieldwork enhances their understanding of specific course contents. Fieldwork also comprises work-study and industry placements which are linked to the training field of the students.

Field visits: This entails activities conducted outside of the classroom designed to expose the students to real work situations. Students, together with the lecturer, observe actual job activities, products or services in process in industries, factories or farms to gain real experiences of what was demonstrated in the classroom. Field visits are productive learning techniques exposing students to the actual work environment (International Labour Organisation [ILO], 2009). This is like the principal on-the-job extension training methods that HEI can use to provide opportunities for students to imagine themselves in the world of work. This type of experiential learning should be effectively organised by identifying learning content and problems that students want to be exposed to. During the visits, students take notes of their observations and experience. After the visit, both lecturer and students conduct a summary session of key issues observed; the notes and experiences are also discussed.

Placement: This type of experiential learning incorporates a series of structured work experiences in which students perform work in an industry or organisation approved by his/her institution (O'Shea,

2014). The main activities entail career exploration and practice to develop specific skills related to future employment.

Practicum work: A relative of the internship, practicum denotes the experience by which career competencies are developed in a workplace environment, with the purpose of meeting the requirements of professional registration of a scientific body. The terms "professional practice placement, clinical placement or professional placement are also used to describe practicum work" (O'Shea, 2014, p.18). Some academic programmes often require work experience combined with learning content and assessment methods based on criteria and professional competencies defined by the accrediting body.

Project-based learning (PBL): PBL is about using relevant projects as a vehicle to help students gain hands-on skills in a particular lesson. In PBL, students acquire knowledge and skills by working for an extended period to explore and respond to an authentic, engaging and complex question, problem, or challenge (Boss and Larmer,2018). Efstratia (2014, p.1257) argues that "the core idea of Project Based Learning is that real-world problems capture students' interest and provoke serious thinking as the students acquire and apply new knowledge in a problem-solving context." Identified projects are directed to students' learning needs and goals and include important project design elements – key knowledge, understanding and skills. PBL can occur both inside and outside classrooms.

On-the-job training (OJT): This is a practical approach to learning that helps a student gain new skills and competencies required for a job in a real working environment. This type of experiential learning occurs whenever an experienced professional shows an inexperienced trainee how to do a job. It takes place at the workplace during the performance of a job (Matsuo, 2014), usually involving one-on-one instructions. There is a strong link between training and practice as the skills acquired through OJT are learned more effectively (O'Shea, 2014).

8.1.2 Classroom-Based Experiential Learning

Having analysed field-based learning approaches or off-campus approaches of experiential learning, this sub-section focuses on classroom-based experiential learning. However, some scholars in the education field have cast doubt on the contribution of classroom-based practice to experiential learning (Green et al., 2017; Perini, *et al.*, 2018). The reason is that it may not always help students to understand the content of knowledge and apply it in the world of work because of its passive nature, yet the active and practical nature of the learning process tends to facilitate a deep understanding of knowledge and skills taught in class. Nevertheless, there is a similarity between motivating and engaging students in the classroom; both actions help them to understand better the content of the lessons and gain practical knowledge and skills.

The four stages of experiential learning (see Figure 1) can also occur in classrooms if lecturers select suitable activities to help students to understand the content of knowledge and apply it. Huang and Jiang (2020) note that classroom-based experiential learning helps students to understand the content of knowledge and apply it in the world of work because it is not passive in nature, but rather active and practical. In the context of higher education, classroom-based experiential approaches help students to participate in practical activities by applying the theoretical knowledge learnt from textbooks to their real-life situations (Villarroel, *et al.*, 2020). Therefore, the classroom-based experiential approach complements field-based experiential learning.

Some of the learning approaches that lecturers can utilise as classroom-based experiential learning are cooperative learning, role-playing, game-based learning, case-based learning, simulations, presentations, and various types of small-group work. The lecturer selects a relevant type of learning approach based on course objectives and learning outcomes. The disadvantage of these approaches is that they require time for preparation and execution, whether inside or outside the classroom.

Small-group work: This approach consists of asking five to seven students to work together in solving a real-life problem, relevant subject matter, or question to discuss. The discussion should be carried out by the students with reference to their past experiences, reaction and values (PRIA, 2014). The purpose is to use this group discussion in sharing students' experiences and lead to new knowledge and new insights. Killen (2015) points out that one of the reasons for using small-group work is to improve the students' problem-solving skills and help them discover that there are multiple solutions to problems. Finding multiple solutions to a problem helps develop students' depth of understanding and increases the possibility that they could resolve future problems in the workplace from various perspectives. In this case, small-group work provides rich opportunities for assigning students' initiative, creativity and flexibility and communication skills.

Cooperative learning: Though small-group learning has similarities with cooperative learning, this variety is more than just group work; students work together in small groups to help one another achieve a common learning goal through reflection. Killen (2015, p.229) notes that cooperative learning is based on "the belief that students can achieve more by working collaboratively than by working alone or passively receiving information from lecturers." As a type of experiential learning, cooperative learning requires that group members be reflective students. Each group must analyse the outcome it has achieved, how well the group functions and how well individuals were able to learn. Cooperative learning is useful in higher education because it is an effective strategy to help students to develop the core skills needed in their future workplace, such as critical thinking and problem-solving collaboratively. It helps students to see that there can be multiple solutions to complex problems.

Role-playing: This is a learning approach in which one or more students are required to "act" apart, referred to as performance activities. In most cases, role-playing involves just a few students (the actors), while most of the class observes and analyses the interactions between the actors. Whichever form of roleplay is used, Killen (2015) advises that the lecturer is responsible for planning, organising, facilitating and monitoring the role-playing, and for guiding the follow-up discussions so that it can generate a learning experience.

Game-based learning (GBL): Within experiential learning, GBL consists of not just creating games for students to play; it is designing learning activities that progressively introduce concepts and guide students towards an end goal (Sera and Wheeler, 2017). GBL provides an exciting, interactive, personalised, and challenging learning environment while enabling students to experience and apply knowledge gained from textbooks. Furthermore, such games deepen the students' understanding of the lesson content so they can solve more complex problems and develop creative and critical thinking (Perini, *et al.,* 2018). Using GBL activities can be an effective teaching method to teach students when content is application-based; this helps students practice with a more hands-on approach or review previous content covered by presentation teaching methods (Thompson, *et al.* 2020). It is a practice-based experience that provides higher education students with real-life training opportunities to apply taught knowledge and develop personal attributes and professional skills (Lucas, 2018).

Case-based learning: This consists of a story from a real-life experience in which an individual or group must decide or solve a problem. In this type of experiential learning, students develop skills in reflective judgment and analytical thinking by reading and discussing complex, real-life scenarios (Bates, 2019). Many real situations often confront workers with unique problems. Case study learning can replicate these situations and improve students' ability to deal with them. Case study learning incorporates many of the features of small-group work and cooperative learning. The case study approach is often used when a lecturer aims to offer students opportunities to grapple with issues, problems, and dilemmas in ways that are engaging, challenging and productive (Killen, 2015). The purpose of adopting case-based learning in the classroom is to prepare students for real-world experience in the future workplace.

There are two types of case studies - live cases and classroom-confined cases. The former engages students with the community (executed outside the campus) and its features are close to those of project-based learning (Georgiou, 2008). The sub-section focuses on the latter type of case study (classroom-confined case) and uses demonstration cases and problem cases. The case-based learning by demonstration, as the term indicates, demonstrates real-world practice and concerns with the practical application of concepts, theories and processes. If the case studies are to be included in Kolb's Experiential Learning Cycle (see Figure 1 above) for the purpose of promoting classroom experiential learning, they must be included as practical means to attain some degree of concrete experience.

8.2 Using Experiential Learning to Promote a Real-Life Environment in Higher Education

Using EL may be difficult for novice lecturers because it calls for proper preparation, a deep knowledge of facilitation methods in higher education and course content. There are several steps to consider when in integrating an experiential component into a course. Depending on the subject, the following steps are important to consider:

1. *Analysis of student population and determination of their needs*: The programme level of students, whether undergraduate or postgraduate, will determine the appropriateness of experiential activities that the students should participate in. The number of students in class is also another factor to consider when choosing the type of practical activities because it can both restrain and provide opportunities for the use of experiential learning. Students have needs to acquire specific skills that are needed in the labour market. Therefore, the lecturer should also be guided by the course outcomes which highlight the types of knowledge and practical skills that students need to acquire and use in the workplace after graduating.
2. *Writing learning objectives and learning outcomes*: Learning objectives set yardsticks for practical learning activities, while outcomes set the final product to be obtained from those objectives (Greene, (2011). In other words, it is important to consider course aims and learning objectives before considering an experiential learning approach for your course.
3. *Identifying resources for experiential learning:* The decision to use experiential learning lies in the availability of resources. The resources offer students well-structured activities and discipline content, that enables them to work successfully through the components of the tasks set. Resources can be in the form of laboratories, workshops, workplace placements, etc.
4. *Selecting suitable activities for students and course content*: This step entails learning activities that are appropriate for the course content and help achieve "the cognitive development needs of the students" (Cantor, 1995, p. 81). In this step, the lecturing ought to determine the aspects of course content that can be blended with experiential learning. However, the activities should be directed to meet course objectives and instructional goals. It is important to stress that the activities should complement the programme curriculum (Cantor, 1995). Activities should facilitate a learning experience that helps students to achieve the envisioned outcomes of the course. Some

of the questions a lecturer may ask in this stage are: What knowledge and skills do you want the students to demonstrate upon completion of the activities? What are the main learning objectives for the activities or tasks? What are activities that will help students achieve the level of Bloom's Taxonomy (remember, understand, apply, analyse, evaluate, create)?

5. *Designing experiential learning within the lesson plan*: When making decisions about experiential learning, there will always be a need for determining a suitable timetable. Experiential learning should be designed to include a balance of theory and practical activities within reasonable time limits. Too much of either component will cause the theory lesson or experiential learning activities to either grow boring or become overly burdensome. It is important that the lecturer keep a balance between the theoretical and practical components of a lesson.

9. THE IMPLICATIONS OF EXPERIENTIAL LEARNING FOR HIGHER EDUCATION

From each section of this chapter, the core element is that EL occurs when students are directly in connection with the realities of the activities they are studying; it encompasses a direct encounter with the phenomenon being studied rather than just thinking about it. In other words, all types of experiential learning help students to make connections between experience and theory (Earnest *et al.*, 2016), facilitating the transition between undergraduate education and the world of work (Friedman and Goldbaum, 2016; Villarroel, 2020). Furthermore, due to the use of stages of the learning cycle (concrete, reflective, abstract and active), the content is expanded, allowing the achievement of significant learning (Calvert, *et al.,* 2016). In this connection, the present chapter demonstrates the results of the contribution of experiential learning to self-directed learning approaches in higher education

The implications for experiential learning making curricula meaningful to promote a real-life, interactive environment in higher education consists of students connecting between experience and theory. To achieve this objective, experiential learning presents opportunities for students to actively participate in practical learning processes. This endeavour has implications for the design of curricula that have components for critical reflection and reflective practice that enhance the relevant skills and competencies for the real world of the workplace (Kong, 2021). Therefore, HEIs also need to ensure that all lecturers have the skills to develop students' skills in reflective practice. Higher education will need to develop evidence-based frameworks to guide lecturers to take a systematic approach to designing experiential learning activities.

Experiential learning supports the development of new workforce skills. Another implication of ELT to this higher education is that lecturers are urged to design course materials that help students acquire practical skills to solve the socioeconomic problems of society (Kalyuga *et al.*, 2010). Universities need to prepare students to be leaders and contributing members of local, national and global communities, who apply their acquired skills to real-world problems (Bartkle, 2015; Huang and Jiang, 2020). Over the course of a degree programme, students should develop "21st-century skills" in addition to core knowledge in their field of study (Bartkle, 2015). So, ELT points to higher education teaching practice that involves teaching-learning approaches aiming at empowering students to perform tasks that are close to those existing in the work environment. Therefore, Teaching-learning approaches must involve both types of EL, namely, field-based experiential learning and classroom-based experiential learning (see section 6.1 above).

10. CONCLUSION

The chapter addresses the problem of graduates who remain unemployed after graduation by exploring the possibilities of implementing the core principles of ELT into the teaching and learning processes of students in higher education. The chapter described how the design and implementation of experiential learning in the course can influence the perception of the students about the ability to apply the knowledge and skills to solve real-life problems or utilise them in labour market environments. The first objective was to determine how experiential learning theory makes higher education learning an active process helping students to apply the information in a practical situation. In applying ELT phases, lecturers should give students more responsibility to create knowledge achieve the needs and objectives of learning and enable them to utilise the acquired skills in the labour market. They ought to use EL in the form of field-based learning and classroom-based learning. It is important to create a variety of out-of-classroom activities including project-based learning, work-based training and other activities which enable students to gain work experiences.

The second objective was to determine how experiential learning makes curricula meaningful promoting a real-life, interactive environment in higher education. Using experiential learning promotes skills and competencies to solve a real-life problem and facilitation a school-to-work transition. Incorporating experiential learning into a degree curriculum is a good strategy to provide students with opportunities to apply theory into practice in a real-world work setting. Experiential learning makes curriculum meaningful by promoting real-life, interactive learning in higher education. Indeed, experiential learning presents opportunities for students to have a personal role in actively participating in their practical learning process. To achieve this purpose, the design of the curriculum contains components of critical reflection and practical experience that develop the relevant skills and competencies of the students.

The third objective was to determine what experiential learning contributes in the design of course materials. The field-based learning and classroom-based learning will help the lecturer select relevant materials for each component of learning. It is important to select various materials appropriate for classroom activities and out-of-classroom activities such as project-based learning, and work-based training. It is noteworthy to reiterate the contribution of EL in terms of helping students develop practical skills and competencies for future workplaces through concrete experience.

Therefore, the major implication of the ELT to this higher education is that lecturers should design course materials that help solve the socioeconomic problems of students. The theory points to higher education teaching practice involving teaching-learning approaches that aim at helping students perform tasks as close to those on the job as possible. The experiential learning approach will entail simulation exercises, skill-practice exercises and field projects.

The study contributes to the knowledge of lecturer professional development by focusing on students' learning outcomes. Effective teaching should lead to improved student achievement using outcomes that matter to their future success. The critical point is that student progress is the yardstick by which lecturer quality should be evaluated. As mentioned above, adult learning is effective in the 'real world' interactive environment.

REFERENCES

Bates, T. (2019). *Teaching in a Digital Age: Guidelines for designing and learning* (2nd ed.). Tony Bates Associates LTD.

Boss, S., & Larmer, J. (2018). *Project-based teaching: How to create rigorous and engaging learning experiences*. Buck Institute for Education.

Bowen, G. A. (2009). Document Analysis as a Qualitative Research Method. *Qualitative Research Journal*, *9*(2), 27–40. doi:10.3316/QRJ0902027

Butler, M. G., Church, K. S., & Spencer, A. W. (2019). Do, reflect, think, apply: Experiential education in accounting. *Journal of Accounting Education*, *48*, 12–21. doi:10.1016/j.jaccedu.2019.05.001

Calvert, F., Crowe, T. & Grenyer, B. (2016). Dialogical reflexivity in supervision: An experiential learning process for enhancing and relational competencies. *Journal of Supervision in Psychotherapy and Mental Health, 35*(1), 1-21. . doi:10.1080/07325223.2015.1135840

Cantor, J. A. (1995). ASHEERIC Higher Education Report: Vol. 7. Experiential Learning in Higher Education. Academic Press.

Cooper, L., Orrell, J., & Bowden, M. (2010). *Work integrated learning: A guide to effective practice*. Routledge. doi:10.4324/9780203854501

Creswell, J. W. (2014). *Research design. Qualitative, quantitative and mixed method approaches* (4th ed.). Sage.

DeGiancomo, J. A. (2002). Experiential learning in higher education. *Forestry Chronicle*, *78*(2), 245–247. doi:10.5558/tfc78245-2

Derakhshan, A., Coombe, C., Arabmofrad, A., & Taghizadeh, M. (2020). Investigating the effects of English language teachers' professional identity and autonomy in their success. *Issue Lang. Teach*, *9*, 1–28. doi:10.22054/ilt.2020.52263.496

Earnest, D., Rosenbusch, K., Wallace-Williams, D., & Keim, A. (2016). Study abroad in Psychology: Increasing cultural competencies through experiential learning. *Teaching of Psychology*, *43*(1), 75–79. doi:10.1177/0098628315620889

Efstratia, D. (2014). Experiential education through project-based learning. *Procedia: Social and Behavioral Sciences*, *152*, 1256–1260. doi:10.1016/j.sbspro.2014.09.362

Friedman, F., & Goldbaum, C. (2016). Experiential learning: Developing insights about working with older adults. *Clinical Social Work Journal*, *44*(2), 186–197. doi:10.1007/s10615-016-0583-4

Georgiou, I., Zahn, C., & Meira, B. J. (2008). A systemic framework for case-based classroom experiential learning. *Systems Research and Behavioral Science*, *25*(6), 807–819. doi:10.1002/sres.858

Girvan, C., Conneely, C., & Tangney, B. (2016). Extending experiential learning in teacher professional development. *Teaching and Teacher Education*, *58*, 129–139. doi:10.1016/j.tate.2016.04.009

Green, R. A., Conlon, E. G., & Morrissey, S. M. (2017). Task values and self-efficacy beliefs of undergraduate psychology students. *Australian Journal of Psychology, 69*(2), 112–120. doi:10.1111/ajpy.12125

Greene, H. (2011). Freshmen Marketing: A first-year Experience with Experiential Learning. *Marketing Education Review, 21*(1), 79–88. doi:10.2753/MER1052-8008210111

Guo, F., Yao, M., Wang, C., Yang, W., & Zong, X. (2016). The effects of service-learning on student problem solving: The mediating role of classroom engagement. *Teaching of Psychology, 43*(1), 16–21. doi:10.1177/0098628315620064

Harvey, M., Coulson, D., & McMaugh, A. (2016). Towards a theory of the Ecology of Reflection: Reflective practice for experiential learning in higher education. *Journal of University Teaching & Learning Practice, 13*(2), 1–20. doi:10.53761/1.13.2.2

Helyer, R., & Lee, D. (2014). The role of work experience in the future employability of higher education graduates. *Higher Education Quarterly, 68*(3), 348–372. doi:10.1111/hequ.12055

Herod, L. (2012). *Adult learning: From theory to practice.* Available at http://en.copian.ca/library/learning/adult_learning/adult_learning.pdf

Huang, R., & Jiang, L. (2021). Authentic assessment in Chinese secondary English classrooms: Teachers' perception and practice. *Educational Studies, 47*(6), 633–646. doi:10.1080/03055698.2020.1719387

International Labour Organisation (ILO). (2009). *Rural Skills Training: A Generic Manual on Training for Rural Economic Empowerment (TREE).* ILO.

Kalyuga, S., Renkl, A., & Pass, F. (2010). Facilitating flexible problem-solving: Cognitive load perspective. *Educational Psychology Review, 22*(2), 175–186. doi:10.1007/s10648-010-9132-9

Kendall, J. C., Duley, T. C., Little, T. S., Permaul, J. S., & Rubin, S. (1986). *Strengthening experiential education within institution.* National Society for Internship and Experiential Education.

Killen, R. (2015). Teaching strategies for quality teaching and learning. Cape Town: Juta and Company (Pty) Ltd.

Kolb, A. Y., & Kolb, D. A. (2005). Learning styles and learning spaces: Enhancing experiential learning in higher education. *Academy of Management Learning & Education, 4*(2), 193–212. doi:10.5465/amle.2005.17268566

Kolb, A. Y., & Kolb, D. A. (2011). Experiential learning theory: A dynamic, holistic approach to management learning, education and development. In *The SAGE Handbook of Management Learning*. Education and Development. doi:10.4135/9780857021038.n3

Kolb, D. (1984). *Experiential learning theory: Experience as source of learning and development.* Prentice-Hall.

Kolb, D., Boyatzis, R. E., & Mainemelis, C. (1999). *Experiential learning theory: Previous research and new directions*. Available at https://learningfromexperience.com/media/2010/08/experiential- learning-theory.pdf

Kong, Y. (2021). The role of experiential learning on students' motivation and classroom engagement. *Frontiers in Psychology, 12*, 771272. doi:10.3389/fpsyg.2021.771272 PMID:34744950

Lantu, D. C., Suharto, Y., Fachira, I., Permatasari, A., & Anggadwita, G. (2022). Experiential learning model: Improving entrepreneurial values through internship program at start-ups. *Higher Education. Skills and Work-Based Learning*, *12*(1), 107–125. doi:10.1108/HESWBL-01-2021-0014

Leedy, P. D., & Ormrod, J. E. (2014). *Practical research.* Pearson.

Lewis, L. H., & Williams, C. J. (1994). In L. Jackson & R. S. Caffarella (Eds.), *Experiential Learning: A New Approach* (pp. 5–16). Jossey-Bass.

Lucas, C., Williams, K., Tudball, J., & Walpola, R. L. (2018). Community, hospital and industry preceptor perceptions of their role in experiential placements - The need for standardization of preceptor responsibilities and evaluations on students. *Currents in Pharmacy Teaching & Learning*, *10*(11), 1447–1455. doi:10.1016/j.cptl.2018.08.002 PMID:30514534

Matsuo, M. (2014). Instructional skills for on-the-job training and experiential learning: An empirical study of Japanese firms. *International Journal of Training and Development*, *18*(4), 225–240. doi:10.1111/ijtd.12035

O'Shea, A. (2014). Models of WIL. In S. Ferns (Ed.), *Work integrated learning in the curriculum. Higher Education Research and Development Society of Australia guide* (pp. 7–14). Australia Collaboration Education Network Ltd.

PRIA. (2014). *A manual for participatory training methodology in development.* New Delhi: PRIA.

Ritter, B. A., Small, E. E., Mortimer, J. W., & Doll, J. L. (2018). Designing management curriculum for workplace readiness: Developing students' soft skills. *Journal of Management Education*, *42*(1), 80–103. doi:10.1177/1052562917703679

Sam, C., & Van Der Sijde, P. (2014). Understanding the concept of the entrepreneurial university from the perspective of higher education models. *Higher Education*, *68*(6), 891–908. doi:10.1007/s10734-014-9750-0

Sampietro, M. (2019). Executive learning and development. In Handbook of Vocational Education and Training: Developments in the Changing World of Work. Springer. doi:10.1007/978-3-319-94532-3

Sarantakos, S. (2005). *Social Research* (3rd ed.). Palgrave Macmillan. doi:10.1007/978-0-230-20901-5

Sera, L., & Wheeler, E. (2017). Game on: The Gamification of the Pharmacy Classroom. *Currents in Pharmacy Teaching & Learning*, *9*(1), 155–159. doi:10.1016/j.cptl.2016.08.046 PMID:29180148

Slavich, G. M., & Zimbardo, G. Z. (2012). Transformational teaching: Theoretical underpinnings, basic principles, and core methods. *Educational Psychology Review*, *24*(4), 569–608. doi:10.1007/s10648-012-9199-6 PMID:23162369

Thompson, M., Kanmaz, T., Cheryl, C., & Tanzer, K. (2020). *Recommendations for Participation in Experiential Education Activities during a National, Regional or Local Crisis.* Available at https://www.acpe-accredit.org/wp-content/ uploads/Experiential- Education-ACPE-Proposal-Response-Final.pdf

Villarroel, V., Bernavente, M., Chuecas, M. J., & Bruna, D. (2020). Experiential learning in higher education. A student-centred teaching method that improves perceived learning. *Journal of University Teaching & Learning Practice*, *17*(5), 1–16. doi:10.53761/1.17.5.8

Wang, Y., Derakhshan, A., & Zhang, L. J. (2021). Researching and practicing positive psychology in second/foreign language learning and teaching: The past, current status and future directions. *Front. Psychol., 12*(731721). doi:10.3389/fpsyg.2021.731721

Xie, F., & Derakhshan, A. (2021). A conceptual review of positive teacher interpersonal communication behaviours in the instructional context. *Frontiers in Psychology*, *12*, 1–10. doi:10.3389/fpsyg.2021.708490 PMID:34335424

Chapter 6
Application of Experiential, Inquiry-Based, Problem-Based, and Project-Based Learning in Sustainable Education

Ranjit Singha
https://orcid.org/0000-0002-3541-8752
Christ University, India

Surjit Singha
https://orcid.org/0000-0002-5730-8677
Kristu Jayanti College (Autonomous), India

ABSTRACT

This chapter explores integrating pedagogical approaches for sustainable teaching and learning, emphasizing the capacity to meet present requirements without compromising the future. It highlights the merits of experiential, inquiry-based, problem-based, and project-based methods in sustainable practices in education. Experiential learning emphasizes practical application and reflection, whereas inquiry-based learning promotes inquiry and exploration. Problem-based learning immerses students in real-world sustainability challenges and interdisciplinary collaboration, whereas project-based learning enables students to take on leadership roles. Integrating these techniques offers a variety of options for addressing complex environmental issues. Future obstacles include integrating technology and ensuring equitable access. Integrated educational practices require learner-centred approaches, collaboration, and continuous feedback, empowering students to become proactive sustainability advocates and promoting positive change for a sustainable future.

DOI: 10.4018/978-1-6684-9859-0.ch006

1. INTRODUCTION

Sustainable refers to the ability to fulfil current needs so that it does not hinder the capacity of future generations to meet their requirements. It includes environmental, social, and economic dimensions. In other words, sustainable endeavours to balance ecological health, social equity, and economic prosperity so that human activities and the natural world can coexist harmoniously. As global challenges such as climate change, biodiversity loss, social inequality, and resource depletion intensify, adopting sustainable practices becomes essential for the survival and prosperity of humanity and the planet. Educating current and future generations about sustainability is necessary to form responsible and well-informed global citizens who can positively contribute to creating a more sustainable future. Effective education for sustainable practice necessitates innovative pedagogical strategies beyond conventional teaching techniques (Lim et al., 2022; Jeronen, 2022; Salovaara et al., 2019). This chapter examines several essential approaches that have effectively engaged students and nurtured a comprehensive understanding of sustainable issues. These methods include: Experiential learning involves hands-on experiences that enable students to engage with situations and challenges from the real world actively. By engaging in sustainable-related activities, students acquire practical skills, critical thinking skills, and a heightened awareness of their actions' environmental and social consequences.

Inquiry-based learning encourages students to query and investigate sustainable issues independently. Learners develop a deeper comprehension of complex problems through investigation and analysis, enabling them to propose inventive solutions and take ownership of their learning journey. Problem-based learning focuses on presenting students with authentic, sustainable issues. Students develop a strong sense of responsibility and purpose through collaborative problem-solving and applying theoretical knowledge to real-world challenges. Project-based learning entails the development of long-term, hands-on, sustainable-related initiatives. These initiatives inspire a sense of empowerment and accomplishment by encouraging students to collaborate, implement interdisciplinary knowledge, and innovate while confronting real-world problems (Smith et al., 2022; Herranen et al., 2020; Jonassen & Hung, 2012; Lam, 2012). This chapter sets the groundwork for investigating pedagogical approaches that can effectively promote teaching and learning for sustainable practice. By delineating and emphasizing the significance of sustainable practice, we establish the context for why such educational endeavours are crucial in the modern world. Experiential, inquiry-based, problem-based, and project-based learning provides educators with valuable tools and insights. By scrutinizing the advantages and implementations of each approach, readers will comprehend how these methods can effectively promote sustainable education. This chapter provides examples of successful deployments, emphasizes best practices, and discusses potential obstacles and their solutions. After reading this chapter, readers will gain the knowledge and inspiration to incorporate these pedagogical approaches into their teaching practices, nurturing a culture of sustainable practices and empowering future generations to become proactive agents of positive change.

This Chapter explores the pedagogical underpinnings of problem-based, project-based, inquiry-based, and experiential learning, all grounded in seminal educational theories. Experiential learning, a methodology that emphasizes direct, hands-on experiences within a learning cycle consisting of concrete experiences, reflective observation, abstract conceptualization, and active experimentation, is examined in light of the insights of educational theorist David Kolb. Experiential learning is a highly pertinent component of sustainable education, as it has been demonstrated to be a highly effective instrument that stimulates students to confront tangible sustainability issues while nurturing a deep understanding of these challenges. It is emphasized that inquiry-based learning, which is predicated on the notion that students construct knowledge by actively exploring, fosters problem-solving and critical thinking abilities, particularly in the context of sustainability. Constructivist in nature, problem-based learning is proposed as an instructional approach that encourages students to collectively tackle complex sustainability challenges, thereby cultivating critical thinking skills and the ability to generate workable solution proposals. Furthermore, the discourse revolves around project-based learning, which is grounded in constructivism and situated learning. Its focal point is on authentic, intricate assignments that foster ingenuity, collaboration, and tangible resolutions to real-world sustainability dilemmas. The primary objective of this chapter is to examine and investigate the productive implementation of these pedagogical methods in the context of sustainable education. Through an analysis of the theoretical foundations that support these methodologies, this chapter aims to offer educators valuable perspectives that can inform the development and execution of inventive pedagogical approaches. The extent of this study includes an investigation into the theoretical underpinnings, an exploration of the practical implementations of sustainable education, an evaluation via case studies, a discourse on the advantages and difficulties associated with these approaches, and the provision of suggestions to assist educators in incorporating them into their practices for the purpose of enriching sustainability education and providing students with meaningful learning experiences.

2. EXPERIENTIAL LEARNING FOR SUSTAINABLE

Experiential learning is an educational method that emphasizes direct experience and reflection. Experiential learning is founded on the idea that learners acquire knowledge, skills, and values most effectively when engaging in meaningful experiences. Active participation in real-world activities allows students to interact directly with sustainable-related challenges and solutions (Bergmann et al., 2021). In experiential learning, Hands-On Experience refers to the direct engagement of students in practical, real-world actions pertinent to the subject being studied, in this instance, sustainable. Instead of relying exclusively on traditional lectures and theoretical concepts, students actively engage with the material by participating in immersive practical experiences. These experiences may include field excursions, simulations, investigations, community service initiatives, or hands-on seminars. Hands-on experiences in sustainable education may also involve conducting environmental surveys, participating in conservation efforts, devising sustainable projects, or engaging with local communities to resolve sustainable challenges (O'Brien et al., 2021). The value of hands-on experience resides in its capacity to foster a deeper and more meaningful comprehension of sustainable principles (Ferreira et al., 2006). By actively engaging with the material, students can observe the real-world effects of sustainable issues and investigate potential solutions. This practical knowledge reinforces the theoretical concepts taught in class and enables students to comprehend the complexity of sustainable challenges.

In addition to fostering a sense of empowerment and agency in learners, Thuneberg et al. (2018) hands-on experiences enable a sense of autonomy. Kalamas et al. (2017) When students actively participate in sustainable-related problem-solving and decision-making processes, they feel more connected to the issues and become proactive agents of change in their communities. This experiential approach fosters a sense of responsibility and encourages students to assume responsibility for their learning and its impact on the world. Moreover, Othman Amiruddin (2010) states that experiential learning stimulates multiple senses and various learning styles. Visual, aural, and kinesthetic learners can all benefit from experiential learning opportunities, as active engagement improves information retention and recall by allowing learners to experience concepts tangibly. In sustainable education, hands-on activities strengthen the relationship between students and the natural world. It cultivates a sense of responsibility for environmental stewardship, sustainable practises, and a deeper appreciation for the natural world. Educators can create transformative learning experiences beyond textbooks and lectures by incorporating hands-on activities into sustainable education. Learners are encouraged to think critically, employ theoretical knowledge in practical contexts, and partake actively in creating a more sustainable future. Ultimately, experiential learning enriches the learning journey and equips students with the skills and motivation to address real-world sustainable challenges effectively.

Learners engage in reflective practises following the experiential activity (Brackenreg, 2004), critically analyzing their experiences, observations, and emotions. This reflection promotes a more profound comprehension and comprehension of sustainable issues. Following the hands-on experience, reflection is an essential component of experiential learning. It entails students pondering their experiences, thoughts, emotions, and observations and analyzing their learning journey critically. Learners are encouraged to make connections between their experiences and the theoretical knowledge they have acquired through reflection.

Students reflect through experiential learning, remembering the tasks, difficulties, and results of their real-world encounters. Throughout these encounters, they identify vital exchanges, events, and insights. The appraisal of the values, abilities, and knowledge they have acquired from the experiencing activities is a part of this reflective activity. The students evaluate how these experiences have affected their comprehension of sustainable ideas and their ability to put theoretical knowledge into practice. Critical thinking is developed in students through reflection, which is essential. It challenges them to think critically, consider other viewpoints, and investigate the underlying causes and ramifications of the sustainable concerns they came upon. In addition, students consider their attitudes, personal growth, and emotional reactions within the framework of the experiential exercise. These self-examinations could reveal areas where the individual needs to improve, such as empathy, awareness, or a stronger sense of duty towards sustainable practice. Through introspection, learners establish connections between their encounters and more comprehensive frameworks and personal facets. To enhance their overall learning experience, they consider how to apply the knowledge they have gained to future circumstances and real-world sustainable concerns.

Reflection is used in experiential learning as a compass to help shape future goals and pinpoint areas that require improvement. It turns encounters into chances for deep understanding through focused questions and group discussions. Employing introspection, students assimilate knowledge, acquiring a more profound comprehension of sustainable practice and developing into proactive worldwide citizens dedicated to promoting constructive transformation. According to Chiang and Xu (2021), experiential learning promotes active engagement, empowering students to take charge of their education and make links between theory and practice. Within this approach, learners' dynamic and participatory role in their educational journey is called active engagement. Through collaborative projects, problem-solving exercises, and hands-on activities, students actively engage with the subject matter instead of passively acquiring knowledge. Students who actively participate in their education can better bridge the gap between theoretical knowledge and practical applications, taking ownership of their learning process and developing a deeper and more meaningful understanding of the material. Students use interactive exercises that require critical thinking and problem-solving abilities in this type of learning. The process is engaging and enjoyable because this active engagement piques interest and motivates learning. Direct experience and learning by active participation are highly valued in experiential learning. To help students see the practical applications of their study ideas, educators urge them to investigate, try, and make judgments in authentic settings. Using this method, teachers provide questions to their pupils relevant to the world and call for using their knowledge and abilities to solve. Making decisions and solving problems helps students develop their critical thinking abilities and prepares them for real-world, practical problem-solving in various contexts.

Students' intrinsic motivation is sparked by actively participating in experiential learning activities. They become more fully involved in the learning process because they are sincerely interested in the material and want to make a difference. Students might also investigate subjects that are meaningful to them personally through experiential learning. Students are more likely to remember and use the knowledge they learn if it is connected to their interests and experiences. Coordination and teamwork are required for many experiential learning activities as students collaborate in groups and share ideas and viewpoints. This improves their communication abilities and cultivates a sense of group accountability for tackling environmental issues. Reflection serves as a bridge between experiential knowledge and theoretical comprehension, complementing active engagement in experiential learning. Students are positioned as change agents ready to apply their knowledge and abilities to real-world situations because their agency and ownership propel a deeper exploration of sustainable principles. This all-encompassing method fosters vital life skills such as critical thinking, problem-solving, teamwork, and a sense of responsibility for a more sustainable future, in addition to helping students get a better grasp of sustainable practices. According to Yang (2012), experiential learning promotes students' engagement with sustainable issues while fostering their creative and problem-solving abilities. According to Kreber (2001) and Montesinos et al. (2023), problem-solving in experiential learning entails posing real-world sustainable issues or challenges to students, who must then apply their knowledge, abilities, and imagination to find answers. This idea seeks to develop students' capacity for critical thought, in-depth analysis of complex problems, and creative problem-solving in the face of sustainable issues. Experience learning activities that are thoughtfully created to reflect the sustainable challenges that communities and organizations confront accurately. These activities enable learners to proactively address problems sustainable by assuring their relevance and significance, thereby creating meaningful learning experiences that resonate with real-world challenges. Students are exposed to the complexity of real-world issues through experiential learning, which motivates them to think about multiple possible solutions. Students learn to bridge the gap between theory and practice by applying the theoretical knowledge they have learned in the classroom to real-world circumstances during problem-solving activities. This helps students retain and better comprehend sustainable principles.

When students solve problems in the context of experiential learning, they are encouraged to employ critical thinking skills as they investigate the underlying causes of sustainable issues, consider the possible outcomes of different solutions, and evaluate the viability of each suggested solution. This technique fosters a comprehensive awareness of the issues and improves their analytical abilities. Collaborating to solve problems frequently requires students to work in groups and use their peers' many viewpoints and experiences. Through experiential learning, students are encouraged to use an interdisciplinary approach, combining their knowledge from several fields to create well-rounded answers. Additionally, experiential learning encourages creativity, motivating students to investigate novel solutions to sustainable-related problems. Students are likelier to take chances and try novel concepts when given a secure and supportive environment. After participating in problem-solving exercises, students evaluate their experiences and results, identifying areas for development and refining their strategies. This iterative procedure encourages continuous improvement and learning. Through problem-solving in experiential learning, students gain the confidence and independence to take the lead as change agents. It is vital to sustainable education because it gives students the hands-on experience they need to solve real-world problems. Making ethically conscious and accountable decisions promotes informed and long-lasting decision-making. Integrating problem-solving techniques into hands-on learning fosters a dedication to constructive change and the advancement of society. Students participate in the practical development and maintenance of a sustainable school garden while learning essential lessons about organic farming, biodiversity conservation, and decomposition techniques. They also travel to nearby ecosystems—forests, wetlands, and coastlines—for educational purposes to observe and study natural processes and ecological relationships. Students also have a deeper understanding of sustainable energy sources through designing, constructing, and managing renewable energy systems, like small wind turbines and solar panels. In addition to these efforts, students actively recognize sustainable issues and, working with local communities, put solutions like recycling and waste management plans into action. This all-encompassing approach gives them a thorough awareness of sustainable practices and environmental stewardship.

According to Martindale et al. (2023), experiential learning provides students with a deep comprehension of the subject by enabling them to grasp sustainable ideas concretely and relevantly. Muzaffar et al. (2023) note that this technique increases the relevance and practicality of the learning experience by applying theoretical information to real-world circumstances. Nir and Musial (2020) have observed that active involvement in sustainable challenges might elicit affective reactions, increasing awareness and empathy for environmental and social concerns. Experiential learning is essential to fostering a sincere and long-lasting commitment to sustainability.

Experiential learning frequently necessitates additional resources, such as time, money, and access to suitable locations, which might provide difficulties for some educational establishments. Evaluations of experience learning may also be more complex than traditional assessments, necessitating creative and thoughtful approaches to evaluation. Furthermore, some sustainable-related activities could have inherent risks, which means strict safety measures and close supervision are needed to guarantee participants' safety. Consequently, even if experiential learning has many advantages, educational institutions must seriously consider the practical and safety issues that come with it.

Educators should clearly define the learning objectives and outcomes for each activity and match them with sustainable concepts and abilities to optimize the impact of experiential learning. To make the experiences more meaningful, it is imperative to choose experiential activities that directly address regional sustainable issues or global challenges. Before, during, and following these exercises, educators should provide students with support and materials to help with understanding and introspection. It is essential to incorporate structured reflection sessions into the educational process so that students may apply sustainable principles to their experiences and critically assess them. Adding local communities, environmental organizations, or business experts to experiential projects can increase their impact and genuineness. During these exercises, promoting teamwork improves problem-solving abilities and social learning. Prioritizing participant safety above all else will always be necessary. Strict safety protocols should be implemented to reduce hazards and provide a safe learning environment for all parties. By adhering to these recommendations, educators can maximize the potential of experiential learning to motivate and enable students, transforming them into engaged supporters of sustainable Practices and agents of constructive social change in their communities.

3. INQUIRY-BASED LEARNING FOR SUSTAINABLE

The core of inquiry-based learning as an instructional strategy is students' active exploration and investigation of issues and queries. Giving children the freedom to create questions, gather data, evaluate the evidence, and make conclusions promotes curiosity, critical thinking, and problem-solving abilities. The notion that students should formulate questions and problems to guide their study and investigation is at the core of this educational philosophy (Wen et al., 2023). Questioning is fundamental to inquiry-based learning because it fosters critical thinking, a deeper engagement with the material, and piques curiosity. With inquiry-based learning, students can create questions using their innate interests and curiosity. Students are empowered and motivated to actively search for answers and solutions because of this autonomy in the learning process, which fosters higher-order thinking abilities like analysis, assessment, and synthesis. Students reinforce their commitment to finding meaningful answers by thinking critically about the relevance and significance of the selected themes. They also frequently connect the questions they generate and their personal and community life. Students are encouraged to investigate a variety of subjects through inquiry-based learning, which leads to a greater comprehension of sustainable issues and a diversity of viewpoints that enhances class discussions. Due to this process, students are continuously refining their inquiry, learning new information, and digging deeper into their questions. Developing questions gives students agency in their learning process by enabling them to look for solutions independently, conduct research, and make connections to what they already know.

Within inquiry-based learning, practical questions foster relevance and application. Students who develop their research and critical thinking abilities become curious knowledge consumers who promote an atmosphere of inquiry and improved understanding. This strategy uses questions as a spark for learning and better comprehension, encouraging lifelong learning and preparing people to become autonomous problem solvers and thinkers regarding environmental issues. To find answers to their concerns, students actively participate in the research process by performing surveys, experiments, interviews, and data analysis (Chin & Chia, 2004). One of the main components of inquiry-based learning is active investigation, which calls for students to participate in the research process to answer their questions actively (Spronken-Smith et al., 2013). Instead of being passive consumers of knowledge, learners become active researchers who perform surveys, experiments, interviews, and data analysis to solve issues and find answers to concerns. Active participation in research activities, such as data collecting, phenomenon observation, and variable manipulation, is required by inquiry-based learning. Through direct engagement with the subject matter, this hands-on method allows students to gain more profound knowledge of sustainable ideas. Students connect their learning experiences to fundamental concerns by actively investigating and acquiring information from real-world circumstances, which makes the knowledge they develop incredibly relevant. They regularly work with others to plan and carry out experiments or research projects to foster collaborative learning. This cooperative endeavour enables the sharing of concepts, information, and viewpoints, improving cooperation and communication. Additionally, students gain problem-solving abilities when they critically assess data, make connections between seemingly unrelated bits of information, and draw conclusions from their investigations backed by evidence.

Active exploration encourages students to be passionate and committed to the subject matter by giving them the tools to take responsibility for their education and learning. A thorough understanding of the research process replaces mindless data memorizing when active investigation is strongly emphasized. Students gain transferable research skills, such as gathering, analyzing, and evaluating data. A cyclical inquiry process is encouraged by inquiry-based learning, which fosters ongoing learning and discovery.

Students get better at analyzing material, become more adept at critical thinking, and comprehend the nuances of sustainable issues. This practical inquiry also fosters accountability and stewardship, inspiring students to take the initiative as leaders for long-term change. Theobald and Ramsbotham (2019) emphasize that inquiry-based learning strongly emphasizes the application of evidence-based reasoning and critically examining information to draw well-informed conclusions. This method uses reliable and pertinent data to support any findings, recommendations, or claims made throughout the investigation. Pupils are urged to evaluate material critically, draw connections between facts and ideas, and make evidence-based judgments. They use various methods to gather data, including surveys, observations, experiments, and interviews. Then, they carefully examine the data for associations, patterns, and trends. To guarantee the correctness and dependability of their evidence, students also assess the authenticity and trustworthiness of the information sources they come across while conducting research. They do this by differentiating between trustworthy and untrustworthy sources. They then integrate theoretical and empirical knowledge to improve their understanding of sustainable principles by relating the data they gather to the theoretical concepts and principles covered in class.

Students are also encouraged to consider various views and points of view using evidence-based reasoning. They investigate the problem from multiple perspectives to have a thorough grasp and openly explain how they arrived at their findings based on the available data. This technique makes it easier to communicate effectively and allows people to evaluate the soundness of their arguments. Students participate in an inquiry cycle as they assess the facts and make judgments. They might rethink their choices or develop new inquiries in light of their discoveries, encouraging ongoing learning and growth. Evidence-based reasoning encourages moral research techniques, critical thinking, and well-informed decision-making in inquiry-based learning. Students acquire a scientific viewpoint, questioning their preconceptions and looking for trustworthy data to solve sustainable-related problems. This method educates students to be astute information consumers and knowledgeable global citizens, contributing to a sustainable future.

Learners take responsibility for their learning voyage by making decisions and directing their inquiries based on their interests and prior experiences. Autonomy and ownership are essential inquiry-based learning principles that empower students to take charge of their learning journey. This method encourages students to be self-directed, selecting the topics they wish to investigate and assuming responsibility for their learning (Levy et al., 2009). Autonomy and ownership foster a sense of agency and motivation as learners actively participate in decision-making and inquiry direction. Students in inquiry-based learning can choose the issues and queries they want to research. Their decisions are frequently shaped by their passions and life experiences, which forges a profound and fulfilling bond with sustainable concerns. Through these self-selected investigations, students can make connections between the course material and their own lives and communities, increasing their commitment to learning and deepening their comprehension of sustainable ideas. Students who choose their own learning objectives, research strategies, and study direction exercise this autonomy in designing and managing their investigations. Their independence fosters their sense of accountability and ownership for their educational journey. Effective time management is also necessary because students need to set aside time for research, gathering data, analyzing it, and reflecting. This helps them acquire critical time management abilities and a stronger sense of accountability. By reflecting on their learning process and progress, learners develop metacognitive abilities through the reflective practice ingrained in this approach. These skills allow learners to assess their areas of strength and improvement. Their capacity to draw lessons from their experiences and modify their tactics as necessary is improved by this reflective process. By letting students follow their interests, inquiry-based learning enables customized learning experiences catered to each student's needs, preferences, and learning styles. Students with this freedom and ownership are more equipped to solve problems on their initiative and promote sustainability. Motivated by their curiosity, they investigate issues with a spirit of investigation, emulating the ideas of lifelong learning. These profoundly fulfilling and transforming encounters encourage deep participation, agency, and inquiry, all advancing sustainable education.

Students participate in various practical tasks as a component of their learning journey through experience. They compute their carbon footprints and investigate ways to lower the carbon emissions from both their own and their communities. Field surveys serve as their classrooms as they discover and catalogue local plant and animal species and explore the importance of biodiversity conservation. Students use nearby rivers and lakes as laboratories to investigate water quality, assess the effects of contamination, and think through possible remediation strategies. Students also explore the intricate world of social injustices connected to environmental problems, like the consequences of climate change on marginalized groups or the availability of clean water. These investigations are all a part of inquiry-based learning, which involves students actively and helps them retain and comprehend sustainable ideas and concepts. Using scrutinizing data, formulating conclusions, and putting forth solutions to sustainable problems, learners cultivate the mental faculties necessary to effect constructive change. In addition, the inquiry method fosters curiosity and encourages independent study outside of the classroom, enabling students to tackle real-world sustainable issues and improve their educational experience. Inquiry-based learning, emphasizing investigation and reflection, may require more time than traditional lecture-based approaches.

Additionally, some students might need extra support in formulating inquiry questions and conducting independent research, as individual inquiries can vary in scope and methodology, making evaluation challenging. To navigate these challenges effectively and create a rich inquiry-based learning experience, several strategies can be employed. First, it is crucial to clearly define the problem or inquiry question, ensuring it is relevant, manageable, and aligned with sustainable objectives. Providing scaffolding and guidance to help students develop appropriate research methodologies and navigate the inquiry process can be immensely beneficial. Access to relevant resources, including books, articles, online databases, and field experts, is essential. Furthermore, encouraging peer collaboration, where students can discuss their investigations, share findings, and provide constructive criticism, enriches the learning experience. Opportunities for review throughout the inquiry process allow students to assess their progress, identify obstacles, and understand their learning outcomes. Finally, it allows students to present their research and findings to the larger school or community to celebrate their accomplishments. It nurtures a sense of pride and a commitment to sustainability. By implementing these strategies, educators can create a stimulating and effective inquiry-based learning environment that empowers students to investigate sustainable issues, develop problem-solving skills, and become informed and engaged global citizens.

4. PROBLEM-BASED LEARNING FOR SUSTAINABLE

Problem-based learning is an educational strategy that presents students with sustainable-related real-world problems. Students work collaboratively to investigate and resolve these issues, utilizing their existing knowledge and acquiring new skills (Lehmann et al., 2008). In problem-based learning, the core principle revolves around authentic and meaningful problems that reflect real-world sustainable challenges, thus promoting student engagement and motivation. These problems are carefully selected to mirror the complexities and dynamics of actual sustainable challenges communities, organizations, and society face. Learners can readily relate to and find value in these authentic dilemmas, as they are grounded in real-world situations. By working on issues with tangible, practical implications, students can observe the real-world ramifications of their education. The authenticity of these problems significantly heightens student engagement and motivation, as learners are naturally drawn to issues directly relevant to their lives and the broader world. Authentic cases often involve multiple disciplines and facets of sustainable practices. They encourage students to draw on knowledge and perspectives from various fields as they investigate these problems, promoting interdisciplinary learning. Addressing these authentic issues necessitates critical thinking and problem-solving skills. Students must analyze the complexities of the problems, consider multiple perspectives, and develop viable solutions. Original problems frequently call for collaborative efforts, with students working in teams to share ideas, merge knowledge, and devise comprehensive solutions. This collaborative approach fosters effective coordination and communication skills. Students integrate their existing knowledge with newly acquired skills and concepts when solving authentic problems, enhancing their comprehension of sustainable practices as a multifaceted and interconnected concept. Solving genuine problems in Problem-based learning leads to significant and practical learning outcomes. It equips students with essential skills and a deeper understanding of sustainable issues, preparing them for real-world challenges. The authenticity of these problems instils a sense of purpose and accountability in students as agents of positive change. Active participation empowers proactive problem-solvers to navigate ambiguity and effectively address sustainable challenges. Education is seamlessly connected with the natural world through these carefully crafted dilemmas, fostering practical skills, critical thinking, and a commitment to sustainable practices. These challenges are powerful tools to prepare learners to tackle real-life complexities and become proactive agents of positive change.

According to Grabinger and Dunlap (2002) and Yu and Zin (2023), students actively engage in problem-solving, investigation, analysis, and decision-making to address the sustainable concerns provided. This promotes critical thinking and the practical application of knowledge. A fundamental tenet of problem-based learning is active learning, which emphasizes students' active participation in the learning process. In Problem-based learning, students tackle real-world sustainable issues through practical problem-solving, investigation, analysis, and decision-making. This method promotes critical thinking, in-depth understanding, and the practical application of knowledge in real-world situations. Students must perform research and inquiry to obtain pertinent information and data about sustainable challenges within the framework of Problem-based learning's active learning. They pick up skills like data analysis, finding reliable sources, and drawing conclusions backed up by facts. Students also synthesize and evaluate the data they have acquired throughout their investigation. They combine many data sources, spot patterns and trends, and create a thorough grasp of the current sustainable problem. Their diversified strategy helps them to tackle complex sustainable issues successfully. Through sharing ideas, viewpoints, and specialized expertise, student teams working together to address sustainable concerns are encouraged to learn actively through problem-based learning. This cooperative setting promotes effective communication, teamwork, and the sharing of different perspectives. Students in Problem-based learning participate in decision-making processes based on their examination and assessment of viable solutions, which motivates them to think about the moral, social, and environmental ramifications of their decisions. Problem-based learning requires active learning, which requires students to apply their prior knowledge and skills to real-world situations. Students learn more about the subject and its applicability in real life by converting abstract ideas into practical solutions for real-world issues. They go through a process of self-reflection and iterative learning as they solve problems, assess their development, pinpoint areas that need work, and hone their strategies in response to criticism and fresh data. Giving pupils a sense of ownership over the problems they are solving is one of the main benefits of active learning in Problem-based learning. Being personally invested in creating workable answers fosters a more profound commitment to learning because of this sense of ownership. Problem-based learning also uses interviews, fieldwork, and simulations to offer valuable perspectives on sustainable practices. It builds dynamic learning environments that encourage critical thinking and active problem-solving among students. In line with constructivist ideas, students actively create a long-lasting understanding of sustainable topics.

Additionally, as observed by Sroufe and Ramos (2015) and Stentoft (2017), Problem-based learning frequently entails integrating information from many disciplines, reflecting the complexity of sustainable concerns and supporting multidisciplinary education. A key component of Problem-based learning is the interdisciplinary approach, which combines viewpoints and knowledge from different sectors to address sustainable issues. Problem-based learning fosters in students an awareness of the complex and interwoven nature of sustainable concerns by encouraging them to investigate and comprehend these issues from various perspectives. Because problem-based learning takes an interdisciplinary approach, it helps students understand sustainable challenges comprehensively by considering the social, environmental, economic, and cultural elements of problems. It invites students to examine these many components' interactions and influences. Problem-based learning facilitates the interchange of diverse viewpoints and knowledge by encouraging students from different academic fields to work together on a common sustainable issue. Students are urged to incorporate ideas and information from other areas into their approach to problem-solving, finding links and intersections pertinent to sustainability. This kind of interdisciplinary teaching fosters systems thinking, which is the capacity to understand how different components of a system are interdependent and interconnected. It is essential to approach complex sustainable concerns from a systemic perspective. Problem-based learning regularly crosses conventional disciplinary lines, encouraging students to engage in transdisciplinary learning experiences where they investigate subject areas that are not limited to a single discipline. This is similar to real-world scenarios where addressing sustainable issues requires knowledge and insight from various fields. Problem-based learning involves students critically analyzing sustainable matters from different disciplinary angles, weighing the benefits and downsides of multiple theories and approaches, and making more informed decisions. Problem-based learning's multidisciplinary approach encourages lifelong learning by giving students the tools to adapt and apply their knowledge in various situations. It captures the complexities of reality and enables flexible problem-solving in cooperative groups to tackle global concerns. Students thereby transcend departmental barriers and get an appreciation for interconnection, which promotes sustainable and constructive change. This method deepens understanding and motivates creative, all-encompassing thinkers committed to a sustainable future. Under the direction of teachers, Problem-based learning students take an active and autonomous part in their educational process, directing their approaches to problem-solving and formulating their learning goals. According to Peterson (2004), a fundamental tenet of Problem-based learning is the student-centred approach, which puts the students at the centre of the learning process.

Under inquiry-based and problem-based learning approaches, students can take charge of their education. Students are free to choose the topics of study, hobbies, and learning goals they want to pursue concerning the sustainable issue being presented. This freedom to choose allows for customized learning pathways that meet the needs and preferences of students with varying learning styles. It also promotes a sense of ownership and responsibility for education. Instead of being traditional teachers, educators in problem-based learning play the role of facilitators. They offer direction, materials, and assistance to foster an atmosphere that encourages student inquiry, exploration, and discovery. Since students are encouraged to find meaningful answers, the student-centred approach promotes active involvement and improves learning and comprehension. They take charge of their education, actively participate in it, track their development, pinpoint their strengths, and, through recurring evaluations, address their areas for growth. Through developing lifelong learning abilities like critical thinking, flexibility, and self-evaluation, this method prepares pupils for learning outside of the classroom. Through active participation in solution-finding and tackling intricate sustainable issues, students feel accountable for their tangible consequences. This method aligns with constructivist ideas, giving students the self-reliance and drive to address sustainable-related problems. It also cultivates curiosity, creativity, and resilience to encourage a love of learning and a commitment to a sustainable future.

In summary, students can contribute to a more sustainable world through inquiry-based and problem-based learning. Students collaborate to design plans for sustainable urban development within the collaborative framework of these learning methodologies. They address topics like green spaces, public transit, and infrastructure for renewable energy to construct socially and environmentally inclusive cities. They look into problems with trash management in their neighbourhood and suggest long-term fixes, such as recycling plans, efforts to promote decomposition, and waste minimization techniques. Students conduct research and create plans for communities to adjust to the consequences of climate change while considering ways to lessen risks and boost resilience. In addition, they assess the risks to regional biodiversity and, working together, create conservation strategies for ecological corridors, habitat preservation, and community education initiatives. A "problem-based learning" strategy actively involves students in real-world sustainable challenges, strengthening their comprehension of the subjects and encouraging the application of information in real-world contexts. Using this approach, students examine and suggest solutions to challenging sustainable issues, honing their critical thinking, problem-solving, and decision-making abilities. In addition to encouraging teamwork and clear communication, problem-based learning reflects the multidisciplinary character of actual sustainable programs. Students' capacity to solve problems independently gives them the confidence to take charge of their education and act as proactive, sustainable agents. Teachers may find that putting problem-based learning into practice requires many resources since it takes time and energy to create real-world challenges, encourage student inquiry, and organize group projects. Educators are responsible for leading and encouraging learning so that students fulfil their learning goals. Because there are many different ways to solve a problem, assessing results can be difficult. Educators should select sustainable challenges pertinent to the local context to ensure authenticity and significance and match them with learning objectives when creating compelling problem-based learning experiences. They should assist students in defining issues, establishing goals, and locating pertinent resources. Encouraging students to use their knowledge from several fields to create holistic answers is crucial. Giving students the chance to receive feedback and evaluations regularly enables them to track their development, overcome challenges, and perform better. Furthermore, it fosters accountability and recognition when students can showcase their answers to a larger group of people, including the school community or nearby stakeholders. With the help of these techniques, teachers may create engaging, problem-based learning opportunities that provide students with the information, drive, and abilities to take the initiative to affect positive change and contribute significantly to sustainable development.

5. PROJECT-BASED LEARNING FOR SUSTAINABLE

A teaching strategy known as project-based learning (PBL) involves students in lengthy, in-depth projects that deal with real-world issues. This approach strongly emphasizes teamwork, active exploration, and applying knowledge and abilities to generate tangible outcomes (Rahmawati et al., 2022; Liao et al., 2023). Project-based learning initiatives are grounded in real-world problems, connecting education to practical applications and fostering a feeling of significance and direction. Through problem-based learning, students are encouraged to collaborate in groups, simulating real-world situations in which interdisciplinary teamwork is essential for resolving challenging sustainable issues. Through research, critical thinking, and problem-solving, students pose questions, look into concerns, and consider potential solutions, all contributing to their deep understanding of sustainable challenges. Project-based learning usually takes place over a more extended period, allowing students to study the subject matter more thoroughly. This helps them explore the topic thoroughly and develop comprehensive solutions. Students use their knowledge and abilities to produce concrete results, refining their comprehension and skills throughout the project by commenting on their experiences and learning. Through developing a strong link between education and practical application, project-based learning empowers students to become change agents capable of effectively addressing sustainable issues. However, students are connected to sustainable challenges through problem-based learning, which gives the learning process significance and relevance. Students gain essential cognitive abilities by dissecting intricate issues, looking into possible fixes, and weighing the consequences. Coordination and cooperation are fostered by problem-based learning, which mimics the teamwork required to solve real-world sustainable challenges. Real-world issues can inspire students to actively solve problems and advocate for sustainability outside of the classroom.

On the other hand, creating problem-based learning activities could require a lot of time, resources, and expertise. It can be challenging for teachers to support problem-based learning because they must balance mentoring students and letting them conduct independent research. Accurately assessing problem-based learning can be difficult because students' methods and solutions to the same topic vary greatly. Teachers can use several techniques to design a problem-based learning program that is both successful and sustainable. One of these tactics is to develop real-world, clearly defined sustainable problems that fit the student's learning objectives and context. It is also critical to give students the support and direction they need and ensure they have access to tools and experts who can help them solve problems. Encouraging students to synthesize information from several fields of study fosters a comprehensive comprehension of sustainable concerns. For students to grow, they must have frequent opportunities to evaluate their progress, pinpoint challenges, and modify their approaches to problem-solving. Setting up events where students can showcase their solutions to the greater community will help them feel proud of their accomplishments. To improve the design and execution of problem-based learning activities for sustainable practices and, eventually, produce a more significant and impactful educational experience, instructors and students must provide feedback. By putting these techniques into practice, teachers may create problem-based learning opportunities that enable students to take the initiative to contribute to sustainable practices and develop into responsible global citizens.

6. INTEGRATING PEDAGOGICAL APPROACHES FOR SUSTAINABLE EDUCATION

Integrating multiple pedagogical approaches in sustainable education increases the efficacy and profundity of students' learning experiences (Fiel'ardh et al., 2023). Educators can create a more comprehensive and holistic learning environment by combining courses with their respective assets and benefits. Integrating instructional strategies facilitates: Students are exposed to various learning experiences that accommodate different learning styles and preferences, enhancing their comprehension of sustainable concepts. Students engage with real-world issues by incorporating experiential, inquiry-based, problem-based, and project-based learning, making the learning process more relevant and meaningful. Different approaches emphasize distinct abilities, such as critical thinking, collaboration, communication, and creativity, equipping students with many skills required to address sustainable challenges. Integrating multiple approaches enables students to take an active role in their learning, fostering a sense of ownership, agency, and responsibility for their education and its global impact. Through inquiry-based learning, students collaborate to identify sustainable issues in their community. The participants then engage in experiential and project-based activities to resolve these issues, implementing solutions and reflecting on their impact. Students are presented with a real-world dilemma in sustainable practices and use inquiry-based learning to investigate its complexities. The students then design and prototype sustainable solutions through problem- and project-based learning. The purpose of field excursions for students is to observe ecosystems, investigate environmental issues, and collect data. They analyze their findings through inquiry-based learning and propose sustainable management strategies through problem-based learning. Students can develop a deeper and more extensive comprehension of sustainable concepts and their practical applications through integration. As they participate in interactive and hands-on activities, diverse instructional strategies keep students engaged and motivated. Integrating approaches allows students to acquire various skills, fostering well-rounded growth and readiness for future challenges. Integration of multiple processes necessitates meticulous planning and coordination among educators to ensure a smooth flow from one activity to the next. Combining various techniques may require additional time, which educators must effectively manage within the curriculum.

Creating assessments aligned with integrated learning experiences can be difficult and requires deliberate evaluation strategies. Ensure that the integration of pedagogical approaches is consistent with the desired learning outcomes and sustainable objectives. Plan the order of activities so that they follow a logical progression and build upon one another, resulting in a coherent and interconnected learning journey. Allow for flexibility and adaptability in the learning process, recognizing that students may advance at different rates and have diverse learning requirements. Encourage collaboration among educators from various disciplines to develop a unified strategy that capitalizes on their expertise. Keep the focus on the students' interests, conditions, and voices, allowing them to determine the course of their educational experiences. Include regular student reflection and feedback opportunities to enhance the integrated approach continuously. By integrating experiential learning, inquiry-based learning, problem-based learning, and project-based learning with care, educators can create a dynamic and effective sustainable education that equips students with the knowledge, skills, and motivation to become agents of positive change in the world.

7. ASSESSING LEARNING OUTCOMES FOR SUSTAINABLE EDUCATION

A thorough evaluation of student's knowledge, abilities, attitudes, and behaviours about sustainable ideas and activities is necessary to determine sustainable education's effectiveness. For instructors to monitor student progress, evaluate the effectiveness of their educational practices, and pinpoint areas for development, this evaluation is crucial. It is also essential to ensure that students are equipped to deal with sustainable issues in the real world and support sustainable development and lifestyle. Teachers use assessment techniques, including standard exams and tests and practical exercises like sustainable projects, problem-solving activities, and presentations. To show their growth and learning path, students frequently put together portfolios of their work that include written comments, research papers, and project outcomes. While student observations in experiential or project-based learning activities offer insights into their real-world application of sustainable principles, peer evaluations and self-assessments encourage metacognition and reflection. Students' opinions, values, and viewpoints regarding environmental issues are measured through surveys and interviews.

Aligning the assessment with the program's or course's targeted learning outcomes and sustainable goals is essential to its effectiveness. Different aspects of sustainable learning, such as knowledge, skills, and values, are captured by different evaluation methodologies. Assessment techniques should reflect the challenges associated with sustainable practices to enable students to apply their knowledge in real-world scenarios. Uniform assessment is more accessible and transparent for students with clear evaluation criteria and rubrics. For students to grow, they need timely and helpful feedback that helps them evaluate their performance and draw links between their achievements and sustainable principles. One way to assess student sustainable projects is to look at factors like how innovative they are, how well they solve problems, and how much they consider the effects on the environment and society. It is possible to evaluate research publications on sustainable issues based on their critical thinking, evidence-based reasoning, and depth of study. Students' presentations on sustainable topics are assessed based on their ability to communicate complicated ideas and their clarity and persuasiveness effectively. Surveys can evaluate students' commitment to sustainable behaviours by measuring their views and values on sustainability. Lastly, the efficiency with which students collaborate on sustainable projects can be a basis for assessing their leadership, teamwork, and communication abilities. Through the implementation of various assessment methodologies that are in line with the principles of sustainable education, educators can acquire a thorough comprehension of the learning outcomes of their students. This strategy encourages the growth of aware, proactive, and socially conscious global citizens dedicated to sustainability.

DISCUSSION

Quantitative and qualitative methods can be employed to assess the effectiveness of pedagogical approaches in sustainable education. These methods include pre-and post-assessments to measure changes in students' sustainable knowledge and skills, surveys to gauge shifts in attitudes and values, rubrics and grading criteria for project evaluation, observations and reflections for practical application, self-assessments and peer evaluations to promote self-reflection, action research, case studies, interviews for deeper insights, tracking long-term impacts, and qualitative analysis of written reflections and narratives. Employing this array of assessment methods allows for a comprehensive grasp of how pedagogical approaches contribute to advancing sustainable education and cultivating students' dedication to sustainable living and development.

However, due to the complexity of the learning outcomes, measuring the effectiveness of pedagogical approaches in the context of sustainable education presents several challenges. Education for sustainable practices seeks to cultivate knowledge, skills, attitudes, and actions related to sustainability. A combination of quantitative and qualitative evaluation techniques is required to resolve these obstacles. Quantitative methods, such as pre- and post-assessments and surveys, provide valuable numerical data for monitoring changes in students' knowledge, skills, attitudes, and values regarding sustainable practices. These data points provide a structured method for measuring the impact of pedagogical approaches, allowing for distinct comparisons between before and after. However, relying solely on quantitative data may need to pay more attention to the deeper, more nuanced aspects of sustainable education, such as shifts in students' perspectives and ability to employ knowledge in real-world contexts. In contrast, qualitative methodologies provide insights into students' experiences, personal development, and practical application of sustainable concepts. Approaches such as interviews, case studies, and written reflections delve into the qualitative dimensions of sustainable education, revealing the students' transformative journey. These strategies facilitate the holistic development of students as global citizens committed to sustainable living and development. The complexity of sustainable education, which endeavours to cultivate not just cognitive but also affective and behavioural domains, necessitates a blended approach. By combining quantitative and qualitative evaluation techniques, researchers can portray a more complete picture of the efficacy of pedagogical approaches in sustainable education. This multidimensional assessment is consistent with the holistic objectives of sustainable education, in which knowledge, skills, attitudes, and actions all contribute to the intended learning outcomes. A common constraint observed in a significant portion of the current body of research is the tendency to solely utilize either qualitative or quantitative methodology to obtain results, with a prevailing inclination towards quantitative methods. The disproportionate focus on a singular process is a notable limitation in numerous research outcomes. It is expected that in the future, scholars will progressively investigate and embrace blended approaches as a means to alleviate this constraint. This evolutionary process has the potential to yield a more comprehensive and holistic understanding of the efficacy of educational methods in the field of sustainable education.

CONCLUSION

In this chapter, we have covered the significance of pedagogical approaches in sustainable education. We started by describing the importance of sustainable practices for addressing global problems. Following that, we looked at project-based, problem-based, inquiry-based, and experiential learning as the four primary pedagogical approaches. Experience learning emphasizes practical experiences and reflective practice, which raises students' knowledge of sustainable concerns. Active inquiry, critical thinking, and problem-solving are all encouraged in inquiry-based learning. Problem-based learning emphasizes the relevance of interdisciplinary teamwork and practical, sustainable mobility challenges. In the end, project-based learning shows how valuable, long-term projects may support students in growing a sense of accountability for sustainable practices.

The combination of these pedagogical approaches has the potential to improve sustainable education. Still, several challenges need to be addressed, such as resource constraints, efficient time management, and the need to identify trustworthy assessment methods. While technological advancements offer opportunities for innovation and the creation of virtual learning environments, it is imperative to ensure equitable access to high-quality, sustainable education for all. Sustainable educators should use a learner-centred approach that encourages students' curiosity, initiative, and feelings of responsibility. The interdisciplinary learning process can be improved when instructors from several areas collaborate. For teachers and students, continual progress is facilitated by regular feedback and self-reflection. Research on sustainable education should assess the value of different pedagogical approaches and the impact of integration on student learning results. Researching innovative uses of technology and the internet can lead to more widely available and adaptable environmental education. We may cultivate a cohort of astute, empathetic, and competent individuals committed to constructing a sustainable future using educational approaches that prioritize active learning, collaboration, and practical application. These teaching strategies will be essential in the future to build a more sustainable world where students become agents of change equipped with the skills, knowledge, and passion to address the most pressing problems of the day.

REFERENCES

Bergmann, M., Schäpke, N., Marg, O., Stelzer, F., Lang, D. J., Bossert, M., & Sußmann, N. (2021). Transdisciplinary sustainability research in real-world labs: Success factors and methods for change. *Sustainability Science*, *16*(2), 541–564. doi:10.1007/s11625-020-00886-8

Brackenreg, J. (2004). Issues in reflection and debriefing: How nurse educators' structure experiential activities. *Nurse Education in Practice*, *4*(4), 264–270. doi:10.1016/j.nepr.2004.01.005 PMID:19038168

Chiang, C., Wells, P. K., & Xu, G. (2021). How does experiential learning encourage active learning in auditing education? *Journal of Accounting Education*, *54*, 100713. doi:10.1016/j.jaccedu.2020.100713

Chin, C., & Chia, L. G. (2004). Problem-based learning: Using students' questions to drive knowledge construction. *Science Education*, *88*(5), 707–727. doi:10.1002/sce.10144

Costes-Onishi, P., & Kwek, D. (2023). Technical skills vs meaning-making: Teacher competencies and strength of inquiry-based learning in aesthetic inquiry. *Teaching and Teacher Education*, *130*, 104152. doi:10.1016/j.tate.2023.104152

Ferreira, A. J. D., Lopes, M. A. R., & Morais, J. P. F. (2006). Environmental management and audit schemes implementation as an educational tool for sustainability. *Journal of Cleaner Production*, *14*(9-11), 973–982. doi:10.1016/j.jclepro.2006.01.003

Fiel'ardh, K., Fardhani, I., & Fujii, H. (2023). Integrating perspectives from Education for Sustainable Development to foster plant awareness among trainee science teachers: A mixed methods study. *Sustainability (Basel)*, *15*(9), 7395. doi:10.3390/su15097395

Grabinger, S., & Dunlap, J. C. (2002, October). Problem-based learning as an example of active learning and student engagement: Invited talk. In *International Conference on Advances in Information Systems* (pp. 375–384). Springer Berlin Heidelberg. https://link.springer.com/chapter/10.1007/3-540-36077-8_39

Herranen, J., Tolppanen, S., Vesterinen, V., & Aksela, M. (2020). Challenges and tensions in collaborative planning of a student-led course on sustainability education. *Nordic Studies in Science Education, 16*(1), 18–36. doi:10.5617/nordina.6583

Jeronen, E. (2022). Sustainable Education. In Springer eBooks (pp. 1–10). doi:10.1007/978-3-030-02006-4_237-1

Jonassen, D. H., & Hung, W. (2012). Problem-Based learning. In Springer eBooks (pp. 2687–2690). doi:10.1007/978-1-4419-1428-6_210

Kalamas Hedden, M., Worthy, R., Akins, E., Slinger-Friedman, V., & Paul, R. C. (2017). *Teaching sustainability using an active learning constructivist approach: Discipline-specific case studies in higher education*. doi:10.3390/su9081320

Kooli, C. (2023). Chatbots in education and research: A critical examination of ethical implications and solutions. *Sustainability (Basel), 15*(7), 5614. doi:10.3390/su15075614

Kreber, C. (2001). Learning experientially through case studies? A conceptual analysis. *Teaching in Higher Education, 6*(2), 217–228. doi:10.1080/13562510120045203

Lam, S. (2012). Project-Based learning. In Springer eBooks (pp. 2707–2709). doi:10.1007/978-1-4419-1428-6_1095

Lehmann, M., Christensen, P., Du, X., & Thrane, M. (2008). Problem-oriented and project-based learning (POPBL) as an innovative learning strategy for sustainable development in engineering education. *European Journal of Engineering Education, 33*(3), 283–295. doi:10.1080/03043790802088566

Levy, P., Aiyegbayo, O., & Little, S. (2009). Designing for inquiry-based learning with the Learning Activity Management System. *Journal of Computer Assisted Learning, 25*(3), 238–251. doi:10.1111/j.1365-2729.2008.00309.x

Liao, Y. C., & Ringler, M. (2023). Backward design: Integrating active learning into undergraduate computer science courses. *Cogent Education, 10*(1), 2204055. doi:10.1080/2331186X.2023.2204055

Lim, C. K., Haufiku, M. S., Tan, K. L., Ahmed, M. F., & Ng, T. F. (2022). Systematic Review of Education Sustainable Development in Higher Education Institutions. *Sustainability (Basel), 14*(20), 13241. doi:10.3390/su142013241

Martindale, L., Cannone, C., Niet, T., Hodgkins, R., Alexander, K., & Howells, M. (2023). Empowering Tomorrow's Problem Solvers: Nexus Thinking and CLEWs Modelling as a Pedagogical Approach to Wicked Problems. *Energies, 16*(14), 5539. doi:10.3390/en16145539

Montesinos, L., Salinas-Navarro, D. E., & Santos-Diaz, A. (2023). Transdisciplinary experiential learning in biomedical engineering education for healthcare systems improvement. *BMC Medical Education, 23*(1), 207. doi:10.1186/s12909-023-04171-x PMID:37013525

Muzaffar, H., Nikolaus, C. J., & Nickols-Richardson, S. M. (2023). Students' reflections on an experiential learning rotation with the peer education about the weight steadiness club program. *Pedagogy in Health Promotion, 9*(2), 138–149. doi:10.1177/2373379920960647

Nir, E., & Musial, J. (2020). The power of experiential learning in emotional courtroom spaces. *Journal of Criminal Justice Education, 31*(4), 542–562. doi:10.1080/10511253.2020.1817515

O'Brien, W., Doré, N., Campbell-Templeman, K., Lowcay, D., & Derakhti, M. (2021). Living labs as an opportunity for experiential learning in building engineering education. *Advanced Engineering Informatics, 50*, 101440. doi:10.1016/j.aei.2021.101440

Othman, N., & Amiruddin, M. H. (2010). Different perspectives of learning styles from the VARK model. *Procedia: Social and Behavioral Sciences, 7*, 652–660. doi:10.1016/j.sbspro.2010.10.088

Peterson, T. O. (2004). Are you considering trying problem-based learning?: Three critical success factors for implementation. *Journal of Management Education, 28*(5), 630–647. doi:10.1177/1052562904267543

Rahmawati, Y., Taylor, E., Taylor, P. C., Ridwan, A., & Mardiah, A. (2022). Students' engagement in education as sustainability: Implementing an ethical dilemma-STEAM teaching model in chemistry learning. *Sustainability (Basel), 14*(6), 3554. doi:10.3390/su14063554

Salovaara, J. J., Soini, K., & Pietikäinen, J. (2019). Sustainability science in education: Analysis of master's programmes' curricula. *Sustainability Science, 15*(3), 901–915. doi:10.1007/s11625-019-00745-1

Smith, K., Maynard, N., Berry, A., Stephenson, T., Spiteri, T., Corrigan, D., Mansfield, J., Ellerton, P., & Smith, T. C. (2022). Principles of Problem-Based Learning (PBL) in STEM Education: Using expert wisdom and research to frame educational practice. *Education Sciences, 12*(10), 728. doi:10.3390/educsci12100728

Spronken-Smith, R., Bullard, J. O., Ray, W., Roberts, C., & Keiffer, A. (2013). Where might dunes be on Mars? Engaging students through inquiry-based learning in geography. In *Active Learning and Student Engagement* (pp. 72–87). Routledge.

Sroufe, R., & Ramos, D. P. (2015). Leveraging collaborative, thematic problem-based learning to integrate curricula. *Decision Sciences Journal of Innovative Education, 13*(2), 151–176. doi:10.1111/dsji.12063

Stentoft, D. (2017). From saying to doing interdisciplinary learning: Is problem-based learning the answer? *Active Learning in Higher Education, 18*(1), 51–61. doi:10.1177/1469787417693510

Theobald, K. A., & Ramsbotham, J. (2019). Inquiry-based learning and clinical reasoning scaffolds: An action research project to support undergraduate students' learning to 'think like a nurse'. *Nurse Education in Practice, 38*, 59–65. doi:10.1016/j.nepr.2019.05.018 PMID:31176910

Thuneberg, H. M., Salmi, H. S., & Bogner, F. X. (2018). How creativity, autonomy and visual reasoning contribute to cognitive learning in a STEAM hands-on inquiry-based math module. *Thinking Skills and Creativity, 29*, 153–160. doi:10.1016/j.tsc.2018.07.003

Ward, K., & Hartley, J. (2006). Using a virtual learning environment to address one problem with problem-based learning. *Nurse Education in Practice, 6*(4), 185–191. doi:10.1016/j.nepr.2005.12.003 PMID:19040876

Wen, Y., Wu, L., He, S., Ng, N. H. E., Teo, B. C., Looi, C. K., & Cai, Y. (2023). Integrating augmented reality into inquiry-based learning approach in primary science classrooms. *Educational Technology Research and Development, 71*(4), 1–21. doi:10.1007/s11423-023-10235-y PMID:37359487

Yang, Y. T. C. (2012). Building virtual cities, inspiring intelligent citizens: Digital games for developing students' problem-solving and learning motivation. *Computers & Education*, *59*(2), 365–377. doi:10.1016/j.compedu.2012.01.012

Yu, L., & Zin, Z. M. (2023, May). The critical thinking-oriented adaptations of problem-based learning models: a systematic review. In *Frontiers in Education* (Vol. 8, p. 1139987). Frontiers. doi:10.3389/feduc.2023.1139987

KEY TERMS AND DEFINITIONS

Experiential Learning: A hands-on, learning-by-doing approach that deepens understanding through real-world experiences.

Future Directions: Anticipated advancements and developments in sustainable education and pedagogical approaches.

Inquiry-Based Learning: Active learning involves questioning, exploring, and critical thinking to develop problem-solving skills.

Integrated Approaches: Combining multiple pedagogical strategies to enhance the effectiveness of sustainable education.

Pedagogical Approaches: Diverse strategies and methods employed by educators to facilitate effective teaching and learning.

Problem-Based Learning: Students actively solve real-world problems, fostering collaboration and interdisciplinary thinking.

Project-Based Learning: Students engage in long-term, real-world projects to apply knowledge and cultivate a sense of responsibility.

Sustainable Education: Educational efforts focused on promoting environmental, social, and economic sustainable principles and practices.

Chapter 7
Environment Education Through Experiential Learning: A Case of Pauri Garhwal, Uttarakhand

Atul Bamrara
https://orcid.org/0000-0002-8390-961X
Department of School Education, Government of Uttarakhand, India

Poonam Bamrara
Department of School Education, Government of Uttarakhand, India

ABSTRACT

Environmental education (EE) has become an essential component of the curriculum from primary to higher education. It connects the students to the world around us, making them aware about both natural and built environments. It heaves awareness of issues impacting the environment upon which we all depend, as well as actions we can take to improve and sustain it. Environmental education has a lot of advantages to students, teachers, schools, and the society as a whole. The curriculum attempts to incorporate the components of environmental education either in language, science, mathematics, or a complete subject as well. The teachers can deliver such knowledge by bringing nature into the classroom, take students outside to learn, or find impromptu teachable moments on an eco-trail. The present study explores the awareness level of students using new ecological paradigm (NEP) scale after and before implementation of an EE program. Further, it differentiates their environmental awareness with respect to their demography.

INTRODUCTION

Environmental Education is the teaching of how natural environments function and how we as humans effect the environment through our behavior. The study was constructed through qualitative research by engaging students in outdoor activities, nature based curriculum, and relating the children's daily behavior to the effects they have on the environment. Significance of research within the Environmental Education

DOI: 10.4018/978-1-6684-9859-0.ch007

field comes in a wide range: building a connection between youth and nature, managing ADHD, increasing focus and self discipline, and enhancing cognitive abilities. Environmental education is widely used to increase awareness of conservation issues. The growing knowledge will improve attitudes towards the environment. Often, environmental education is aimed at children with the assumption that this can also impact adults through intergenerational transfer of knowledge and attitudes. However, there are few detailed studies evaluating the effectiveness of environmental education on changing knowledge and attitudes, and whether any changes do transfer between generations. We evaluate the effect of a school education program entitled "Ek-Prithvi" run by WWF India. We assess changes in the knowledge and attitudes of participating children in Ek-Prithvi Program. The main motto behind the said program is to develop the skills of Leadership, Deep Thinking, Effective Communication, Decision Making, Problem Solving, and Participation among school students.

REVIEW OF LITERATURE

Classroom instruction may contribute to, but may sometimes not be sufficient to increase environmental knowledge of the children, develop environmental awareness and sense of responsibility (Erdogan, 2009). In this regard, out-of class activities which mainly involve field trips, site visits and so on enable the pupils to communicate with the nature directly and to comprehend the various aspects of the nature (Palmberg & Kuru, 2000). Outdoors is much effective settings for addressing to learn about environmental issues (Martin, 2003). Outdoor education activities also provide in-depth understanding to grasp the relationship between living and non-living things, and also cause and effect relationship within the nature (Erdogan et al, 2010). Furthermore, participation in field trips, outdoor and nature-related activities could enhance participants' appreciation of nature, conservation behaviors (Bogner, 2002; Sia, Hungerford and Tomera, 1985), knowledge of environmental issues, environmental responsibility and sensitivity, and also action skills (Palmberg & Kuru, 2000). This is because of the fact that the more the people engage in environmental activities and outdoor education programs, the more they feel interconnected with the nature which results in more tendencies to protect the environment. This is in line with the claims of Bogner (1998) indicating that EE programs provided first-hand experiences and participatory interaction. Thus, outdoor education as an informal method of teaching and learning provides the pupils first-hands observation and direct experiences (Lee, 1984). Field studies which are undertaken out-of-class provides the most effective way to study and learn about the environmental issues (Neal, 1994) and increase curiosity. In a model proposed by Dresner and Gill (1994), it is indicated that increased interests and curiosity about nature stimulate to learn about environmental issues, which turn into motivation to take environmentally responsible actions.

Today's youth will become tomorrow's leaders and as their form of guidance it is our responsibility to provide them with educational experiences, for them to make knowledgeable decisions regarding the environment. Many people believe that environmental education is one of the most important factors for preventing environmental problems (Ozden, 2008). In order to make formative decisions, students must not only become an expert on the subject but also develop a connection to the environment. Judith Cheng and Martha Monroe designed a study that explicitly shows the impact of youth building a connection to the environment. Cheng and Monroe measured children's affective attitude towards the natural environment after participants completed a mandatory EE program. The study consisted of four specific research questions regarding the participants overall connection toward nature. The survey results showed

4 dimensions of children's connection to nature: enjoyment of nature, empathy for creatures, sense of oneness, and sense of responsibility. Cheng and Monroe found that the children's connection influences their intention to participate in mature based activities in the future. Their perceived family value toward nature and perceived control positively influenced their interest in environmentally friendly behavior (Cheng & Monroe, 2010). The work of Donald Burgess and Jolie Mayer- Smith also explored the connection between children and the environment. They investigated children's experience and perception during an outdoor EE program.

OBJECTIVES

- To study the awareness of students before implementation of Ek-Prithvi Program
- To study the awareness of students after implementation of Ek-Prithvi Program

RESEARCH METHODOLOGY

The present study explores the awareness level of students before and after implementation of Ek–Prithvi Program in Government Model Primary School and Government Inter College Ganga Bhogpur Malla. For the purpose, a representative sample of 57 Primary School Children and 43 Upper primary School Children has been interviewed/ observed using New Ecological Paradigm Scale (NEP Scale for Children) for Children consisting of 10 questions assessing three subscales that contribute to environmental world view including rights of nature, eco-crisis, and human exceptionalism. The collected data has been analyzed using SPSS to generate appropriate statistic.

ANALYSIS AND DISCUSSION

Plants And Animals Have as Much Right as People to Live

Before implementation of the program, it has been explored that 28% students of 7-10 age group and 9% students of 11-14 age group opined neutrally towards "Plants and animals have as much right as people to live", whereas 25% students of 7-10 age group and 19% students of 11-14 age group disagreed towards "Plants and animals have as much right as people to live". Only 12% students were agreed to the statement.

After implementation of the program, it has been explored that only 3% students of 7-10 age group and 1% students of 11-14 age group opined neutrally towards "Plants and animals have as much right as people to live", whereas 12% students of 7-10 age group and 0% students of 11-14 age group disagreed towards "Plants and animals have as much right as people to live". Further, 21% students of 7-10 age group and 22% students of 11-14 age group were agreed and 21% students of 7-10 age group and 20% students of 11-14 age group were strongly agreed towards "Plants and animals have as much right as people to live".

Table 1. Plants and animals have as much right as people to live (students' opinion before the program)

Before Program			Plants and Animals Have as Much Right as People to Live					Total
			Strongly Disagree	Disagree	Neutral	Agree	Strongly Agree	
Age (in Years)	7-10	Count	4	25	28	0	0	57
		% of Total	4.0%	25.0%	28.0%	0.0%	0.0%	57.0%
	11-14	Count	3	19	9	10	2	43
		% of Total	3.0%	19.0%	9.0%	10.0%	2.0%	43.0%
Total		Count	7	44	37	10	2	100
		% of Total	7.0%	44.0%	37.0%	10.0%	2.0%	100.0%

Table 2. Plants and animals have as much right as people to live (students' opinion after the program)

After Program			Plants and Animals Have as Much Right as People to Live					Total
			Strongly Disagree	Disagree	Neutral	Agree	Strongly Agree	
Age (in Years)	7-10	Count	0	12	3	21	21	57
		% of Total	0%	12.0%	3.0%	21.0%	21.0%	57.0%
	11-14	Count	0	0	1	22	20	43
		% of Total	0%	0%	1%	22.0%	20.0%	43.0%
Total		Count	0	12	4	43	41	100
		% of Total	0%	12.0%	4.0%	43.0%	41.0%	100.0%

There Are Too Many People on Earth

Before implementation of the program, it has been explored that 32% students of 7-10 age group and 15% students of 11-14 age group opined neutrally towards "There are too many people on Earth", whereas 13% students of 7-10 age group and 13% students of 11-14 age group strongly disagreed towards "There are too many people on Earth". Only 12% students were strongly/ agreed to the statement.

After implementation of the program, it has been explored that only 13% students of 7-10 age group and 3% students of 11-14 age group opined neutrally towards "There are too many people on Earth", whereas 2% students of 7-10 age group and 0% students of 11-14 age group disagreed towards "There are too many people on Earth". Further, 25% students of 7-10 age group and 21% students of 11-14 age groups were agreed and 14% students of 7-10 age group and 19% students of 11-14 age groups were strongly agreed towards "There are too many people on Earth".

People Are Clever Enough to Keep from Ruining Earth

Before implementation of the program, it has been explored that 35% students of 7-10 age group and 22% students of 11-14 age group opined neutrally towards "People are clever enough to keep from ruining Earth", whereas 13% students of 7-10 age group and 6% students of 11-14 age group strongly

Table 3. There are too many people on Earth (students' opinion before the program)

Before Program			There Are Too Many People on Earth					Total
			Strongly Disagree	Disagree	Neutral	Agree	Strongly Agree	
Age (in Years)	7-10	Count	13	3	32	5	4	57
		% of Total	13.0%	3.0%	32.0%	5.0%	4.0%	57.0%
	11-14	Count	13	12	15	0	3	43
		% of Total	13.0%	12.0%	15.0%	0%	3.0%	43.0%
Total		Count	26	15	47	5	7	100
		% of Total	26.0%	15.0%	47.0%	5.0%	7.0%	100.0%

Table 4. There are too many people on Earth (students' opinion after the program)

After Program			There Are Too Many People on Earth					Total
			Strongly Disagree	Disagree	Neutral	Agree	Strongly Agree	
Age (in Years)	7-10	Count	3	2	13	25	14	57
		% of Total	3.0%	2.0%	13.0%	25.0%	14.0%	57.0%
	11-14	Count	0	0	3	21	19	43
		% of Total	0%	0%	3.0%	21.0%	19.0%	43.0%
Total		Count	3	2	16	46	33	100
		% of Total	3.0%	2.0%	16.0%	46.0%	33.0%	100.0%

Table 5. People are clever enough to keep from ruining Earth (students' opinion before the program)

Before Program			People Are Clever Enough to Keep From Ruining Earth					Total
			Strongly Disagree	Disagree	Neutral	Agree	Strongly Agree	
Age (in Years)	7-10	Count	3	13	35	3	3	57
		% of Total	3.0%	13.0%	35.0%	3.0%	3.0%	57.0%
	11-14	Count	5	6	22	5	5	43
		% of Total	5.0%	6.0%	22.0%	5.0%	5.0%	43.0%
Total		Count	8	19	57	8	8	100
		% of Total	8.0%	19.0%	57.0%	8.0%	8.0%	100.0%

disagreed towards "People are clever enough to keep from ruining Earth". Only 16% students were strongly/ agreed to the statement.

After implementation of the program, it has been explored that only 12% students of 7-10 age group and 3% students of 11-14 age group opined neutrally towards "People are clever enough to keep from ruining Earth", whereas 6% students of 7-10 age group and 8% students of 11-14 age group disagreed towards "People are clever enough to keep from ruining Earth". Further, 26% students of 7-10 age group

Table 6. People are clever enough to keep from ruining Earth (students' opinion after the program)

After Program			People Are Clever Enough to Keep From Ruining Earth					Total
			Strongly Disagree	Disagree	Neutral	Agree	Strongly Agree	
Age (in Years)	7-10	Count	6	6	12	26	7	57
		% of Total	6.0%	6.0%	12.0%	26.0%	7.0%	57.0%
	11-14	Count	2	8	3	14	16	43
		% of Total	2.0%	8.0%	3.0%	14.0%	16.0%	43.0%
Total		Count	8	14	15	40	23	100
		% of Total	8.0%	14.0%	15.0%	40.0%	23.0%	100.0%

and 14% students of 11-14 age groups were agreed and 7% students of 7-10 age group and 16% students of 11-14 age groups were strongly agreed towards "People are clever enough to keep from ruining Earth".

People Must Obey the Laws of Nature

Before implementation of the program, it has been explored that 37% students of 7-10 age group and 32% students of 11-14 age group opined neutrally towards "People must obey the laws of nature", whereas 5% students of 7-10 age group and 0% students of 11-14 age group strongly disagreed towards "People must obey the laws of nature". Only 21% students were strongly/ agreed to the statement.

After implementation of the program, it has been explored that only 6% students of 7-10 age group and 3% students of 11-14 age group opined neutrally towards "People must obey the laws of nature", whereas 4% students of 7-10 age group and 2% students of 11-14 age group disagreed towards "People must obey the laws of nature". Further, 23% students of 7-10 age group and 26% students of 11-14 age groups were agreed and 16% students of 7-10 age group and 11% students of 11-14 age groups were strongly agreed towards "People must obey the laws of nature".

Table 7. People must obey the laws of nature (students' opinion before the program)

Before Program			People Must Obey the Laws of Nature					Total
			Strongly Disagree	Disagree	Neutral	Agree	Strongly Agree	
Age (in Years)	7-10	Count	5	5	37	4	6	57
		% of Total	5.0%	5.0%	37.0%	4.0%	6.0%	57.0%
	11-14	Count	0	0	32	8	3	43
		% of Total	0%	05	32.0%	8.0%	3.0%	43.0%
Total		Count	5	5	69	12	9	100
		% of Total	5.0%	5.0%	69.0%	12.0%	9.0%	100.0%

Table 8. People must obey the laws of nature (students' opinion after the program)

After Program			People Must Obey the Laws of Nature					Total
			Strongly Disagree	Disagree	Neutral	Agree	Strongly Agree	
Age (in Years)	7-10	Count	8	4	6	23	16	57
		% of Total	8.0%	4.0%	6.05	23.0%	16.0%	57.0%
	11-14	Count	1	2	3	26	11	43
		% of Total	1.0%	2.0%	3.0%	26.0%	11.0%	43.0%
Total		Count	9	6	9	49	27	100
		% of Total	9.0%	6.0%	9.0%	49.0%	27.0%	100.0%

When People Mess With Nature It Has Bad Results

Before implementation of the program, it has been explored that 37% students of 7-10 age group and 32% students of 11-14 age group opined neutrally towards "When people mess with nature it has bad results", whereas 5% students of 7-10 age group and 0% students of 11-14 age group strongly disagreed towards "When people mess with nature it has bad results". Only 21% students were strongly/ agreed to the statement.

After implementation of the program, it has been explored that only 3% students of 7-10 age group and 1% students of 11-14 age group opined neutrally towards "When people mess with nature it has bad results", whereas 15% students of 7-10 age group and 0% students of 11-14 age group disagreed towards "When people mess with nature it has bad results". Further, 19% students of 7-10 age group and 21% students of 11-14 age groups were agreed and 18% students of 7-10 age group and 20% students of 11-14 age groups were strongly agreed towards "When people mess with nature it has bad results".

Nature Is Strong Enough to Handle the Bad Effects of Our Modern Lifestyles

Before implementation of the program, it has been explored that 33% students of 7-10 age group and 19% students of 11-14 age group opined neutrally towards "Nature is strong enough to handle the bad

Table 9. People mess with nature (students' opinion before the program)

Before Program			When People Mess With Nature					Total
			Strongly Disagree	Disagree	Neutral	Agree	Strongly Agree	
Age (in Years)	7-10	Count	3	5	37	6	6	57
		% of Total	3.0%	5.0%	37.0%	6.0%	6.0%	57.0%
	11-14	Count	0	0	32	8	3	43
		% of Total	0%	0%	32.0%	8.0%	3.0%	43.0%
Total		Count	3	5	69	14	9	100
		% of Total	3.0%	5.0%	69.0%	14.0%	9.0%	100.0%

Table 10. People mess with nature (students' opinion after the program)

After Program			When People Mess With Nature					Total
			Strongly Disagree	Disagree	Neutral	Agree	Strongly Agree	
Age (in Years)	7-10	Count	2	15	3	19	18	57
		% of Total	2%	15.0%	3.0%	19.0%	18.0%	57.0%
	11-14	Count	1	0	1	21	20	43
		% of Total	1%	0%	1%	21.0%	20.0%	43.0%
Total		Count	3	15	4	40	38	100
		% of Total	3.0%	15.0%	4.0%	40.0%	38.0%	100.0%

Table 11. Nature is strong enough to handle the bad effects of our modern lifestyles (students' opinion before the program)

Before Program			Nature Is Strong Enough to Handle the Bad Effects of Our Modern Lifestyles					Total
			Strongly Disagree	Disagree	Neutral	Agree	Strongly Agree	
Age (in Years)	7-10	Count	3	13	33	5	3	57
		% of Total	3.0%	13.0%	33.0%	5.0%	3.0%	57.0%
	11-14	Count	5	6	19	8	5	43
		% of Total	5.0%	6.0%	19.0%	8.0%	5.0%	43.0%
Total		Count	8	19	52	13	8	100
		% of Total	8.0%	19.0%	52.0%	13.0%	8.0%	100.0%

effects of our modern lifestyles", whereas 13% students of 7-10 age group and 6% students of 11-14 age group strongly disagreed towards "Nature is strong enough to handle the bad effects of our modern lifestyles". Only 21% students were strongly/ agreed to the statement.

After implementation of the program, it has been explored that only 12% students of 7-10 age group and 3% students of 11-14 age group opined neutrally towards "Nature is strong enough to handle the bad effects of our modern lifestyles", whereas 5% students of 7-10 age group and 8% students of 11-14 age group disagreed towards "Nature is strong enough to handle the bad effects of our modern lifestyles". Further, 28% students of 7-10 age group and 14% students of 11-14 age groups were agreed and 7% students of 7-10 age group and 16% students of 11-14 age groups were strongly agreed towards "Nature is strong enough to handle the bad effects of our modern lifestyles".

People Are Supposed to Rule Over the Rest of Nature

Before implementation of the program, it has been explored that 35% students of 7-10 age group and 22% students of 11-14 age group opined neutrally towards "People are supposed to rule over the rest of nature", whereas 8% students of 7-10 age group and 8% students of 11-14 age group strongly disagreed

Table 12. Nature is strong enough to handle the bad effects of our modern lifestyles (students' opinion after the program)

After Program			Nature Is Strong Enough to Handle the Bad Effects of Our Modern Lifestyles					Total
			Strongly Disagree	Disagree	Neutral	Agree	Strongly Agree	
Age (in Years)	7-10	Count	5	5	12	28	7	57
		% of Total	5.0%	5.0%	12.0%	28.0%	7.0%	57.0%
	11-14	Count	2	8	3	14	16	43
		% of Total	2.0%	8.0%	3.0%	14.0%	16.0%	43.0%
Total		Count	7	13	15	42	23	100
		% of Total	7.0%	13.0%	15.0%	42.0%	23.0%	100.0%

Table 13. People are supposed to rule over the rest of nature (students' opinion before the program)

Before Program			People Are Supposed to Rule Over the Rest of Nature					Total
			Strongly Disagree	Disagree	Neutral	Agree	Strongly Agree	
Age (in Years)	7-10	Count	3	3	35	13	3	57
		% of Total	3.0%	3.0%	35.0%	13.0%	3.0%	57.0%
	11-14	Count	5	5	22	6	5	43
		% of Total	5.0%	5.0%	22.0%	6.0%	5.0%	43.0%
Total		Count	8	8	57	19	8	100
		% of Total	8.0%	8.0%	57.0%	19.0%	8.0%	100.0%

towards "People are supposed to rule over the rest of nature". Only 27% students were strongly/ agreed to the statement.

After implementation of the program, it has been explored that only 12% students of 7-10 age group and 3% students of 11-14 age group opined neutrally towards "People are supposed to rule over the rest of nature", whereas 6% students of 7-10 age group and 8% students of 11-14 age group disagreed towards "People are supposed to rule over the rest of nature". Further, 26% students of 7-10 age group and 14% students of 11-14 age groups were agreed and 7% students of 7-10 age group and 16% students of 11-14 age groups were strongly agreed towards "People are supposed to rule over the rest of nature".

People Are Treating Nature

Before implementation of the program, it has been explored that 37% students of 7-10 age group and 32% students of 11-14 age group opined neutrally towards "People are treating nature badly", whereas 5% students of 7-10 age group and 0% students of 11-14 age group strongly disagreed towards "People are treating nature badly". Only 21% students were strongly/ agreed to the statement.

After implementation of the program, it has been explored that only 6% students of 7-10 age group and 3% students of 11-14 age group opined neutrally towards "People are treating nature badly", whereas

Table 14. People are supposed to rule over the rest of nature (students' opinion after the program)

After Program			People Are Supposed to Rule Over the Rest of Nature					Total
			Strongly Disagree	Disagree	Neutral	Agree	Strongly Agree	
Age (in Years)	7-10	Count	6	6	12	26	7	57
		% of Total	6.0%	6.0%	12.0%	26.0%	7.0%	57.0%
	11-14	Count	2	8	3	14	16	43
		% of Total	2.0%	8.0%	3.0%	14.0%	16.0%	43.0%
Total		Count	8	14	15	40	23	100
		% of Total	8.0%	14.0%	15.0%	40.0%	23.0%	100.0%

Table 15. People are treating nature badly (students' opinion before the program)

Before Program			People Are Treating Nature Badly					Total
			Strongly Disagree	Disagree	Neutral	Agree	Strongly Agree	
Age (in Years)	7-10	Count	5	5	37	4	6	57
		% of Total	5.0%	5.0%	37.0%	4.0%	6.0%	57.0%
	11-14	Count	0	0	32	8	3	43
		% of Total	0%	0%	32.0%	8.0%	3.0%	43.0%
Total		Count	5	5	69	12	9	100
		% of Total	5.0%	5.0%	69.0%	12.0%	9.0%	100.0%

4% students of 7-10 age group and 2% students of 11-14 age group disagreed towards "People are treating nature badly". Further, 23% students of 7-10 age group and 26% students of 11-14 age groups were agreed and 16% students of 7-10 age group and 11% students of 11-14 age groups were strongly agreed towards "People are treating nature badly".

People Will Someday Know Enough About How Nature Works to Be Able to Control It

Before implementation of the program, it has been explored that 37% students of 7-10 age group and 32% students of 11-14 age group opined neutrally towards "People will someday know enough about how nature works to be able to control it", whereas 5% students of 7-10 age group and 0% students of 11-14 age group strongly disagreed towards "People will someday know enough about how nature works to be able to control it". Only 23% students were strongly/ agreed to the statement.

After implementation of the program, it has been explored that only 3% students of 7-10 age group and 1% students of 11-14 age group opined neutrally towards "People will someday know enough about how nature works to be able to control it", whereas 15% students of 7-10 age group and 2% students of 11-14 age group disagreed towards "People will someday know enough about how nature works to be able to control it". Further, 19% students of 7-10 age group and 19% students of 11-14 age groups were

Table 16. People are treating nature badly (students' opinion after the program)

After Program			People Are Treating Nature Badly					Total
			Strongly Disagree	Disagree	Neutral	Agree	Strongly Agree	
Age (in Years)	7-10	Count	8	4	6	23	16	57
		% of Total	8.0%	4.0%	6.05	23.0%	16.0%	57.0%
	11-14	Count	1	2	3	26	11	43
		% of Total	1.0%	2.0%	3.0%	26.0%	11.0%	43.0%
Total		Count	9	6	9	49	27	100
		% of Total	9.0%	6.0%	9.0%	49.0%	27.0%	100.0%

Table 17. People will someday know enough about how nature works to be able to control it (students' opinion before the program)

Before Program			People Will Someday Know Enough About How Nature Works to Be Able to Control It					Total
			Strongly Disagree	Disagree	Neutral	Agree	Strongly Agree	
Age (in Years)	7-10	Count	3	5	37	6	6	57
		% of Total	3.0%	5.0%	37.0%	6.0%	6.0%	57.0%
	11-14	Count	0	0	32	8	3	43
		% of Total	0%	0%	32.0%	8.0%	3.0%	43.0%
Total		Count	3	5	69	14	9	100
		% of Total	3.0%	5.0%	69.0%	14.0%	9.0%	100.0%

Table 18. People will someday know enough about how nature works to be able to control it (students' opinion after the program)

After ProgramStrongly Disagree	People Will Someday Know Enough About How Nature Works to Be Able to Control It				Total
	Disagree	Neutral	Agree	Strongly Agree	

agreed and 18% students of 7-10 age group and 19% students of 11-14 age groups were strongly agreed towards "People will someday know enough about how nature works to be able to control it".

If Things Don't Change, We Will Have a Big Disaster in the Environment Soon

Before implementation of the program, it has been explored that 33% students of 7-10 age group and 17% students of 11-14 age group opined neutrally towards "If things don't change, we will have a big disaster in the environment soon", whereas 13% students of 7-10 age group and 6% students of 11-14 age group strongly disagreed towards "If things don't change, we will have a big disaster in the environment soon". Only 21% students were strongly/ agreed to the statement.

Table 19. If things don't change, we will have a big disaster in the environment soon (students' opinion before the program)

Before Program			If Things Don't Change, We Will Have a Big Disaster in the Environment Soon					Total
			Strongly Disagree	Disagree	Neutral	Agree	Strongly Agree	
Age (in Years)	7-10	Count	3	13	33	5	3	57
		% of Total	3.0%	13.0%	33.0%	5.0%	3.0%	57.0%
	11-14	Count	7	6	17	8	5	43
		% of Total	5.0%	6.0%	19.0%	8.0%	5.0%	43.0%
Total		Count	10	19	50	13	8	100
		% of Total	10.0%	19.0%	50.0%	13.0%	8.0%	100.0%

Table 20. If things don't change, we will have a big disaster in the environment soon (students' opinion after the program)

After Program			If Things Don't Change, We Will Have a big Disaster in the Environment Soon					Total
			Strongly Disagree	Disagree	Neutral	Agree	Strongly Agree	
Age (in Years)	7-10	Count	5	6	11	28	7	57
		% of Total	5.0%	5.0%	12.0%	28.0%	7.0%	57.0%
	11-14	Count	2	8	3	14	16	43
		% of Total	2.0%	8.0%	3.0%	14.0%	16.0%	43.0%
Total		Count	7	14	14	42	23	100
		% of Total	7.0%	14.0%	14.0%	42.0%	23.0%	100.0%

After implementation of the program, it has been explored that only 11% students of 7-10 age group and 3% students of 11-14 age group opined neutrally towards "If things don't change, we will have a big disaster in the environment soon", whereas 6% students of 7-10 age group and 8% students of 11-14 age group disagreed towards "If things don't change, we will have a big disaster in the environment soon". Further, 28% students of 7-10 age group and 14% students of 11-14 age groups were agreed and 7% students of 7-10 age group and 16% students of 11-14 age groups were strongly agreed towards "If things don't change, we will have a big disaster in the environment soon".

CONCLUSION

In this study we explored that there was very low awareness of nature and environmental issues at large based on NEP Scale for Children. After implementation of Ek-Prithvi program, children in the School who received environmental education had developed their environmental awareness via eco-trails; expert interaction and the academic program run by WWF India with the help of school teachers. The study concludes that Environmental education programs can have a lasting effect, certainly on knowledge, but engagement of the research and NGO community is needed to build the capacity of teachers in rural areas to stir up their pupils about ecology and conservation issues.

REFERENCES

Cheng, J. C. H., & Monroe, M. C. (2012). Connection to nature: Children's affective attitude toward nature. *Environment and Behavior*, *44*(1), 31–49. doi:10.1177/0013916510385082

Dresner, M., & Gill, M. (1994). Environmental education at summer nature camp. *The Journal of Environmental Education*, *25*(3), 35–41. doi:10.1080/00958964.1994.9941956

Erdogan, M. (2009). *Fifth grade students' environmental literacy and the factors affecting students' environmentally responsible behaviors* [Unpublished doctoral dissertation]. Middle East Technical University, Turkey.

Erdogan, M., Erentay, N., Balaban, D., Çakır, U., Şen, C., Aydogan, B., & Nechita, A. (2010). Expanding the horizons through field trips: Developing global action plan for saving endangered species and threatened environments. Academic Press.

Lee, C. L. (1984). *Outdoor Education Activities for Elementary School Students*. Academic Press.

Martin, S. C. (2003). The influence of outdoor schoolyard experiences on students' environmental knowledge, attitudes, behaviors, and comfort levels. *Journal of Elementary Science Education*, *15*(2), 51–63. doi:10.1007/BF03173843

Ozden, M. (2008). Environmental awareness and attitudes of student teachers: An empirical research. *International Research in Geographical and Environmental Education*, *17*(1), 40–55. doi:10.2167/irgee227.0

Palmberg, I. E., & Kuru, J. (2000). Outdoor activities as a basis for environmental responsibility. *The Journal of Environmental Education*, *31*(4), 32–36. doi:10.1080/00958960009598649

Palmer, J. A., & Neal, P. (2003). *The handbook of environmental education*. Routledge.

Chapter 8
Transformative Social Emotional Learning for a Sustainable EFL Education:
A Key to Celebrating Diversity and Promoting Equity

Gülşah Öz
Aksaray University, Turkey

ABSTRACT

Social emotional learning (SEL) can be defined as a process in which individuals make responsible decisions and choices by managing their emotions, setting goals, establishing positive relationships with empathy towards others. This chapter aims to discuss the current SEL literature through a transformative lens of social justice and equity for a sustainable educational future. With this purpose in mind, this chapter reviews the following issues: fundamentals of SEL along with core components, the crucial role of SEL in celebrating diversity, promoting equity in EFL classes through SEL, and benefits of SEL integrated foreign language learning and teaching. Moreover, applications of transformative SEL integration methods and strategies are addressed. Based on the current review, some implications for educators on how to integrate SEL key components into their teaching contexts and suggestions for further research to enhance equity and sustainability in the field of foreign language education are yielded.

INTRODUCTION

Do all students have equal educational opportunities in our classes? Are our classroom environments positive, inclusive, and culturally responsive enough to foster learning and teaching process for all students from diverse backgrounds? These are currently hot questions that aspire educators and researchers at all fields of education in recent years. In addition, digital instruction has become a crucial component of education all over the world as a result of the rapid transition from traditional face-to-face education to

DOI: 10.4018/978-1-6684-9859-0.ch008

online learning and teaching platforms brought on by the Covid-19 (Köroğlu & Öz, 2023). Therefore, in responding to promoting equity and excellence in educational contexts either traditional or online throughout the world, a variety of innovative pedagogical approaches have been proposed and implemented, and Social Emotional Learning (SEL) is the one that has received increasingly great attention over the past decades. More specifically, with the paradigm shift from cognitive aspect to social emotional aspect of learning, scholarly attention to SEL has increased (Dix et al., 2012). SEL helps children and adults make responsible decisions and choices by managing their emotions, setting goals, establishing positive relationships with empathy towards others (CASEL, 2022). Furthermore, SEL "advances educational equity and excellence through authentic school-family-community partnerships to establish learning environments and experiences that feature trusting and collaborative relationships, rigorous and meaningful curriculum and instruction, and ongoing evaluation" (ibid). Additionally, besides mitigating physical and emotional negative responses such as anxiety, depression, or dropping out, SEL has the potential to enhance learners' academic success in different disciplines (Panayiotou et al., 2019). In this respect, a growing body of relevant research revealed that SEL promotes academic achievement as well as ensures educating democratic citizens with positive interpersonal relationships with others (Durlak, Weissber, Dymnicki, Taylor, & Schellinger, 2011; Sklad, Diekstra, De Ritter, & Ben, 2012).

Recently, the term of transformative SEL has emerged grounding SEL, more specifically, at the center of equity and providing equal optimal social and educational conditions for individuals regardless of their diverse backgrounds (Jagers et al., 2019). As the researcher introduces, transformative SEL is associated with the concept of justice-oriented citizenship (ibid.). On the other hand, in many non-English speaking nations, learning English as a foreign language (EFL) is mandated and considered a necessary ability for global competency (Villegas, 2017). Furthermore, the growing global population of English language learners has made equity in the context of English instruction a crucial issue that needs to be appropriately addressed (Murray, 2020). Thus, in the context of education's shift toward equity, EFL instruction is strongly tied to educational equity and has a big impact on the global effort to reduce social and educational disparities. Offering each student equal opportunity can help to create equity in EFL instruction (Lachance et al., 2019) as well as meeting diverse student communities' educational demands (Leiva et al., 2021). In other words, teachers are the key players in guaranteeing justice and fairness in the educational process in such a transforming environment (Crawford-Garrett, 2017). Therefore, to create a more democratic, interactive, equal, professional, and fair future in English language instruction, EFL teachers need to have accurate conceptions of equity and implement pertinent equitable pedagogy in their real classrooms (Yazan& Rudolph, 2018). All things considered, EFL teachers should stress the value of equal opportunities for every learner while also demonstrating a strong dedication to the process of creating those opportunities to provide education for everyone (Guo et al., 2019).

Besides current equity-based issues and problems faced in the field of foreign language learning and teaching, its multifaceted nature can create some challenges and require teachers to possess additional knowledge and skills to overcome possible difficulties that they might encounter in their diverse classroom settings. The reason behind this stems from the fact that unlike other academic areas, language learning can be highly socially and emotionally sensitive. Therefore, affective, and social aspects of language learning and teaching play crucial roles in education of all learners from different cultural backgrounds. Given the difficult cognitive burden of translating and the fear of utilizing an L2 for public or real communication, L2 learners are highly prone to feel unpleasant feelings like anxiety and shame (MacIntyre & Gregersen, 2012; Shao et al., 2013). Within this concern, studies have shown that unpleasant feelings like boredom, anxiety, and sadness can have a negative impact on learning a foreign language (Dewaele

& MacIntyre, 2014; MacIntyre et al., 2019). Furthermore, since it has been demonstrated that both good and negative emotions have an impact on behavior, current psychology research has begun to emphasize the value of being aware of one's emotions since it acts as a monitor to reduce impulsive emotional reactions (Ludwig et al., 2020). According to Bai et al. (2021), individuals can improve the way they control their learning behaviors by maximizing the good and limiting the negative impacts suggesting that self-awareness plays an important role in enhancing English language learning. In this regard, understanding how students build their SEL is especially important since excellent emotion control and social competencies will boost both their general wellbeing and academic success (Hoffmann et al., 2020). Similarly, in Dowdy et al. (2011)' s study, teachers' assessment indicated that students with poor English proficiency had higher learning difficulties, fewer adaptive abilities, and comparable degrees of internalizing and externalizing problems as compared to their more skilled English-speaking peers. Taken together, it is evident that learners' social and emotional skills play a paramount effect on their academic skills and overall psychological wellbeing and resilience in the process of foreign language learning.

In the light of the relevant studies presented above, it is obvious that schools today face a difficulty in educating pupils with far more complicated social-emotional experiences who come from ethnic and multilingual environments. In addition, students from diverse social and cultural backgrounds frequently find it difficult to fit in and be accepted by their environment. In order to overcome these obstacles, social-emotional skills are now essential for both instructors and pupils (Suganda et al., 2021). Therefore, in todays' multicultural globalized world, as we all live in a diverse society that requires individuals to be responsible and respectful towards one another by caring and celebrating differences, it is crucial to enhance justice among citizens, in turn students in all fields of education. To this end, the role of SEL through a transformative lens play a significant role in fostering educational equity in todays' classrooms. In other words, transformative SEL is paramount to enhance educational equity by empowering children and adults and addressing various forms of inequity, which contributes to establishing justice-oriented educational and social communities. This case is the same for foreign language learning and teaching process throughout the world. Similar to the society outside, EFL classrooms also host students as individuals from different backgrounds i.e., gender, ethnicity, sexual orientation, culture, education, religion, social status, perspectives, beliefs and so on. From this point of view, SEL seems to serve as a lever to ensure equity in education and promote cognitive, social, and emotional developments among learners, in turn fostering academic success at all disciplines, in foreign language learning as well. The significance of the current study lies in the fact that this chapter aims to discuss the current SEL literature through a transformative lens of social justice and equity for a sustainable educational future and propose that SEL/TSEL as an equity-based pedagogical approach can be benefited in EFL classrooms to ensure more equal, inclusive, and socially, emotionally, and culturally responsive environment for all language learners. The chapter begins by introducing theoretical background of SEL along with its five core components and discuss the crucial role of SEL in celebrating diversity and promoting equity in EFL classes, followed by benefits of SEL integrated foreign language learning and teaching. Furthermore, sample implementations of transformative SEL classroom practices in EFL classes are delineated to shed light on the ways of teachers, instructors, curriculum, and program designers and researchers in the field of foreign language learning and teaching.

BACKGROUND

Social and Emotional Learning (SEL)

The phrase "social and emotional learning" (SEL) first appeared at a 1994 Fetzer Institute conference when researchers, teachers, and child advocates participated in a range of education-based programs designed to promote the healthy development of young children (Borowski, 2019). The process by which both children and adults acquire the abilities, information, and attitudes required to comprehend and control their emotions, build, and maintain healthy relationships, set and accomplish constructive goals, and demonstrate empathy for others is known as social and emotional learning, according to Borowski (2019). Within this concern, SEL was presented as a framework to address young people's needs and assisting in the alignment and coordination of educational initiatives. To enhance students'' social, emotional, and intellectual learning, the framework of SEL is designed to serve as a roadmap for educators, families, and communities. Regarding this, the Collaborative for Academic, Social, and Emotional Learning (CASEL) provides a cohesive framework that clarifies the ways in which communities, families, and educators can collaborate to support SEL. The framework is comprehensive and includes the following three aspects knowledge, skills, and attitudes as well as focuses on systemic approaches to enhance individuals' intrapersonal, interpersonal, and cognitive competence in a variety of contexts. The CASEL framework along with its core competences are applied globally from early childhood through high school, both within and outside of the classroom.

Five Core Components of SEL

CASEL's framework presents five overarching and interactive competence areas namely *self-awareness*, *self-management*, *social awareness*, *relationship skills*, and *responsible decision-making* giving instances of each. From early childhood to adulthood, these five essential elements can be utilized and taught in a range of cultural contexts and developmental stages.

Self-awareness is known as the ability to acknowledge one's own emotions, goals, beliefs, and values as well as how they influence behavior (Borowski, 2019). Additionally, it refers to the capacity to evaluate one's strengths and shortcomings confidently and optimistically. In this sense, a person's ability to identify emotions, accurately perceive oneself, identify strengths, have confidence in oneself, and believe in oneself are all components of self-awareness. *Self-management*, on the other hand, is the ability to regulate feelings, ideas, and actions in a variety of situations (Borowski, 2019). Stress management, impulse control, self-motivation, self-discipline, organizational abilities, and creating and achieving academic and personal objectives are crucial aspects of self-management. According to Borowski (2019), the ability to comprehend and accept the opinions of others, especially those from different origins and cultures, is referred to as *social awareness*. Additionally, understanding moral and ethical norms, recognizing differences as well as similarities among people, and, most importantly, having respect for them are all components of social awareness. The capacity to build and preserve satisfying relationships with a variety of individuals and groups is known as *relationship skills*. These skills include the following abilities namely effective communication, attentive listening, teamwork, resistance to unwarranted social pressure, constructive dispute resolution, and asking for and providing assistance when required (Borowski, 2019). The ability to make wise decisions regarding one's behavior and relationships with others based on moral principles, safety considerations, and social norms, as defined by Borowski (2019),

Figure 1. Interactive CASEL wheel

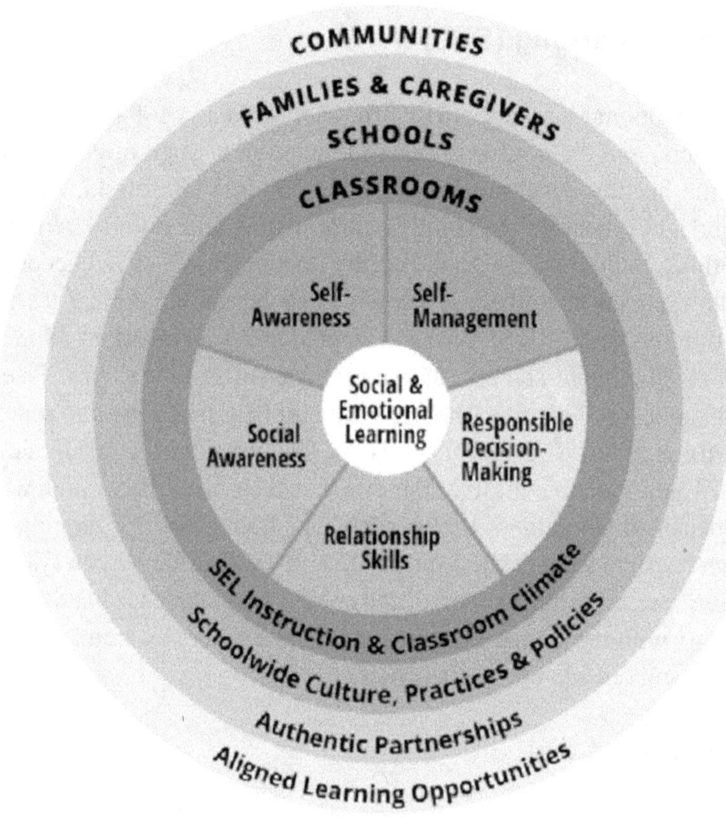

refers to *responsible decision-making*, which necessitates abilities including problem-solving, situation analysis, evaluation, reflection, and ethical responsibility.

As indicated in Figure 1, academic, social, and emotional learning competencies are formed at four levels: classrooms, schools, families and caregivers, and community levels. Developing SEL, at classroom level, entails teaching and demonstrating social and emotional competencies while giving students chances to practice and apply the competencies in a variety of contexts (Weissberg et al. 2015). Moreover, the researchers claim, "teacher practices that provide students with emotional support and create opportunities for students' voice, autonomy, and mastery experiences promote student engagement in the educational process." (p.8). Programs that foster positive school environments and bullying prevention methods are examples of SEL strategies used in schools; these include policies, procedures, and structures that give kids opportunities to solve problems. Partnerships between the family and the community can also aid in social and emotional learning that is taught in schools by giving students more chances to practice these skills outside of the classroom.

Transformative Social Emotional Learning (TSEL)

SEL, emerged as a field about thirty years ago, has exploded in popularity in recent years particularly as a transformative educational approach among educators and researchers, and it has been extensively

Table 1. SEL competencies and skills

Competencies	Skills
Self-awareness	Identifying emotions Accurate self-perception Recognizing strengths Self-confidence Self-efficacy
Self-management	Impulse control Stress management Self-discipline Self-motivation Goal setting Organizational skills
Social awareness	Perspective taking Empathy Appreciating diversity Respect for others
Relationship skills	Communication Social engagement Relationship building Teamwork
Responsible decision-making	Identifying problems Analyzing situations Solving problems Evaluating Reflecting Ethical responsibility

valued and practiced at all fields of education throughout the world with a common goal of advancing equity and social justice in the society. In this sense, transformative social emotional learning (TSEL) is a concept that has just lately gained attention as a means of grounding SEL in a focus on equity and advancing conditions that allow for the best possible human development regardless of background or circumstances (Jagers et al., 2019). More specifically, TSEL refers to "a process whereby students and teachers build strong, respectful relationships founded on an appreciation of similarities and differences, learn to critically examine root causes of inequity, and develop collaborative solutions to community and societal well-being" (Jagers et al., 2018, p. 3). Within this perspective, CASEL updated the definition of SEL in 2020, highlighting the importance of educational equity through real partnerships at the school-family-community level in order to jointly design challenging and meaningful curriculum and instruction that foster positive social environments (Niemi, 2020).

This shift in the meaning of "SEL" is in line with contextual and embodied learning theories, which place an emphasis on learning by engagement with real-world situations, contexts, and learner cultures. (Liew, Erbeli et al., 2020). In the light of the findings of Jagers et al. (2018), CASEL (2022) expanded its SEL framework to TSEL that emphasizes the abilities necessary to promote democratic, equitable, and inclusive communities as well as the abilities for personal success, interpersonal relationships, and community building. This framework purposely analyzes SEL via a culturally sensitive, equity-focused perspective because it recognizes that social and emotional capabilities vary by culture. According to Jagers et al. (2019), TSEL places a strong emphasis on achieving educational excellence and equity, with the objective of creating fair learning settings that result in fair outcomes for students that face margin-

Table 2. TSEL focal constructs

SEL Core Competencies	TSEL Equity Elaboration
Self-awareness	Identity
Self-management	Agency
Social awareness	Belonging
Relationship skills	Collaborative problem solving
Responsible decision-making	Curiosity

alization. Through TSEL, both students and adults acquire the social and emotional skills necessary to participate in the classroom and community, with an emphasis on rights and responsibilities for fostering caring and just learning environments with emphasis on *identity*, *agency*, *belonging*, *collaborative problem solving*, and *curiosity* CASEL (2022).

Many educators and academics are familiar with the interactive CASEL main five competency categories, which are self-awareness, self-management, social awareness, relationship skills, and responsible decision making. These areas have been around for a while; however, when equity takes center stage, the importance of TSEL and its main concepts—identity, agency, belonging, collaborative problem-solving, and curiosity—are highlighted (Jagers et al., 2021). In this regard, these focal constructs that are added to the framework after revision of the SEL definition are explicated respectively. *Identity* denotes perceptions and understandings connected to complex individual and societal group identities (typically explored in terms of intersectionality as well as positionality), suggesting introspection and respect for oneself. *Agency* grants the ability to favorably affect this psychological and social reality. It exhibits optimism and initiative. *Belonging* alludes to the sense of community and trust required to actively participate in creating an egalitarian, prosperous local community, and a functioning civil society, in turn raises one's sense of value. In local, national, and global community contexts, *collaborative problem solving* recognizes and supports the collective rights and responsibilities of full citizenship, which is an essential part of efforts to advance excellence and equity. *Curiosity*, on the other hand, reflects the strong desire to constantly uncover and gather knowledge of oneself in connection to others and the external world. It places a high priority on making well-informed choices based on inquiries conducted with an open mind and care that lead to societal analysis and self-development.

THE CRUCIAL ROLE OF TSEL IN CELEBRATING DIVERSITY AND PROMOTING EQUITY

In this multicultural globalized world, classrooms are hosting students with diverse cultural, linguistic, and socio-economic backgrounds. For this reason, it is paramount to accept and respect all students creating positive and inclusive classroom environments at all fields of education throughout the world. On the other hand, educational equity can be ensured through giving every student the opportunity to access to academic resources and support they need on the way of success. In this sense, as stressed in revised definition of SEL and CASEL framework itself including core competences, the crucial role of TSEL in celebrating diversity and promoting equity cannot be ignored since it may prove to be a useful instrument in creating communities that are inclusive, fair, caring, and supportive of everyone's ability to reach their greatest potential. A systemic application of SEL, from a transformative standpoint, necessitates and promotes an atmosphere in which every person feels appreciated, valued, and validated for their distinct interests, skills, social identities, cultural norms, and backgrounds. In this regard, it is evident that TSEL can play a paramount role in promoting equal educational opportunities for all learners regardless of their backgrounds.

To create learning environments that are inclusive, responsive, and respectful of all students, it is imperative to consider the relationship between educational justice and learners' social and emotional development (Ramirez et al., 2021). Furthermore, SEL and educational equity are closely related, suggesting that SEL can support educational equity objectives by helping all students feel recognized, accepted, and competent in the classroom; in addition, a conscious focus on equity enhances SEL practice by ensuring that SEL is relevant, approachable, and beneficial for all individuals (Ramirez et al., 2021). Additionally, adults need to become aware of their own prejudices and expectations, learn about the cultures and origins of students, and address and question any unequal practices and/or policies that may impede or hurt children's development in order to effectively support SEL for all students. Therefore, CASEL's work promotes cooperation among schools, families, and communities in order to stress what is best for their children and in their communities. SEL improves educational equity and excellence within this framework by *supporting authentic school-family-community partnerships*, *fostering trusting and collaborative relationships*, *promoting rigorous and meaningful curriculum and instruction*, and *applying ongoing evaluation of policies, practices, and outcomes* (CASEL, 2022), which are explicated below in detail:

1. Supporting authentic school-family-community partnerships: In order to best serve all learners in their particular local setting, CASEL encourages real school-family-community partnerships in which children, families, educators, and community partners work together to develop, implement, and continuously improve approaches and programs. Through SEL, individuals can build the connections and abilities needed for productive cross-perspective dialogue and teamwork.

2. Fostering trusting and collaborative relationships: Teachers can better understand each student's developmental needs and unique strengths with the aid of CASEL, as well as learn more about their backgrounds, cultures, and learning preferences. Teachers will be better able to construct inclusive educational settings in the classroom with this knowledge. SEL teaches and encourages the development of empathy, perspective-taking, and active listening skills in both adults and children in order to foster deeper relationships.

3. Promoting rigorous and meaningful curriculum and instruction: To help students develop the intellectual, social, and emotional skills they need to achieve their objectives and contribute to their communities, CASEL encourages rigorous curriculum and instruction that sets high standards and expectations for all students. Evidence-based SEL programs and methods suggest that all children should regularly be given the chance to learn and apply skills that are appropriate for their developmental stage.

4. Applying ongoing evaluation of policies, practices, and outcomes: To guarantee that every student receives fair treatment, has access to secure learning settings, engages in excellent learning opportunities and programs, and succeeds in their academic, social, and emotional goals, CASEL continuously assesses policies, practices, and outcomes. Focusing efforts on achieving just and desirable results is made easier by implementing SEL with a research-based framework and a dedication to ongoing improvement.

Given the ways of how SEL celebrates diversity and promotes equity within the framework of CASEL, it is also paramount to take into account TSEL focal constructs and SEL core components along with skills in addressing how CASEL is dedicated to advancing equity and celebrating diversity in all fields of education. As understood from key components and constructs, SEL enhance individuals' self-awareness and social awareness allowing them to be aware of their own and others' biases, emotions, thoughts, and beliefs, which in turn can enhance their understanding and empathy towards themselves and other individuals so that they can become more open-minded people who can appreciate and respect diversity. By fostering the development of responsible decision-making and self-management abilities, such as controlling one's emotions and thoughts in the face of different situations and making responsible decisions, SEL can encourage individuals to think over the potential consequences of their decisions on themselves as well as others. Moreover, thanks to relationship skills components of SEL, individuals can have positive relationships with others regardless of their differences and improve their communication, collaboration, negotiation, and cooperation skills as well. In addition, with an emphasis on identity, agency, belonging, collaborative problem solving, and curiosity to establish equal and just learning environments, TSEL can enhance participation in civic and school life in order to redistribute power and advancc social justice and equity.

TRANSFORMATIVE SOCIAL EMOTIONAL LEARNING IN EFL EDUCATION

In the field of foreign language learning and teaching, like any other domains of education, individuals' not only cognitive dimensions but also social and emotional aspects are crucial to obtain positive outcomes on the sides of both learners and teachers. It is well known that learning a foreign language is a highly complex process that has physical, cognitive and emotional impacts on individuals while they are learning the language (Öz & Şahinkarakaş, 2023). That is why, integration of TSEL approach to foreign language teaching can help to enhance foreign language learning process providing a much more just and positive learning environment among students and their peers as individuals of the multicultural classrooms of todays' society. Like other disciplines, foreign language learning should encourage inclusive education through practical applications (Köroğlu et al., 2023). Putting equity at the center of education and accepting diversity of learners considering the crucial roles of social and emotional factors, TSEL can enhance foreign language teachers' way of understanding diversity and equity, and teaching language for all students regardless of their diverse backgrounds. As stressed in the relevant SEL literature, SEL integrated education has been linked favorably to a variety of learner outcomes, including wellbeing, active citizenship, academic success, and so on (Bisquerra, 2003; Bridgeland et al., 2013; CASEL, 2020; Durlak et al., 2011; Goodman et al., 2015; Roffey, 2017; Schonert-Reichl et al., 2017; Zins, 2004). Along with the above-described benefits, SEL can be seen as the key to the door that permits every learner's voice to be heard in the context of education, as well as foreign language teaching. In this respect, as Burgin et al. (2021) state, in order to uphold the rights of all kids, SEL makes sure that teachers cultivate an environment where diversity is seen as a strength. For this reason, training is necessary to improve instructors' comprehension of and ability to use SEL (Castillo et al., 2013; Roffey & McCarthy, 2013; Shulman & Sherin, 2004; Talvio et al., 2013). In a study, Burgin et al. (2021) provided insight into how a professional development workshop could change educators' ideas on SEL, its elements, and their own involvement in putting it into practice. The study also illustrated the lack of knowledge and planning required for the effective integration of SEL in classroom context. Overall findings revealed that the formulation of goals and objectives for each grade level will be key to the effective implementation of SEL ideas in the classroom since this will allow kids to grow in accordance with their emotional needs and levels, providing them the skills they need to succeed in the future, therefore, additional training is required to equip teachers with the skills to integrate SEL into their lesson plans for effective implementation (Burgin et al., 2021). In this respect, it is paramount for educators to know how to integrate SEL/TSEL into their classroom practices. From this point of view, there are some recent studies related to SEL integration to teaching contexts in general, indicating the ways and strategies of effective practices of SEL in education of all students (Jagers et al., 2018; Jagers et al., 2019; Jones et al., 2018; Lau & Shea, 2022; Ramirez et al., 2021; Surya et al., 2023). More specifically, the number of the studies addressing the integration of SEL/TSEL into foreign language teaching seems to be limited in the related literature (Bai et al., 2021; Lau & Shea, 2022; McLean, 2018; Melani et al., 2020; Mori, 2022; Suganda et al., 2021). Based on the review of these relevant studies regarding both conducted in EFL contexts and in general education fields, the crucial role of SEL and variety of strategies and methods how to implement it efficiently on the ground in foreign language teaching is presented below in detail.

As for SEL relation to EFL, for instance, Bai et al. (2021)'s study sought to profile the level of SEL competence among Hong Kong secondary school students and investigate the relations between SEL competence clusters (self-awareness, self-management, social awareness, and social management) and English learning outcomes. The findings showed that, altogether, 315 grade level 3 students from secondary school had a medium level of SEL competence and that there was a discrepancy between their awareness skills at high level and their management abilities at medium level. Two mediation paths were validated by structural equation modeling (SEM) analyses, underscoring the crucial role that self-management abilities play as a mediator in boosting the impacts of self and social awareness skills on the achievement of English language learning. In a similar vein, Melani et al. (2020) contend that SEL promotes social interaction, which helps second language learners combine their cognitive and emotive components to improve their proficiency in second language, despite the fact that there are still few studies on SLA and SEL. Furthermore, the researchers state that in some cases, the emotional moods of learners may obstruct their ability to acquire L2, and an understanding of how SEL curricula might be integrated into school curricula with ELLs or taught separately is provided by existing research. According to the researchers, SEL-related classroom activities can be divided into two categories: those that are taught apart from the standard curriculum in the classroom and those that are included in the curriculum and taught. In another research, to study SEL and reveal some of the potential issues with its implementation, Mori (2022) aimed to look at the current educational system in Japan, namely the absence of a smooth transition from K–12 to higher education, as well as thinking habits, prejudice against certain cultures, and gender inequality. According to the researcher, this inquiry towards SEL competency introduction is necessary in today's world and there is no specific culture that English belongs to. Moreover, it is no longer necessary to acquire the English language through rote memorization of grammar rules. Additionally, more empathy for one another, teachers, and students has fostered in Japan through English language instruction. Teachers will be better equipped to deal with ongoing change if they can integrate a broader understanding and tolerance of diverse personalities, educational desires, and requirements, in addition to emotional understanding and empowerment. They will also be able to respond to changes without bias, negotiate suitable implementations, and foster meaningful connections. On the other hand, in the framework of social emotional learning (SEL), Suganda et al. (2021), in a case study, examined the English SEL model class at an Indonesian middle school and the EFL teacher's usage of code-switching. The results of their study demonstrated the following: (1) code-switching is the method of instruction used by EFL teachers to support all teaching and learning activities in the context of social-emotional learning; (2) implementing code-switching in EFL classroom discourse is one of the teachers' social-emotional skills; and (3) code-switching has a positive effect on the social-emotional development of both teachers and students. Furthermore, the growth of social and emotional competencies in an EFL learning setting can be greatly aided by teachers who code-switch. In another study, Lau and Shea (2022) aim to empower English learners through social-emotional learning strategies that are culturally sensitive by outlining a discussion of the value of culturally responsive pedagogy (CRP) and successful social-emotional learning curriculum. The researchers recommend the new educational approach CULTURE with a focus on inclusive classroom practices because in order for English language learners to actively reflect on and integrate their perspectives, lived experiences, beliefs, identities, and cultural assets into many facets of their literacy development, it is imperative that teachers provide them with a relevant and equal atmosphere for learning in the classroom. To Lau and Shea (2022), by doing this, students will have the self-assurance they need to successfully manage their own social and emotional development and make a positive contribution to their communities and schools. Furthermore, their recommended pedagogical strategy, CULTURE, is predicated on the concepts of CRP and the notion that instructional techniques that are culturally aware and SEL-integrated can be used as an instrument for systemic change and advancement. They claim that students' development as critically analytical, profoundly empathic, and socially responsible adults can be greatly influenced by literacy classes that are framed through an equality lens. In this sense, it is important to remember that language learners are not meant to be passive consumers of remedial instruction intended to correct language deficits or other behaviors that are considered undesirable. Instead, according to the researchers, language learners ought to be motivated to actively think about and integrate into many aspects of their literacy development their perspectives, lived experiences, values, identities, abilities, and cultural assets. Therefore, a key component of the effective delivery of SEL interventions is educators' comprehension and awareness of the formal and informal ways (such as planned curriculum, and after-school activities) that SEL is integrated into students' school and home environments (Lau & Shea, 2022). On the other hand, McLean (2018) advocates that one strategy for social and emotional learning (SEL) integration in English as a foreign language is to use instructional strategies that promote communication, self-awareness, empathy for others, active listening, teamwork, negotiation, and problem-solving. Additionally, learner-centered, communicative, and cooperative learning and teaching are some instructional techniques that are related to social and emotional learning since they all support peer interaction, participation, negotiation, and introspection, which consequently creates a collaborative learning atmosphere that is necessary to foster the growth of emotional and social competencies.

As the relevant literature is limited in both number and applications of SEL integrated instruction in EFL contexts, the strategies and effective practices identified based on the research results of general education can be utilized and recommended to foreign language teachers in creating a positive classroom environment where all learners are accessed providing equal opportunities and considering social and emotional competencies. Within this concern, more specifically, effective SEL practices and strategies in the literature have been reviewed and presented in this research to support the sustainability of foreign language learning and teaching with the belief of power of transformative SEL in promoting equity among all individuals. For example, project-based learning and youth collaborative action research incorporate youth voice, commitment, and collaborative problem-solving and decision-making as well as opportunities for students to utilize and cultivate transformative SEL skills that support youth autonomy and leadership for social transformation (Jagers et al., 2018; Jagers et al., 2019). In another research, Ramirez et al. (2021) discovered that among the programs they coded, three equitable approaches—equitable problem-solving and critical thinking, equitable narrative, and equitable emotional understanding and expression— were most frequently used. In order to incorporate more equity-oriented approaches into SEL programs, these approaches might serve as good starting points. The research findings of Surya et al. (2023) conducted a thorough examination of social emotional learning (SEL) in paternalistic elementary schools, and their findings indicate that many programs, such as Strong Kids, PATHS, Collaborative Learning, and Social Support, can improve SEL. These programs work well to support the growth of social-emotional skills in a paternalistic nation because they shift the perspective of teachers and reframe their role as equal partners in the implementation of SEL. In a similar vein, Jones et al. (2018) identified three crucial practices in the most recent literature on best practices for SEL implementation. The first best practice is that SEL needs to extend outside the classroom and take into account settings like the home and community where students could feel less supported and protected. On the other hand, it is also true that some student populations may perceive schools as unsafe environments. Using data to track results, identify patterns, and identify "teachable moments" that could be covered in SEL is the second-best practice for implementing SEL. The third implementation best practice is that successfully implemented SEL takes local context into account. In addition to the strategies and methods utilized in implementation of SEL into education, there are some studies paying attention to the relation of SEL to multiculturalism and diversity. Researchers have found that adolescents who attend schools that place a high priority on multicultural understanding and acceptance of variety perform better academically and socially and emotionally (Brand et al., 2003; Chang & Le, 2010; Juvonen et al., 2006; Le et al., 2009). More precisely, studies by Brand et al. (2003), Chang & Le (2010), and Le et al. (2009) found that minoritized kids who believed their school valued cultural diversity showed greater academic achievement, happiness, and social-emotional adaptation. Similarly, Juvonen et al. (2006) reported that the appreciation of variety in the classroom has been connected to good self-worth and social contentment.

CONCLUSION

As an educational framework, social and emotional learning (SEL) attempts to support the growth of key abilities and skills for individuals' social and emotional wellbeing along with its five key components namely self-awareness, self-management, social awareness, relationship skills, and responsible decision-making. When combined, these five components assist individuals in developing the social and emotional competencies required for success in the workplace, in the classroom, and in life. They also support people in maintaining their wellbeing and mental health. In the light of the review of relevant literature, SEL programs can provide a variety of advantages, including enhanced prosocial behavior, less behavioral issues and emotional distress, better long-term results like higher graduation rates, and better mental health. The importance of SEL in education and personal development is being acknowledged more widely because it gives people the skills they need to succeed in a variety of spheres of life. Based on the literature, the crucial role and contributions of SEL into individuals and learning/teaching process to enhance equity and celebrate diversity for a sustainable education and foreign language education can be summarized below.

Through the promotion of understanding, empathy, and respect for people with various histories, cultures, and identities, (SEL) plays a significant part in celebrating diversity. First, SEL promotes self-awareness by teaching learners to recognize their own feelings, prejudices, and beliefs, therefore, they are able to recognize any preconceptions or stereotypes they may have and attempt to overcome them thanks to this self-reflection. As a result, learners become more receptive to embracing and appreciating various perspectives and experiences when they have a greater awareness of themselves. Second, through SEL, individuals develop the ability to take into account various points of view, cultures, and experiences, which improves their capacity to comprehend and value others from various backgrounds so that individuals from different communities can connect with one another and form deep relationships when they learn to empathize. Third, developing strong interpersonal and teamwork skills is a top priority in SEL. With these skills, people can bridge cultural gaps and participate in meaningful conversations. They can also build inclusive societies that value diversity and promote collaboration among individuals working toward shared goals. SEL encourages the development of secure and welcoming environments in educational settings where diversity is not only acknowledged but celebrated. Thanks to SEL core principles, individuals may embrace and value the richness of diversity by improving the SEL talents, which will lead to a more equal and inclusive society.

The analysis of the relevant studies in the literature indicates that SEL can be a powerful tool for fostering equity in EFL classrooms and for developing welcoming learning environments. Instructors can establish a fair and encouraging learning atmosphere where all students can succeed by including SEL into EFL education. In this sense, recent research shows that SEL programs can be taught in schools either separately or as part of the regular curriculum. From this point of view, either planning their classes appropriately or altering their instructing style, language teachers can eliminate the complex or ambiguous circumstances in their teaching contexts (Öz, 2022), furthermore, by incorporating culturally responsive pedagogy with SEL, teachers can offer students a meaningful and inclusive learning environment in the classroom where they are able to actively consider and integrate their values, identities, real-life experiences, perspectives, and cultural assets into many facets of their literacy development. In addition, instructional strategies that promote communication, self-awareness, empathy for others, active listening, teamwork, negotiation, problem-solving, learner-centered, communicative, and cooperative learning, and teaching are some teaching strategies of SEL implementation suggested in the related literature. More specifically, through techniques like project-based learning and youth collaborative action research, which include youth voice, involvement, and collaborative problem-solving and decision-making, transformative SEL skills that support young autonomy and leadership for social change can be offered. Furthermore, the following strategies can be sample for effective SEL integrated teaching for educators i.e., going beyond the classroom and considering locations and situations where students might feel less safe and supported, utilizing data to track results, identify patterns, and pinpoint instructive moments for SEL, and considering local context. Teachers may encourage equity, inclusivity, and social-emotional well-being in their classrooms by integrating TSEL into EFL instruction and putting these strategies into practice because TSEL encourages academic and personal success regardless of ethnicity or level of language proficiency for all pupils. Overall, SEL integration in EFL instruction can significantly influence students' linguistic, social, and emotional growth, resulting in a comprehensive and enriching language learning experience.

Thus, transformative social emotional learning instruction in EFL is a valuable instrument for recognizing diversity and advancing equity. Students' emotional intelligence, critical thinking abilities, resilience, and sense of wellbeing can all be developed through SEL through establishing a warm and secure learning atmosphere. It encourages students to become caring global citizens by providing them with the tools they need to succeed in a diverse and interconnected society. On the other hand, the educational techniques provided here indicate that it takes time and effort to master SEL competencies and skills in practice of foreign language teaching. Therefore, it is obvious that further study is required regarding how transformative SEL along with its principles is conceptualized in different EFL contexts from the perspective of social justice and equity in education. Additionally, more practice-oriented research is necessary to provide teachers with more specific activities and tasks that can be utilized to integrate transformative SEL principles into EFL teaching process.

REFERENCES

Bai, B., Shen, B., & Wang, J. (2021). Impacts of social and emotional learning (SEL) on English learning achievements in Hong Kong secondary schools. *Language Teaching Research*, 13621688211021736. doi:10.1177/13621688211021736

Bisquerra, R. (2003). Educación emocional y competencias básicas para la vida. *Revista de Investigación Educacional*, *21*(1), 7–43.

Borowski, T. (2019). CASEL's framework for systemic social and emotional learning. *Measuring SEL: Using Data to Inspire Practice*, *8*, 1–7.

Brand, S., Felner, R., Shim, M., Seitsinger, A., & Dumas, T. (2003). Middle school improvement and reform: Development and validation of a school-level assessment of climate, cultural pluralism, and school safety. *Journal of Educational Psychology*, *95*(3), 570–588. doi:10.1037/0022-0663.95.3.570

Bridgeland, J., Bruce, M., & Hariharan, A. (2013). *The missing piece: A national teacher survey on how social and emotional learning can empower children and transform schools*. A Report for CASEL. Civic Enterprises.

Burgin, X., Coli, S., & Daniel, M. C. (2021). Exploratory Study of Ecuadorian Teachers' Understanding of Social Emotional Learning: An Examination of Primary School Teachers. *GIST Education and Learning Research Journal*, *22*(22), 125–146. doi:10.26817/16925777.964

Castillo, R., Fernández-Berrocal, P., & Brackett, M. A. (2013). Enhancing teacher effectiveness in Spain: A pilot study of the RULER approach to social and emotional learning. *Journal of Education and Training Studies*, *1*(2), 263–272. doi:10.11114/jets.v1i2.203

Chang, J., & Le, T. N. (2010). Multiculturalism as a dimension of school climate: The impact on the academic achievement of Asian American and Hispanic youth. *Cultural Diversity & Ethnic Minority Psychology*, *16*(4), 485–492. doi:10.1037/a0020654 PMID:21058811

Crawford-Garrett, K. (2017). 'The problem is bigger than us': Grappling with educational inequity in TeachFirst New Zealand. *Teaching and Teacher Education*, *68*, 91–98. doi:10.1016/j.tate.2017.08.010

Dewaele, J.-M., & MacIntyre, P. D. (2014). The two faces of Janus? Anxiety and enjoyment in the foreign language classroom. *Studies in Second Language Learning and Teaching*, *4*(2), 237–274. doi:10.14746/ssllt.2014.4.2.5

Dix, K. L., Slee, P. T., Lawson, M. J., & Keeves, J. P. (2012). Implementation quality of whole school mental health promotion and students' academic performance. *Child and Adolescent Mental Health*, *17*(1), 45–51. doi:10.1111/j.1475-3588.2011.00608.x PMID:22518095

Dowdy, E. T., Dever, B. V., DiStefano, C., & Chin, J. K. (2011). Screening for Emotional and Behavioral Risk among Students with Limited English Proficiency. *School Psychology Quarterly*, *26*(1), 14–26. doi:10.1037/a0022072

Durlak, J. A., Weissberg, R. P., Dymnicki, A. B., Taylor, R. D., & Schellinger, K. (2011). The impact of enhancing students' social and emotional learning: A meta-analysis of school-based universal interventions. *Child Development*, *82*(1), 405–432. doi:10.1111/j.1467-8624.2010.01564.x PMID:21291449

Goodman, A., Joshi, H., Nasim, B., & Tyler, C. (2015). *Social and emotional skills in childhood and their long-term effects on adult life*. Institute of Education.

Guo, Q., Tao, J., & Gao, X. (2019). Language teacher education in system. *System*, *82*, 132–139. doi:10.1016/j.system.2019.04.001

Hoffmann, J. D., Brackett, M. A., Bailey, C. S., & Willner, C. J. (2020). Teaching emotion regulation in schools: Translating research into practice with the RULER approach to social and emotional learning. *Emotion (Washington, D.C.)*, *20*(1), 105–109. doi:10.1037/emo0000649 PMID:31961187

Jagers, R. J., Rivas-Drake, D., & Borowski, T. (2018). *Equity & social and emotional learning: A cultural analysis.* CASEL Assessment Work Group Brief Series.

Jagers, R. J., Rivas-Drake, D., & Williams, B. (2019). Transformative social and emotional learning (SEL): Toward SEL in service of educational equity and excellence. *Educational Psychologist*, *54*(3), 162–184. doi:10.1080/00461520.2019.1623032

Jagers, R. J., Skoog-Hoffman, A., Barthelus, B., & Schlund, J. (2021). Transformative Social Emotional Learning: In Pursuit of Educational Equity and Excellence. *American Educator*, *45*(2), 12.

Jones, S., Bailey, R., Brush, K., & Kahn, J. (2018). *Preparing for effective SEL implementation.* Harvard Graduate School of Education Easel Lab. Available from Wallace Foundation website: https://www.wallacefoundation. org/knowledgecenter/Documents/Preparing-for-Effective-SEL-Implementation. pdf

Juvonen, J., Nishina, A., & Graham, S. (2006). Ethnic diversity and perceptions of safety in urban middle schools. *Psychological Science*, *17*(5), 393–400. doi:10.1111/j.1467-9280.2006.01718.x PMID:16683926

Köroğlu, Z. Ç., & Öz, G. (2023). Pre-Service English Teachers' Speaking Skills Assessment Literacy: Transformation to Digital Assessment. *The Reading Matrix: An International Online Journal, 23*(1).

Köroğlu, Z. Ç., Öz, G., & Yüce, E. (2023). Inclusive Education Practices During the COVID-19 Pandemic From the Eyes of Turkish EFL Student Teachers: A Qualitative Research. In Promoting Diversity, Equity, and Inclusion in Language Learning Environments (pp. 175-192). IGI Global.

Lachance, J. R., Honigsfeld, A., & Harrell, G. (2019). Equal educational opportunity for English learners: ESL teachers' conceptualizations on the importance of academic language. *TESOL Journal*, *10*(2), e00415. doi:10.1002/tesj.415

Lau, W. S., & Shea, M. (2022). Empowering English learners in the classroom through culturally responsive social-emotional teaching practices. *Journal of Multilingual and Multicultural Development*, 1–18. doi:10.1080/01434632.2022.2078337

Le, T. N., Lai, M. H., & Wallen, J. (2009). Multiculturalism and subjective happiness as mediated by cultural and relational variables. *Cultural Diversity & Ethnic Minority Psychology*, *15*(3), 303–313. doi:10.1037/a0015507 PMID:19594259

Leiva, L. R., Miranda, L. P., & Riquelme-Sanderson, M. (2021). Social justice in the preparation of English language teachers. *Mextesol J.*, *45*, 1–14.

Ludwig, V. U., Brown, K. W., & Brewer, J. A. (2020). Self-regulation without force: Can awareness leverage reward to drive behavior change? *Perspectives on Psychological Science*, *15*(6), 1382–1399. doi:10.1177/1745691620931460 PMID:32857672

MacIntyre, P. D., & Gregersen, T. (2012). Affect: The role of language anxiety and other emotions in language learning. In S. Mercer, S. Ryan, & M. Williams (Eds.), *Language learning psychology: Research, theory and pedagogy* (pp. 103–118). Palgrave. doi:10.1057/9781137032829_8

MacIntyre, P. D., Ross, J., & Clément, R. (2019). Emotions are motivating. In M. Lamb (Ed.), *The Palgrave handbook of motivation for language learning* (pp. 183–202). Palgrave Macmillan. doi:10.1007/978-3-030-28380-3_9

McLean, A. (2018). *"Tää oli kiva tunti!": A social and emotional learning intervention in an EFL classroom*. Academic Press.

Melani, B. Z., Roberts, S., & Taylor, J. (2020). Social emotional learning practices in learning English as a second language. *Journal of English Learner Education*, *10*(1), 3.

Mori, L. (2022). Global Journal of Foreign Language Teaching. *Language Teaching*, *12*(4), 267–276.

Murray, D. E. (2020). The world of English language teaching: Creating equity or inequity? *Language Teaching Research*, *24*(1), 60–70. doi:10.1177/1362168818777529

Niemi, K. (2020, December 15). *Niemi: CASEL is updating the most widely recognized definition of social-emotional learning. Here's why.* https://www.the74million.org/article/niemi-casel-is-updating-the-most-widely recognizeddefinition- of-social-emotional-learning-heres-why/

Öz, G. (2022). An Investigation into Turkish EFL Learners' Ambiguity Tolerance. *Turkuaz Uluslararası Türk Dünyası Bilimsel Araştırmalar Dergisi*, *3*(1), 63–70. doi:10.54970/turkuaz.1119010

Öz, G., & Şahinkarakaş, Ş. (2023). The Relationship Between Turkish EFL Learners' Academic Resilience and English Language Achievement. *The Reading Matrix: An International Online Journal*, *23*(1).

Panayiotou, M., Humphrey, N., & Wigelsworth, M. (2019). An empirical basis for linking social and emotional learning to academic performance. *Contemporary Educational Psychology*, *56*, 193–204. doi:10.1016/j.cedpsych.2019.01.009

Ramirez, T., Brush, K., Raisch, N., Bailey, R., & Jones, S. M. (2021). Equity in social emotional learning programs: A content analysis of equitable practices in PreK-5 SEL Programs. *Frontiers in Education*.

Roffey, S., & McCarthy, F. (2013). Circle solutions, a philosophy and pedagogy for learning positive relationships: What promotes and inhibits sustainable outcomes? *International Journal of Emotional Intelligence*, *5*(1), 36–55.

Schonert-Reichl, K. A., Kitil, M. F., & Hanson-Peterson, J. (2017). To reach the students, teach the teachers: A national scan of teacher preparation and social and emotional learning. A report prepared for the Collaborative for Academic, Social, and Emotional Learning (CASEL). University of British Columbia.

Shao, K. Q., Yu, W. H., & Ji, Z. M. (2013). An exploration of Chinese EFL students' emotional intelligence and foreign language anxiety. *Modern Language Journal*, *97*(4), 917–929. doi:10.1111/j.1540-4781.2013.12042.x

Shulman, L. S., & Sherin, M. G. (2004). Fostering communities of teachers as learners: Disciplinary perspectives. *Journal of Curriculum Studies*, *36*(2), 135–140. doi:10.1080/0022027032000135049

Sklad, M., Diekstra, R., De Ritter, M., & Ben, J. (2012). Effectiveness of school-based universal social, emotional, and behavioral programs: Do they enhance students' development in the area of skill, behavior, and adjustment? *Psychology in the Schools*, *49*(9), 892–909. doi:10.1002/pits.21641

Suganda, L. A., Petrus, I., & Zuraida, Z. (2021). EFL Teacher's Code Switching in the Social Emotional Learning Context. *Indonesian Research Journal in Education, 5*(2), 317-344.

Surya, A., Retnawati, H., & Haryanto, H. (2023). Findings and Implications of Social Emotional Learning (SEL) in Paternalistic Culture in Elementary Schools: A Systematic Literature Review. *Pegem Journal of Education and Instruction*, *13*(3), 151–158.

Talvio, M., Lonka, K., Komulainen, E., Kuusela, M., & Lintunen, T. (2013). *Revisiting Gordon´s teacher effectiveness training: An intervention study on teachers´ social and emotional learning*. Academic Press.

Villegas, D. (2017). Colombia's Nationwide EFL policy and the construction of equity in policy documents. *App. J. App. Lang. Stu.*, *11*(4), 57–74. doi:10.17011/apples/urn.201708083437

Yazan, B., & Rudolph, N. (Eds.). (2018). *Criticality, Teacher Identity, and (In)equity in English Language Teaching: Issues and Implications*. Springer. doi:10.1007/978-3-319-72920-6

Zins, J. E. (Ed.). (2004). *Building academic success on social and emotional learning: What does the research say?* Teachers College Press.

Chapter 9
Teachers' Characteristics for a Sustainable Future: From the Perspective of Pre-Service Teachers

Şenol Şen
https://orcid.org/0000-0003-3831-3953
Hacettepe University, Turkey

Senar Temel
Hacettepe University, Turkey

ABSTRACT

The role of teachers in raising awareness and equipping students with the necessary skills to contribute to a sustainable future is crucial. This study aims to identify the descriptions of the characteristics of effective teachers made by pre-service teachers who have just started university. The research was conducted with the participation of pre-service teachers attending a state university. The data were collected through semi-structured interview forms prepared by the researchers. The interview form was comprised of questions about what the effective and qualified teachers' characteristics are and about what characteristics of effective and qualified teachers would have positive reflections into teaching environments. Each interview lasted approximately 15-20 minutes. The data collected were then put to content analysis. The analyses demonstrated that the participants usually describe effective teachers as individuals who have comprehensive knowledge of their branch and who continuously make self-improvement.

INTRODUCTION

The teacher is the main component of the educational system and social development. The teaching profession is a professional job open to continuous development. Teachers' tasks and responsibilities have increased today and society's perspectives of the teaching profession have changed. Teachers can

DOI: 10.4018/978-1-6684-9859-0.ch009

raise awareness and prepare their students for a sustainable future by providing them with resources relevant to their interests, encouraging critical thinking skills and enabling them to collaborate to meet the challenges of a sustainable future (Su et al., 2012). Teacher training policies were revised in many countries as a result of technological and social changes especially to raise the quality of teachers and thus societies embarked on a new quest. The reason for it is that effective teachers should be raised for effective teaching. For this reason, training teachers having various efficacies have become a target for all countries (Çelebi, 2015).

Countries across the world need effective, well-educated and competent teachers (Rushton, Morgan, & Richard, 2007). Even though the effective teacher is a concept, which everyone is aware of and which is known to everybody, it is difficult to make a description of an effective teacher and the characteristics of an effective teacher (Korthagaen, 2004; Pozo-Munoz, Rebolloso-Pacheco, & Fernandez-Ramirez, 2000). Besides, there is no agreed definition of a good teacher in the literature (Kızıltepe, 2002). Shulman (1987), for instance, listed the knowledge that a teacher should have under 7 categories as "(1) content knowledge, (2) general pedagogical knowledge, (3) curriculum knowledge (4) pedagogical content knowledge, (5) knowledge of learners and their characteristics; (6) knowledge of educational contexts (7) knowledge of educational ends, purposes, and values, and their philosophical and historical grounds". Moreover, the description of an effective teacher also changes over time. The perspective concerning effective teachers changed from the description of "ideal teachers" in the 1950s to "satisfying teachers" and to "diversity responsive teachers" in the 1990s (Cruickshank & Haele, 2001). At first, an effective teacher was described as "a good person". An effective teacher was described as a teacher who considers the needs of the society and who raises ideal citizens according to the needs and who set a good role model to their students. Thus, teachers were evaluated rather based on the behaviours they had displayed, but their behaviours in the learning/teaching process were considered as of secondary importance. However, it was pointed out later that describing an effective teacher according to the ideals of society was wrong and that what was important was the behaviours a teacher displayed in the classroom (Borich, 2014).

Characteristics of Effective Teachers

The American Association of School Administrators (AASA) divided the characteristics of effective teachers into two groups as (1) management and teaching techniques and (2) personal properties. AASA generally describes effective teachers as good managers and mentions them as people (1) who maintain discipline by taking preventive precautions, (2) who use systematic but varied teaching techniques, (3) who are knowledgeable about subjects while arranging to teach according to students' needs, (4) who are flexible, enthusiastic and creative, (5) who believe in their capabilities and who have high expectations, (6) who behaves democratically in the classroom, who is warm in their interaction with students and who cares about that, (7) who are easily accessible outside the classroom (Minor, Onwuegbuzie, Witcher, & James, 2002). McDonald (1975) described teachers' characteristics in three categories as (a) personal characteristics, (b) intellectual qualities, (c) approaches and attitudes related to teaching methods. Minor et al. (2002) stated that effective teachers are experts in determining, arranging and presenting the content. Also, effective teachers are creative, they encourage active student participation, they become masters in asking effective questions, they encourage critical and creative thinking and they take time to wait for students' answers to the questions. They also give feedback, monitor students' progress and use both the traditional and the alternative evaluation methods and they are fair in measurement and evaluation. Such teachers keep developing by evaluating their practices and by benefiting from their experiences (Minor et al., 2002). According to Brophy and Alleman (1991), effective teachers make self-evaluations, share knowledge with their colleagues, participate in seminars and have an exchange of views and they make efforts to promote their knowledge and experiences. According to Borich (2014), effective teachers present their lessons in a way that their students can understand. They explain concepts and subjects step by step in a manner that students at different levels can understand, and they use different methods to increase comprehensibility. They consider variation and flexibility important while teaching a lesson. They also use such supportive elements as computers and the internet to have variation. Another behaviour of effective teachers is to provide opportunities for students to learn and to have practice for the reinforcement of what has been learnt. Such teachers make plans on how much time to spend on the learning-teaching process and how to spend the time and they act accordingly. According to Borich, another effective teacher behaviour is to secure students' active participation in the learning-teaching process. Teachers make efforts to increase students' interest in particular. They assure that students understand exactly and correctly, they are taught (Borich, 2014).

The descriptions of effective teachers are mostly related to students' academic performance and the effects of methods and techniques used by teachers on students (Berliner, 1987). Effective teachers present knowledge that will be useful in students' learning in the future, and they expect their students to concentrate so that their attention is not distracted in the classroom and to make efforts to do the challenging assignments (Braga, Paccagnella, & Pellizzari, 2014). Effective and positive student-teacher relations are possible in the classroom on the condition that teachers are tolerant, patient, open-minded, knowledgeable, affectionate and humoristic. Teachers who do not have such characteristics cannot create a positive classroom atmosphere no matter how well equipped they are professionally (Göksoy & Arıcan, 2018). Loving the profession and students underlie the teaching profession. Teachers who love their students can set up warmer and more positive relations with their students (Kıroğlu, 2012).

Korthagaen (2004) states that a framework for the characteristics of effective/good teachers should be drawn for three reasons. The first is the changes in the goals and methods of teacher training occurring throughout the world. Short-term educational programs were made available in teacher training in many countries and as a result, the quality of the programs was questioned. The answer to the complex and challenging question "who is a good teacher?" is usually neglected by politicians- which has become a serious problem in teacher training. The second reason is that we demand that teachers are effective teachers but we do not know what the characteristics of an effective teacher are true. The third is that although teacher education is built on insights from other disciplines, especially psychology, the changes and developments in these disciplines do not affect the basic idea of teacher training.

Sustainable Development and Characteristics of Teachers

Sustainable development is defined in the Brundtland Report (1987) by the World Commission on Environment and Development (WCED) as meeting the needs of the present generation without compromising the ability of future generations to meet their own needs. Sustainable development can also be defined as the use of resources in a way that meets the needs of future generations and the proper management of these resources for well-being (Repetto, 1992; Tietenberg & Lewis, 2006). Teachers have a great responsibility in educating future generations for sustainable future and development. This is because teachers have a significant impact on future generations in other words on students (Barnett, 2003; Day et al., 2007). Therefore, education can be an important key for sustainable development (McKeown et al., 2002). Teachers also play an important role in raising of future generations awareness about sustainable development (Guo et al., 2020). The education of teacher candidates is also important in this context for achieving sustainable development goals (Tekgöz et al., 2010). From this point of view, education for sustainable development can be defined as improving knowledge, values, and skills for effective and active participation in various decision-making processes for both societies and individuals to improve their quality of life without harming our world (Summer et al., 2005). As a result of a study conducted by Günkör (2017), it was emphasized that education positively supports sustainable development by contributing to individuals' adaptation to the new world, being supplied to the market as qualified labor force, increasing their income, developing technology, and improving their quality of life, and even being the source of development.

Education for sustainability is a paradigm that aims to produce lifelong learners who will contribute to a prosperous life and to produce knowledgeable, skilled, and engaged individuals (Fien & Maclean, 2000). The competencies that teachers need to have for a sustainable future can be classified into four dimensions: belief, character, knowledge, and ability, and these dimensions are consistent with sustainable development goals (Guo et al., 2020). Guo and others (2020) have stated that it is very important to integrate sustainable development into teachers' professional development training in order to achieve sustainable development goals. Ferguson and others (2021) emphasized that teachers need to be knowledgeable about sustainable development for a sustainable future and have positive attitudes and perspectives on this issue. Also, Ferguson and others stated that teachers should be aware of sustainability issues and should not shy away from taking responsibility for solving problems.

Albareda-Tiana et al., (2018, 2019) emphasized that it is important for teachers to not only have knowledge about sustainability but also to have competencies in this area. Teachers must also adopt sustainability principles and have a lifestyle in line with these principles (Albareda-Tiana et al., 2019). As stated by Echegoyen-Sanz and Martín-Ezpeleta (2021), effective teachers should be able to convey the importance of sustainability to their students and persuade them to exhibit sustainable behaviors. In addition, teachers must also be aware of sustainability and have environmental consciousness (Echegoyen-Sanz & Martín-Ezpeleta, 2021). Teachers should be encouraging and guiding in developing a sense of attachment to nature in their students (Anđić & Šuperina, 2021). Finally, it is also an important feature for teachers to have appropriate sustainability attitudes and values (Echegoyen-Sanz and Martín-Ezpeleta, 2021).

The Current Study

Education has an important role on improving the quality of people's lives and for the sustainability of the world (Burbules et al., 2020). Education can also be described as one of the main driving forces of sustainable development (Badea et al., 2020). For this reason, a high-quality education is necessary for sustainable development (Lai, & Peng, 2019). For quality education, it is necessary to determine which effective teacher characteristics for sustainable future teachers should have. The characteristics of effective teachers should primarily be known to be able to raise effective teachers. Raising effective teachers is one of the fundamental goals of educational systems because effective teachers are influential in improving students' academic performance and thus in their achievement in their careers (Lopez-Augudo, & Marcenaro-Guitierrez, 2017). Also, students' achievement and teachers' quality are highly correlated (Muller, Katz, & Dance, 1999). Yet, important questions such as "Who/what is an effective teacher? What is the role of personality in being an effective teacher? Are some types of teacher personality more effective than others in the learning/teaching process?" arise at this point (Rushton et al., 2007). Korthagen

(2004) states that answers to the questions "What are the characteristics of a good teacher?" and "How can we help to become a good teacher?" will differ according to the context. Korthagen also says that it is impossible to formulate the characteristics that a effective teacher should have and that there is no need to do such a thing pedagogically. Therefore, because the characteristics of an effective teacher for sustainable education change from society to society and from context to context, this study aimed to reveal the characteristics that effective teachers should have by setting out from the definitions offered by pre-service teachers who are first-year university students in Türkiye. As Ferguson et al. (2021) stated, teachers need encouragement and resource support to effectively transfer sustainable development in education. So, this type of research can help to identify teachers' needs and the topics they need to be trained on. Because teachers' thoughts shape their classroom practices and influence the outcomes of the learning process (Ferguson et al., 2021). Given that teachers play a key role in sustainable development and future, it would be beneficial for teachers, education researchers, and policymakers to identify the characteristics of effective teachers for a sustainable future.

METHOD

Participants

A total of 30 pre-service teachers (26 females and 4 males) who were first-year university students and who registered in the course Introduction to Education were included in the study. Of them 10 were in the Early Childhood Education department while 7 were in the Chemistry Education department, 6 were in the Mathematics Education Department, 4 were in Guidance and Psychological Counselling department, 2 were in Gifted Education department and 1 was in Science Education department. All the participants ranged from 18 to 24 years with an average age of 19.8 (SD=1.56). A purposeful sampling method was used in choosing the participants. Efforts were made to find more satisfactory answers to the questions on which the study focuses by using purposeful sampling (Patton, 1990). All the participants took part in the study based on volunteering. Before the study, the participants were informed of the study content and of evaluating the data collected; and they were also assured that their names would be kept confidential.

Instrument

Semi-structured interviews were held with the 30 participants to find how they described the characteristics of effective teachers for sustainable education. The interview form containing questions about what the effective and qualified teachers' characteristics are and about what characteristics of effective and qualified teachers would have positive reflections into teaching environments was prepared by the researchers. In addition to that, probe questions were also included in the form to collect more data. Expert opinions of two lecturers were consulted for the content validity of the questions in the interview form. Having obtained expert opinion, the interview questions were given to a different group of pre-service teachers to determine the understandability of the questions. In accordance with their views, the final shape was given to the interview form. All of the interviews with the pre-service teachers were conducted one-to-one outside the classroom by one of the researchers. The data were recorded on a voice recorder with the participants' permission to prevent the loss of data. Each interview lasted approximately 15-20 minutes.

Data Analysis

The whole data collected through the interviews were transcribed by one of the researchers and were put to the computer. The data on the computer was exposed to content analysis. Following the analysis, the codes and categories about the qualities of effective teachers for sustainable education were distinguished. After that, relevant themes were distinguished. The qualitative data were analyzed by experts and thus methods to increase internal validity were used. The similarities and differences between codes made by two coders were compared and the interrater reliability was found to be 92%. At least 70% reliability should be attained in qualitative research studies (Miles & Huberman, 1994; Yıldırım & Şimşek, 2011).

RESULTS

The findings obtained following the content analysis of the data are shown in Table 1.

The pre-service teachers' views on the characteristics of effective teachers were divided into six categories (social dimension, the dimension of personal traits, the dimension of personal development, occupational dimension, the dimension of teaching and the dimension of students). An examination of the responses in the category of social dimension showed that 40% of the participants described well-qualified teachers as teachers preparing students for social life while 26.2% described them as individuals guiding society in development and improvement. Some of the statements made by the participants in this category were as in the following:

Teachers shape the future of societies through the students they educate. They guide societies in development and improvement. Teaching is a special profession that is responsible for building the future of a nation and of a government. (P1)

Our teachers should be the individuals who shape our future and who secure that we go up to more advanced levels socially. (P5)

Characteristics such as working to a plan and a schedule (63.3%), being equal and fair (56.6%) and having effective communication skills (53.3%) were the characteristics which were included in another category- the dimension of personal traits and which were put forward by the majority of the pre-service teachers. Some of the statements made by the participants in this category were as in the following:

Well-qualified teachers should come to their classes with a plan and a schedule. They should not be unprepared for their lessons by thinking that they always teach the same lesson. (P13)

Teachers should be fair. They should not have better attitudes towards some students since they like them more; they should treat students equally. (P16)

Working to a plan makes both teachers and students' jobs easier. If teachers plan what assessment they will make and when they will make it and what subjects they will teach, their job becomes easier in allocating time to activities apart from teaching. (P16)

Table 1. The results of the content analysis about the characteristics of effective teachers for sustainable education

Themes	Categories	Codes	f
Qualities of effective teachers for sustainable education	**Social dimension**	Preparing students for social life	12
		Guiding society in development and improvement	8
		The architects of the future of a country	7
		Directing the society	7
		Raising individuals useful to society	4
	Dimension of personal traits	Working to a plan and a schedule	19
		Equal and fair	17
		Having effective communication skills	16
		Patient	12
		Sincere and warm	11
		Self-disciplined	11
		Empathetic	10
		Cheerful	9
		Having general knowledge	9
		Respectful	9
		Skilful at the art of oratory	7
		Self-sacrificing	7
		Idealist	7
		Humanist	7
		Understanding	7
		Dependable	6
		Art lover	6
		Consistent	5
		Affectionate and happy	5
		A good observer	4
		Self-confident	4
		Humoristic	3
		Encouraging	3
		Versatile	3
		Polite	2
	Occupational dimension	Having a comprehensive knowledge of their area	21
		Setting a model with their thoughts and behaviours	16
		Loving their job	15
		Loving learning and teaching	3
		Keeping private life different from occupational life	3

continued on following page

Table 1. continued

Themes	Categories	Codes	f
	Dimension of personal development	Making continuous personal development	21
		Benefitting from/following technology	18
		Keeping up to date	15
		Having exchange of knowledge with colleagues	2
	Dimension of teaching	Knowing the methods and techniques suitable to the subject and their students	12
		Using strategies encouraging students to discover and to think	9
		Making lessons enjoyable	8
		Making different identities a whole in the same environment	6
		Presenting the lessons in good language	6
		Attracting students' attention in classes	6
		Choosing student-centred teaching	6
		Not using test marking as a threat	5
		Being knowledgeable about individual differences and teaching based on multiple intelligences	3
		Getting students to engage in lessons and attracting their attention	3
		Supporting their lesson with audio-visual aids	2
		Identifying students' strengths and weaknesses with different measurement instruments	2
	Dimension of students	Considering individual differences	14
		Dealing with students' problems	9
		Communicating with students' parents	9
		Recognising students	8
		Preserving teacher-student relations	7
		Motivating students	7
		Noticing students who need more interest and support	6
		Considering students valuable	6
		Uncovering students' potentials and supporting them	6
		Listening to students	5
		Teaching students how to use their imagination	4
		Being aware of students' abilities and guiding students	3
		Orientating students into research	3
		Teaching students how to think	2
		Being aware of students' developmental features	2
		Preventing students from being monotype	2
		Appreciating their students' efforts and achievements	2

Characteristics such as having comprehensive knowledge of their area (70%), setting a model with their thoughts and behaviours (53.3%) and loving their job (50%) were mentioned in the category of occupational dimension. Some of the statements made by the participants in this category were as in the following:

In my opinion, another characteristic that a teacher should have is comprehensive knowledge of his area. He should know what he is going to teach very well. If he is not knowledgeable about his branch and if students become aware of it, their respect for their teacher can be diminished. (P2)

An effective teacher should be competent in his area. Such competence is attained by transferring accurate knowledge into students and by being well equipped in his branch. (P3)

The secret of success is to do one's job with love. Teaching also requires one to teach with love and to be the best in the job. If you want to be a good teacher, you should love learning and teaching. (P4)

On examining the statements made in the category of the dimension of personal development, we saw that characteristics such as making continuous personal development (70%), benefiting from/ following technology (60%) and keeping up to date (50%) were mentioned by the majority of the participants as the characteristics of effective teachers. Some of the statements made by the participants in this category were as in the following:

In my opinion, an effective teacher should keep up to date, he should be informed of what is happening in the world and the country. (P11)

A teacher, who should include up-to-date subjects in his classes, should not stick to the past and the classic knowledge only. He should also teach the new things in the world which is changing. (P16)

Such characteristics as knowing the methods and techniques suitable to the subject and his students (40%), using strategies encouraging students to discover and to think (30%) and making lessons enjoyable (26.6%) were some of the teacher characteristics listed by the pre-service teachers in the category of the dimension of teaching. Some of the sample statements were as in the following:

A teacher should employ suitable strategies to teach well. The strategies he/she uses should encourage students to discover and to think critically. (P21)

Determining the strategy of teaching is among the important characteristics of an effective teacher. The strategies used should encourage students to discover and to think. (P3)

The category of the dimension of students included such teacher qualities as considering individual differences (46.6%), dealing with students' problems (30%) and communicating with students' parents (30%). Some of the statements made by the participants in this category were as in the following:

A teacher should have close communication with his students' families. Student's behaviours at school and home are important to a teacher. In this way, he has the opportunity to recognise his students better. (P19)

Each student is different. Therefore, each student is unique and specific. In this case, a teacher should observe his students and approach them in appropriate ways. (P24)

DISCUSSIONS

This paper analyzed pre-service teachers' views on the qualities of effective teachers for sustainable education and made a general evaluation of the results obtained. Although the pre-service teachers described quite a big number of qualities, the number of those who described the qualities was small. They described effective teachers as individuals who guide society in development and improvement. The descriptions of effective teachers as people who direct society and shape the future and raise individuals useful to the society were offered by very few participants.

While the majority of the participants listed such characteristics as working to a plan and a schedule, being equal and fair and having effective communications skills as the personal traits that effective teachers should have; a small number of the participants mentioned characteristics such as being patient, sincere and warm, self-disciplined, self-sacrificing, respectful, idealist, consistent, self-confident, humoristic and polite. It was found that the studies available in the literature also obtained results similar to the ones obtained in this study. Murray (1983), for instance, found that the most favourite teacher personality traits were (1) being collaborative, displaying democratic attitudes, (2) flexibility, (3) having an interest in students' problems, (4) showing interest, (5) being fair and impartial, (6) joking, (7) being patient, (8) having consistent behaviours, (9) being polite and understanding. Pozo-Munoz et al., (2000), on the other hand, found based on participants' statements that teachers should have personality traits such as being good listeners and being accessible, motivating, fair and dependable in teacher-student relations. Also, Badea et al., (2020) confirmed that personal and professional skills of teachers is effective in improving students' sustainable behavior. According to Guo et al., (2020), the development of teachers' professional quality is so important on realizating the goal of sustainable development.

Effective teachers should have comprehensive knowledge of their area, that they should set models for their students with their thoughts and behaviours and that they should love their job were especially emphasized by the pre-service teachers. The participants considered it important for teachers to make continuous self-development, to keep up-to-date and to benefit from technology for their personal development. On the other hand, the category containing the biggest number of participants' views was the dimension of teaching. Whereas the pre-service teachers described effective teachers as individuals who know the methods and techniques suitable to the subject and to their students, who use strategies encouraging students to think and to discover, who attract students' attention in classes, who know individual differences and teach based on multiple intelligences and who get students to engage in lessons and attract their attention; the number of participants mentioning those characteristics was again small. As Albareda-Tiana et al., (2019) stated, when using active teaching-learning strategies related to the sustainable development goals, sustainability skills can be are improved. According to Albareda-Tiana et al., (2018) project oriented learning is effective in the development of sustainability skills. Also, characteristics such as dealing with students' problems, recognizing students, motivating students, listening to students, being aware of students' developmental features, appreciating students' efforts and achievements were among the characteristics listed by some of the participants in the category of the dimension of students. A review of the literature showed that similar results were obtained in the literature. For example, Pozo-Munoz et al (2000) found that characteristic that was considered the most important for ideal teachers was "teaching competency". They also found that characteristics such as being knowledgeable, transferring the knowledge clearly and being competent in teaching were also among the characteristics of ideal teachers. Minor et al. (2002) divided the characteristics of effective teachers into seven categories labelled as student-centred teaching, effective classroom and behaviour management, competent educator, ethical, eager to teach, knowledgeable and professional in the research conducted with the participation of pre-service teachers. In a similar study, Witcher and Onwuegbuzie (1999) divided the characteristics of effective teachers into six categories as (a) student-centred, (b) eagerness to teach, (c) ethics, (d) classroom and behaviour management, (e) teaching methodology, (f) content knowledge according to the percentages of the pre-service teachers' responses. Çakmak (2009) found that top five characteristics listed by pre-service teachers were (1) stating the goal of the lesson, (2) making students active in lessons, (3) loving the job, (4) teaching based on students' interests and abilities, (5) rapport with students. In another study, Wimberly, Faulkner and Moxley (1978) summarized the characteristics of effective teachers described by students as (1) considering students' rights and needs, (2) contributing to students' intellectual development, (3) evaluating students' performance fairly, (4) having comprehensive knowledge of the content, (5) attracting students' attention, (6) having occupational integrity, (7) setting the objectives of the course, (8) presenting the subject clearly.

In conclusion, on considering the findings obtained in general, the pre-service teachers described effective teachers for sustainable education as individuals;

- Who prepare students for society,
- Who guides the society in development and improvement,
- Who work to a plan and schedule, are equal and fair, have effective communication skills,
- Who make continuous development,
- Who keep up to date,
- Who benefits from technology,
- Who set models to their students with their thoughts and behaviours,
- Who loves their job,
- Who know the methods and techniques suitable to the subject and their students,
- Who consider individual differences.

RECOMMENDATIONS

This study did not make a classification and evaluation of pre-service teachers according to their traits. The studies to be conducted in the future could do analyses based on individual differences such as gender, age and the types of school of graduation. For instance, whether or not teacher characteristics described by male and female pre-service teachers change can be investigated.

The characteristics of pre-service teachers at the previous stages of education (such as male, female, experienced, inexperienced, etc.) could be identified and comparisons could be made with the teacher characteristics they describe. Especially analyses on the gender and personal traits of primary school and pre-school teachers can yield different results.

Analyses according to the departments of pre-service teachers attend could be done. In this way, whether or not the characteristics of effective teachers differ according to departments could be investigated.

Longitudinal studies can also be conducted to identify the differences between pre-service teachers and the changes in them.

In a similar vein, the views of teachers at different ages could be obtained and thus a cross-sectional/age study could also be performed.

REFERENCES

Albareda-Tiana, S., García-González, E., Jiménez-Fontana, R., & Solís-Espallargas, C. (2019). Implementing pedagogical approaches for ESD in initial teacher training at Spanish Universities. *Sustainability (Basel), 11*(18), 4927. doi:10.3390/su11184927

Albareda-Tiana, S., Vidal-Raméntol, S., Pujol-Valls, M., & Fernández-Morilla, M. (2018). Holistic approaches to develop sustainability and research competencies in pre-service teacher training. *Sustainability (Basel), 10*(10), 3698. doi:10.3390/su10103698

Anđić, D., & Šuperina, L. (2021). How important is future teachers' "Connectedness to Nature"? Adaptation and validation of the connectedness to nature scale. *Education Sciences, 11*(5), 250. doi:10.3390/educsci11050250

Badea, L., Şerban-Oprescu, G., Dedu, S., & Piroşcă, G. (2020). The impact of education for sustainable development on romanian economics and business students' behavior. *Sustainability (Basel), 12*(19), 8169. doi:10.3390/su12198169

Barnett, W. S. (2003). *Better teachers, better preschools: Student achievement linked to teacher qualifications* (Preschool Policy Matters No. 2). NIEER. Retrieved from https://nieer.org/resources/policy-briefs/2.pdf

Berliner, D. C. (1987). Simple views of effective teaching and a simple theory of classroom instruction. In D. C. Berliner & B. Rosenshine (Eds.), *Talks to teachers* (pp. 93–110). Random House.

Blair, T. R., Rupley, W. H., & Nichols, W. D. (2007). The effective teacher of reading: Considering the "what" and "how" of instruction. *The Reading Teacher, 60*(5), 432–438. doi:10.1598/RT.60.5.3

Borich, G. D. (2014). *Effective teaching methods* (8th ed.). Pearson Prentice Hall.

Braga, M., Paccagnella, M., & Pellizzari, M. (2014). Evaluating students' evaluations of professors. *Economics of Education Review, 41*, 71–88. doi:10.1016/j.econedurev.2014.04.002

Brophy, J., & Alleman, J. (1991). Activities as instructional tools: A framework for analysis and evaluation. *Educational Researcher, 20*(4), 9–23. doi:10.2307/1176471

Brundtland Commission. (1987). *Our Common Future, Chapter 2: Towards Sustainable Development.* Geneva, United Nation, World Commission on Environment and Development (WCED). http://www.un-documents.net/ocf-02.htm#I

Burbules, N., Fan, G., & Repp, P. (2020). Five trends of education and technology in a sustainable future. *Geography and Sustainability*, *1*(2), 93–97. doi:10.1016/j.geosus.2020.05.001

Çakmak, M. (2009). Prospective teachers' thoughts on characteristics of an "effective teacher". *Education in Science*, *34*, 74–82.

Çelebi, N. (2015). Bir meslek olarak öğretmenlik [Teaching as a profession]. In A. Bakioğlu (Ed.), *Eğitim bilimine giriş* [Introduction to Education] (pp. 115–145). Nobel Akademik Yayıncılık.

Cruickshank, D. R., & Haefele, D. (2001). Good teachers, plural. *Educational Leadership*, *58*(5), 26–30.

Day, C., Sammons, P., Stobart, G., Kington, A., & Gu, Q. (2007). *Teachers matter*. Open University Press.

Echegoyen-Sanz, Y., & Martín-Ezpeleta, A. (2021). A holistic approach to education for sustainability: Ecofeminism as a tool to enhance sustainability attitudes in pre-service teachers. *Journal of Teacher Education for Sustainability*, *23*(1), 5–21. doi:10.2478/jtes-2021-0002

Ferguson, T., Roofe, C., & Cook, L. D. (2021). Teachers' perspectives on sustainable development: The implications for education for sustainable development. *Environmental Education Research*, *27*(9), 1343–1359. doi:10.1080/13504622.2021.1921113

Fien, J., & Maclean, R. (2000). Teacher Education for Sustainability. II. Two Teacher Education Projects from Asia and the Pacific. *Journal of Science Education and Technology*, *9*(1), 37–48. doi:10.1023/A:1009420923692

Göksoy, S., & Arıcan, K. (2018). Öğretmenlik mesleği [Teaching profession]. In K. Kıroğlu & C. Elma (Eds.), *Eğitime giriş* [Introduction to Education] (pp. 260–277). Pegem Akademi.

Günkör, C. (2017). Exploration of the relationship between educationand development. *Journal of International Social Sciences Education*, *3*(1), 14–32.

Guo, S. Y., Wu, W. C., & Cai, A. T. (2020, November). A study on promoting teachers' professional quality to achieve sustainable development goals. *IOP Conference Series. Earth and Environmental Science*, *576*(1), 012013. doi:10.1088/1755-1315/576/1/012013

Kıroğlu, K. (2012). Bir meslek olarak öğretmenlik [Teaching as a profession]. In Ö. Demirel & Z. Kaya (Eds.), *Eğitim bilimine giriş* [Introduction to Education] (pp. 343–372). Pegem Akademi.

Kızıltepe, Z. (2002). Good and effective teacher. *Education in Science*, *27*, 10–14.

Korthagen, F. A. (2004). In search of the essence of a good teacher: Towards a more holistic approach in teacher education. *Teaching and Teacher Education*, *20*(1), 77–97. doi:10.1016/j.tate.2003.10.002

Lai, Y.-C., & Peng, L.-H. (2019). Effective Teaching and Activities of Excellent Teachers for the Sustainable Development of Higher Design Education. *Sustainability (Basel)*, *12*(1), 28. doi:10.3390/su12010028

Lopez-Agudo, L. A., & Marcenaro-Gutierrez, O. D. (2017). Engaging children in lessons: The role of efficient and effective teachers. *School Effectiveness and School Improvement*, *28*(4), 650–669. doi:10.1080/09243453.2017.1364272

McDonald, F. (1975). *Research on teaching and its implications for policymaking: Report on phase II of the beginning teacher evaluation study*. Educational Testing Service.

McKeown, R., Hopkins, C. A., Rizi, R., & Chrystalbridge, M. (2002). *Education for Sustainable Development Toolkit*. Energy, Environment and Resources Center, University of Tennessee Knoxville. http://kpe-kardits.kar.sch.gr/Aiforia/esd_toolkit_v2.pdf

McMillan, J. H. (1996). *Educational research: Fundamentals for the consumer*. HarperCollins College Publishers.

Miles, M. B., & Huberman, A. M. (1994). *Qualitative data analysis: An expanded sourcebook*. Sage Publications.

Minor, L. C., Onwuegbuzie, A. J., Witcher, A. E., & James, T. L. (2002). Preservice teachers' educational beliefs and their perceptions of characteristics of effective teachers. *The Journal of Educational Research, 96*(2), 116–127. doi:10.1080/00220670209598798

Muller, C., Katz, S. R., & Dance, L. J. (1999). Investing in teaching and learning: Dynamics of the teacher-student relationship from each actor's perspective. *Urban Education, 34*(3), 292–337. doi:10.1177/0042085999343003

Murray, H. G. (1983). Low-inference classroom teaching behaviours and student ratings of college teaching effectiveness. *Journal of Educational Psychology, 75*(1), 138–149. doi:10.1037/0022-0663.75.1.138

Patton, M. Q. (1990). *Qualitative evaluation and research methods*. SAGE Publications, Inc.

Pozo-Munoz, C., Rebolloso-Pacheco, E., & Fernandez-Ramirez, B. (2000). The "Ideal Teacher". Implications for student evaluation of teacher effectiveness. *Assessment & Evaluation in Higher Education, 25*(3), 253–263. doi:10.1080/02602930050135121

Repetto, R. (1992). Accounting for Environmental Assets. *Scientific American, 266*(6), 94–101. doi:10.1038/scientificamerican0692-94 PMID:1585151

Rushton, S., Morgan, J., & Richard, M. (2007). Teacher's Myers-Briggs personality profiles: Identifying effective teacher personality traits. *Teaching and Teacher Education, 23*(4), 432–441. doi:10.1016/j.tate.2006.12.011

Shulman, L. (1987). Knowledge and teaching: Foundations of the new reform. *Harvard Educational Review, 57*(1), 1–23. doi:10.17763/haer.57.1.j463w79r56455411

Su, Y. H., Feng, L. Y., Yang, C. C., & Chen, T. L. (2012). How teachers support university students' lifelong learning development for sustainable futures: The student's perspective. *Futures, 44*(2), 158–165. doi:10.1016/j.futures.2011.09.008

Summers, M., Childs, A., & Corney, G. (2005). Education for sustainable development in initial teacher training: Issues for interdisciplinary collaboration. *Environmental Education Research, 11*(5), 623–647. doi:10.1080/13504620500169841

Tekgöz, G., Şahin, E., & Ertepinar, H. (2010). Çevre okuryazarlığı, öğretmen adayları ve sürdürülebilir bir gelecek. *Hacettepe Üniversitesi Eğitim Fakültesi Dergisi, 39*, 307–320.

Tietenberg, T., & Lewis, L. (2016). *Environmental and Natural Resource Economics* (10th ed.). Routledge. doi:10.4324/9781315523972

Wimberly, R., Faulkner, G., & Moxley, R. (1978). Dimensions of teacher effectiveness. *Teaching Sociology*, *6*(1), 7–20. doi:10.2307/1317161

Witcher, A., & Onwuegbuzie, A. J. (1999). *Characteristics of effective teachers: perceptions of preservice teachers*. Paper presented at the Annual Meeting of the Mid-South Educational Research Association, Point Clear, AL. Retrieved from ERIC database. (ED438246).

Yıldırım, A., & Şimşek, H. (2011). Sosyal bilimlerde nitel araştırma yöntemleri (8. Baskı) [Qualitative research methods in social sciences (8th Edition)]. Ankara: Seçkin Yayıncılık.

KEY TERMS AND DEFINITIONS

Characteristics of Effective Teachers: These are features such as ensuring the effectiveness of the teaching process, supporting the student in the process, and mastering the subject area.

Continuous Self-Improvement: This involves being aware of developments in the field, being able to constantly renew oneself, and being a lifelong learner.

Effective Teachers: They are elements of the education system that make the education process effective and efficient.

Sustainable Development: This is a type of development aimed at providing a sustainable ecosystem.

Sustainable Education: This is a type of education aimed at raising awareness of sustainable development.

Sustainable Future: This refers to a future that aims to raise awareness in the fields of environment, climate, and sustainable development.

Teaching Profession: This is a profession that trains the manpower needed by society.

Chapter 10
Teacher and Student Attitudes Towards STEM Education:
A Study of SASTRA-Managed STEM Labs

Usha Ajithkumar Malayankandy
https://orcid.org/0000-0001-8555-8406
SASTRA University, India

ABSTRACT

The study aimed at investigating the attitudes of secondary school teachers and students towards STEM approach. The study is a follow up of the social initiative of SASTRA University, India, where STEM labs were set up in rural schools to promote STEM education. Data for this chapter came from 10 schools where the labs were set up in the first phase of the STEM project. The tool of the study was two questionnaires which were validity and reliability verified. Thirty-three secondary school teachers and 136 students were administered the questionnaire after one year of exposure to the lab. Descriptive techniques were used to analyze the data. Results of the study showed positive student and teacher attitude toward STEM education. In light of results of the study, the researcher highly recommends using STEM approach in teaching subjects in schools as being crucial to personal development and citizenship in the 21st century. The study also recommends the 'adoption' of a school, or a cluster of schools, by a local STEM industry/enterprise, university, NGOs, and other public, private bodies.

INTRODUCTION

STEM is an essential and unavoidable element of today's world. STEM's emergence has ushered in a technical, social, economic, and political transformation. To provide all citizens with the competences required to succeed in this new information-based and highly technological society, education in the fields of science, technology, engineering, and mathematics (STEM) is becoming increasingly more important. There are many important things that students have to possess in this era, namely developing inquiry skills and stimulating positive attitudes towards STEM education (Lin et al., 2020). Educators and the school community need to integrate STEM in learning because it can benefit the national economy

DOI: 10.4018/978-1-6684-9859-0.ch010

(Razali, et.al, 2018). STEM integration aims to combine multiple disciplines as a way to engage students in topics of interest while teaching critical thinking and core content knowledge for future success in STEM careers (Wieselmann et al., 2020).

The terminology "STEM education" refers to the integrated instruction and learning associated with the fields of science, technology, engineering, and mathematics, which usually involves instructional programs at all levels of education in both formal and informal classroom settings. STEM has been defined as the connection between disciplines of integrating technology and engineering practices into current mathematics and science lessons (Bybee, 2010). STEM also refers to problem solving and gives the emphasis on drawing the concepts and procedures from mathematics and science to be connected with the engineering design and technology tool (Shaughnessy, 2013). The goal of the convergence of STEM disciplines is to enhance the capabilities of students in the "four C's" of the 21st century: critical thinking, creativity, communication, and collaboration. STEM education aims at training students for the future and to help develop them into problem solvers who can participate in the development of the economy.

This educational approach was designed to enhance the creative, innovative, and analytical thinking ability of students to improve their levels of productivity. The basis of STEM education starts in childhood. From the early grades and through their play interaction and family settings, children interrelate with the universe in ways that will encourage science, technology, engineering, and mathematics-related literacy.

STEM education builds upon curriculum integration theories in two perspectives. One perspective is that STEM education enables teachers to integrate correlated subjects without ignoring the unique characteristics, depth, and rigor of their main discipline (National Research Council, 2011). However, there is a gap between how STEM subjects are taught in schools and the knowledge, skills, and beliefs required for STEM education (Cuadra & Moreno, 2005). Reducing the gap between current instructional practices and the actual skills needed for STEM education is contingent upon the expertise of STEM teachers to successfully transition from the departmentalized model of teaching to an integrated teaching model (Furner & Kumar, 2007). STEM education requires teachers to excel in utilizing natural and active exchanges of knowledge, skills, and beliefs among STEM disciplines. Teachers are the people who are ideally adapted to implement STEM education. They play a significant role in educating the students. High-quality STEM teachers are more successful in providing STEM education effectively and students will enjoy studying STEM according to STEM objectives. Student's anxiety is predicted to decrease if STEM is presented by a quality STEM teacher. The attitude of teachers towards STEM is the foundation on which the knowledge, skills and beliefs can be built. Therefore, it is imperative to find out the teachers' attitude about STEM education.

The field of STEM will be the main occupation in the future (Mariani, et.al 2013). Many countries have broadly adopted STEM education in their own countries as they realize the needs and importance of this field. STEM has evolved exponentially and is supported by 21st-century skills. STEM education can produce skilled and competitive students in the era of globalization. It is important to know the students' attitude towards STEM education.

STEM combines four disciplines and is supported by 21st-century skills (Beheshti, et al., 2018). This STEM education practically replaces traditional teaching methods with a more student-centered teaching approach (Breiner, et al., 2012). Students can acquire new knowledge and solve problems presented by STEM teachers through projects, inquiry approaches, critical and crisis skills, collaboration, and so on, but most importantly, student learning should be in line with their daily life. Students' activities involve social elements, cultural factors, and physical interactions among them in the learning process.

Therefore, educators will have the confidence to believe that STEM education can produce skilled and competitive students in the era of globalization. Therefore, this study will explore teachers and students' attitude towards STEM education.

BACKGROUND

SASTRA University in Thanjavur, Tamil Nadu stays actively involved in various social projects aimed at the upliftment of the community. It has launched the SASTRA-CNR Rao Lab initiative to setup the STEM laboratory facility with DIY kits on Robotics, IoT, 3D Printing, AI and Drones, in schools of Tamil Nadu. This is to strengthen the science and technology-based learning at the school level. As part of this initiative, SASTRA had announced a competitive proposal submission from all schools all over Tamil Nadu. This proposal asked for a detailed presentation on the lab's objective, outcome and need for STEM. More than a hundred schools had submitted their proposals and 25 schools were shortlisted to interact and present their school's vision on STEM. On the basis of the presentation and project proposals by the shortlisted schools at SASTRA Thanjavur Campus on February 28 and March 8, 2020, ten schools were selected, in phase I. The selection took into consideration the objectives and outcome of this laboratory set up in their respective schools.

This study is a follow up of the STEM lab project. It aims to examine how participation in the STEM lab activities over a period of one year has influenced teachers and students' attitude of STEM education.

Benefits of STEM Laboratories

STEM labs are educational spaces that encourage active learning and problem solving. In these labs, students can develop their science, engineering and mathematics skills by using technology to create, collaborate and complete projects. For example, a study by researchers in Italy found that children's preference for spatial toys and spatial sports promotes spatial thinking skills, which contributes to the successful inclusion of engineering thinking in STEM programs based on a project-based and interdisciplinary approach (Moe, Jansen & Pietsch, 2018). Learning spatial reasoning contributes to a successful STEM career in a person's life. Moreover, in the context of STEM technology, the conditions for independent learning, on the one hand, and a friendly and supportive environment, on the other hand, are created, which increases the motivation for independent problem solving (León et al., 2011). An analysis of the results of a study by researchers from China showed that children develop mathematical skills, especially spatial reasoning, which is necessary for future engineers, especially at the age of 5-6, through the approach of STEM. It was found that boys' choice of STEM subjects was based on their interest in the field, while for girls this choice was determined by their confidence in their mathematical abilities (Sakellariou & Fang, 2021). Along the same lines, the results of a study conducted by a group of researchers from Switzerland have shown that by fostering students' interest in mathematics and science, the inherent value of mathematics and science and the likelihood of choosing a STEM career in the future increase (Aeschlimann et al., 2016). STEM forms high motivation in students, and independence in decision making contributes to the formation of appropriate conditions for creativity (Vanykina & Sundukova, 2020). Through STEM education, students can delve deeper into the logic of ongoing phenomena, understand their interrelationships, study the world systematically, and thereby develop curiosity, an engineering style of thinking, the ability to get out of critical situations, develop teamwork skills, and master the basics of management and self-presentation (Lazareva & Marchuk, 2019).

STEM Education and the Students' Academic Performance

Students who attended STEM school have a better academic performance than students who did not attend. A particularly striking difference is the students' prior academic performance in STEM (Wiswall, Stiefel, Ellen, & Boccardo, 2014). Students who attend STEMs have much higher average 8th grade Reading and Mathematics z-scores than those who attend non-STEMs (Wiswall et al., 2014). These significant different on the scores clearly indicate that the average student in a STEM school is markedly different from one in a non-STEM school. Furthermore, research on the impact of a STEM Program on academic achievement of eighth grade students in a South Texas Middle School shows that eighth grade students in the STEM program would outperform the eighth grade students in the non-STEM program on the basis of academic achievement in mathematics, science, and reading. Students who perceived STEM learning approach will perform better in mathematics, science and reading might due to their ability to think critically and creatively. They are being trained to solve problems independently compared to the non-STEM students. STEM students' application of knowledge and skills sets them apart from other students. They do not simply "fill in the blanks." They take their understandings and apply them by creating new solutions to complex problems. In other words, active learning occurs in the classroom.

Integrated STEM Education

The integrated STEM education entails 'an interconnected entity [of disciplines] with a strong collaborative connection to life. This STEM education approach directs teachers to diffuse paradigmatic knowledge, skills, values and language differences and teach the integrated discipline as one cohesive entity. In doing so, teacher and student interactions should take the centre stage to enable them to collaboratively construct new knowledge, skills and beliefs at the intersection of more than one STEM subject area. Driving such interactions in the classrooms necessitates that teachers comprehend STEM content and acquire supportive pedagogical content knowledge specific to their subjects as well as working knowledge in another (Corlu, et. Al, 2014). This integrated approach argues that the 'real-life' application of STEM is naturally integrated. A mathematically rigorous science education argument that disputes the epistemological (paradigmatic) view of mathematics and science as distinct to an extent that they are imposable to integrate illustrates how this model functions (Corlu, et al., 2014). This argument aligns the insightful thinking that desperate epistemological assumptions underlying STEM disciplines detract their integration (Willimas, 2011) and the interdependence relevance of science and mathematics to real life (Başkan et al., 2010). This interdependence of science and mathematics perspective afore their applications into real-world situations. Thus, the application of sciences and mathematics in engineering and technology invalidates their compartmentalized views and brings in an understanding of STEM education as an integrated entity. STEM teaching can occur at the space where two or more STEM subjects such as mathematics and science intersect. Class interactions draw into this space the content and processes such as problem-solving and quantitative reasoning of both mathematics and science. Mathematics used in science or mathematically rigorous science education brings to the attentions of teachers an interdisciplinary understanding of STEM education that 'does not create an independent meta-discipline while preserving the subject-specific knowledge, skills, and attitudes' (Corlu, et al., 2014).

Learning Theories for STEM Education

Constructivist Theory of Learning

Constructivism is the philosophy of learning that proposes learners need to build their own understanding of new ideas. Through the constructivist theory of learning, students will be actively involved, rather than being passive listeners. Students will be able to create organizing principles that they can take with them to other learning settings. Constructivism also gives students ownership of what they learn, since learning is based on students' questions and explorations, and often the students have a hand in designing the assessments as well. Students are also more likely to retain and transfer the new knowledge to real life. Furthermore, constructivism promotes social and communication skills by creating a classroom environment that emphasizes collaboration and exchange of ideas. Students learn how to articulate their ideas clearly as well as to collaborate on tasks effectively by sharing in group projects.

Social Constructivism

Russian psychologist Lev Vygotsky, believed that knowledge was constructed through dialogue and interaction with others. He argued that knowledge is co-constructed in a social environment and that in the process of social interaction, people use language as a tool to construct meaning (Churcher, Downs & Tewksbury, 2014). The ability to learn through dialogue and interaction with others is central to knowledge generation (Churcher et al., 2014). Student-student interactions are encouraged during the STEM lessons so that students can construct their own understanding themselves.

Collaborative Learning

Collaborative learning is based on the idea that learning is a naturally social act in which the participants talk among themselves (Gerlach, 1994). It is through the conversation among themselves that learning occurs. In the collaborative learning environment, the learners are challenged both socially and emotionally as they listen to different perspectives, and are required to articulate and defend their ideas. In so doing, the learners begin to create their own unique conceptual frameworks and not fully rely on teacher in the classroom. For example, by working together in tutorial groups, students develop interpersonal skills and learn how to become good collaborators.

5E (Engagement, Explanation, Exploration, Elaboration, and Evaluation) Learning Model

The 5E Model is based on the constructivist theory to learning, which suggests that people construct knowledge and meaning from experiences The 5E Learning Cycle involves learning something new, or attempting to understand something familiar in greater depth (Acisli, Yalcin, & Turgut, 2011). The 5E model is divided into five sessions including Engagement, Explanation, Exploration, Elaboration and Evaluation (Fazelian, Ebrahim, & Soraghi, 2010). Rodger Bybee, introduced it at the beginning of the 80s (Fazelian et al., 2010)

The 5 stages of Figure 1 are briefly described for the sake of clarity. The *engage* stage provides an opportunity to activate the students' prior knowledge. The *explore* stage consists of a teacher-guided

Figure 1. 5E model
Source: Bybee et al. (2006)

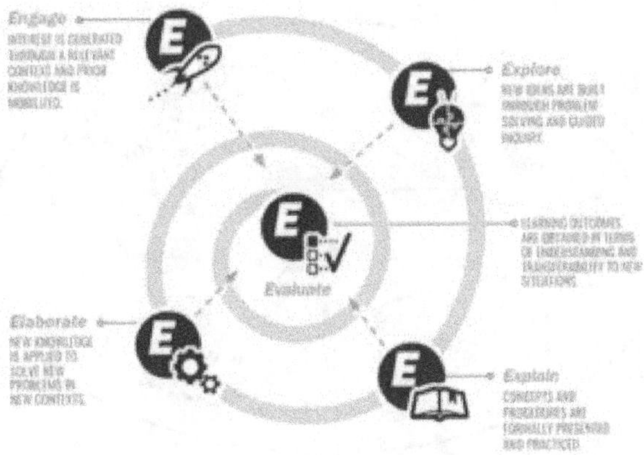

inquiry activity that provides students opportunities to solve the cognitive conflict presented to them in the first stage, and to build new explanations that make sense to them. In the *explain* stage, teachers directly and formally introduce the concepts developed in the *explore* stage, and help students organize their new knowledge in such a way as to facilitate acquisition and the ability to later apply what they know. The *elaborate* stage challenges students to apply what they have learned in solving problems in new contexts. At this stage, the knowledge acquired is shown to be "productive," that is, useful for understanding and addressing situations other than those used to build the concepts and procedures. The final phase *evaluate*, provides an opportunity for students to review and reflect on their own learning, and on their new understanding and skills.

Learning Approaches Used in STEM Education

Learning approaches such as inquiry based learning and project based learning are the most common used approach embedded in STEM education in the schools. Inquiry-based learning is an instructional style based on the idea that learning may be facilitated by giving students the opportunity to explore an idea or question on their own (Reeve, 2015). Inquiry-based learning is a student-centred, active learning approach focusing on questioning, critical thinking, and problem solving. STEM education should emphasize using "inquiry-based learning" (Reeve, 2015). On the other hand, Project-based Learning is another approach that can be embedded in STEM. PBL aims to focus on solving a problem by developing a product or creation. One of the hallmarks of PBL in STEM is authenticity, solving real- world problems as actual teams of scientists and engineers might do in their jobs. Students collaborate with one another as they identify problems and analyze data.

This study uses the STEM education model given in Figure 2. The model delineates the focus of teaching in STEM education. The model links the STEM education to integrated teaching. While the oval STEM shapes indicate the preservation of unique characteristics within each STEM discipline, such as in-depth knowledge, skills, and beliefs, the arrows from the shapes represent the teacher and student-driven interactions. The interactions exist because they are often integral parts of the STEM disciplines, rather than optional. However, the model also hypothesizes that it takes a well-educated teacher with a

Figure 2. STEM education model with a particular focus on mathematics and science

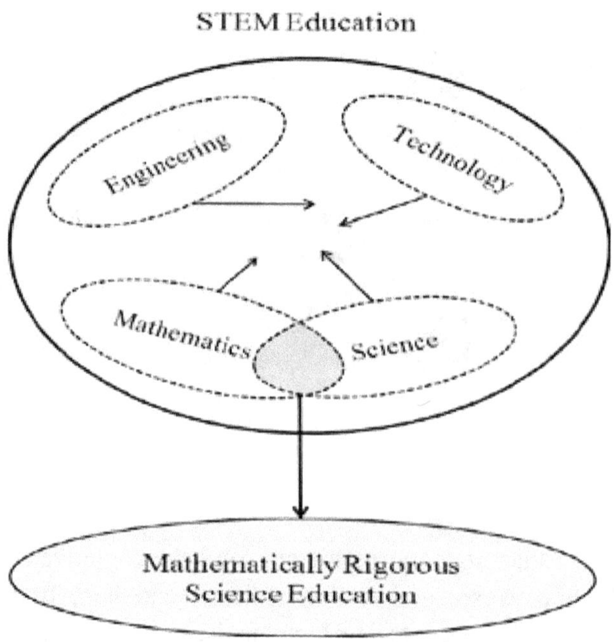

strong integrated teaching knowledge to such interactions actually occur in the classroom settings. The notion of integrated teaching knowledge is defined at the nexus of STEM teacher's expert content and pedagogical content knowledge in their main subject area and working knowledge in another STEM subject, which is mainly developed through participation in professional learning communities (Corlu, 2014). The model is designed with the potential to address all other interactions between STEM subjects; however, the conceptualization of integrated teaching knowledge is beyond the scope of the current paper.

Purpose of the Study

This study aimed to describe teachers and students' attitude towards STEM education. Notably, the two main research questions that drive this study are

1]What are the attitudes of the secondary school teachers towards STEM education?

2] What are the attitudes of secondary school students towards STEM education?

RESEARCH METHODOLOGY

This study adopted a survey research design. The current study used a quantitative approach to answer the objectives of the study. The quantitative method for this study is being used because of the statistical, arithmetical, or a numerical study of facts gathered through surveys, and questionnaires (Babbie,

2010). Survey researches are based on the opinions of the participants about a subject or event or the characteristics such as their interests, skills, abilities and attitudes, and generally the studies in which the sample is larger than other studies (Büyüköztürk et al., 2013). The method used in this investigation is discussed under the following headings: participants, instrument, and the data analysis technique.

Participants

Data for this paper came from the ten schools where the STEM labs were set up in the first phase of the STEM project of SASTRA University. For this study, data was collected from 136 students of grade 9 and 33 teachers handling the STEM lab were distributed the questionnaire. The convenience sampling method was used to determine the study group. Yıldırım and Şimşek, 2008). Convenience sampling is defined as the making sampling of individuals (volunteers) who are in the immediate vicinity, easy to reach and willing to participate in the study (Erkuş, 2009).

Research Instruments

Altakhyneh and Abumusa, (2020), Scale of Trends towards the Use of STEM Approach in Education, was used to collect data about secondary school teachers' attitude towards STEM Education. The questionnaire consists of 19 statements distributed into four cores: the core of desire of applying the strategy, core of developing thinking and problem solving, core of motivation and self-learning skills, and communication and collaboration. It is a Likert five-level scale ranging from 5= Strongly Agree, 4= Agree, 3= Neutral, 2= Disagree and 1=Strongly Disagree The reliability of the tool is as high as 0.83

Benek, and Akcay, B. (2019), STEM Attitude Scale for Secondary School Students was used to collect data from students. There are 33 items of students' attitudes towards STEM and 21st-century skills divided into 4 items for science, 8 items of mathematics, 6 items for engineering, 6 for technology, 5 for Science-Mathematics- Engineering-Technology and 4 for career. Each item consists of a scale of 1-5 which must be filled in by all participants. A reliability coefficient index of 0.78 was obtained using Cronbach's alpha.

Data Analysis Process

All the data collected from the respondents were gathered together to be analysed using Statistical Package for the Social Sciences (SPSS) version 20. The analysis included descriptive analysis. The researchers used descriptive analysis mean and standard deviation to identify the attitude of students and secondary school teachers about STEM education.

The rubric shown in Table 1 was used to interpret the descriptive analysis of data.

Results

Research Question 1. What are the attitudes of secondary school teachers towards STEM education?

To answer the above research question, the descriptive findings are presented in the table below-

As evident in Table 2 secondary school teachers have a high attitude towards STEM education.

In the Desire to Apply Strategy area, all items have high mean score, with the highest for, "I think this strategy develops the learner's motivation to study".

Table 1. Rubric of questionnaire

Range	Level
x ≤2.00	Very low
2.00 <x ≤3.00	Low
3.00 <x ≤4.00	Moderate
4.00 <x ≤4.50	High
4.50 <x ≤5.00	Very High

On the domain Thinking and Problem Solving, the highest mean score is for the item, "STEM educational strategy develops autonomy in thinking".

On the domain, Motivation and Self-learning, all items have high mean score, with "I think this strategy is interesting and fun for learners", scoring highest value.

Table 2. Descriptive statistics of the attitudes of secondary school teachers towards STEM education

S.No	Item	Mean	Description
	Core (1): Desire to Apply Strategy		
1	I think STEM strategy is useful to me in teaching	4.29	High
2	I feel that tools that have been used in the course on the STEM system are useful to me .	4.26	High
3	I gained important new experience for me in teaching through the course	4.17	High
4	I believe that STEM strategy has provided elements of complementarity in the three areas of purpose cognitive, skilful and compassionate	4.26	High
5	I feel a desire to apply the STEM strategy in teaching math and science courses	4.23	High
6	I think this strategy develops the learner's motivation to study	4.34	High
	Core (2): Thinking and Problem Solving		
7	I think this strategy is important for the development of creative thinking	4.43	High
8	I think this strategy is important for the development of critical thinking	4.4	High
9	I think this strategy is important for the development of computational thinking	4.34	High
10	STEM educational strategy develops autonomy in thinking	4.51	Very High
	Core (3): Motivation and Self-learning		
11	I think this strategy is interesting and fun for learners	4.33	High
12	I think this strategy relieves students anxiety in mathematics and science	4.06	High
13	STEM strategy has educational benefit of increasing the self - confidence of learners.	4.26	High
14	The STEM education strategy develops the ability to self-manage and organize	4.14	High
15	The STEM educational strategy takes into account individual differences among students	4.26	High
	Core (4): Collaboration and Communication		
16	I feel that STEM educational strategy fosters Collaborative work among students	4.09	High
17	STEM stereotyping strategy develops the ability to communicate with others	4.06	High
18	STEM strategy can be applied in math and science classes at all stages	3.83	Moderate
19	STEM tools are multiple and varied.	4.11	High

Table 3. Descriptive statistics of attitudes of the secondary school students towards STEM education

S.no	Item	Mean	Description
	Factor (1):Science		
1	I enjoy participating in science-related activities.	4.68	Very High
2	I am interested in science course.	4.54	Very High
3	I am curious about science.	4.55	Very High
4	I enjoy working on science at home.	4.44	High
	Factor (2):Mathematics		
5	Mathematical operations are enjoyable.	4.11	High
6	I am interested in mathematics.	4.34	High
7	Mathematics is a fun lesson.	3.96	Moderate
8	I'm interested in mathematics in extracurricular times.	4.17	High
9	I enjoy solving mathematical problems.	4	High
10	I do mathematical activities in my spare time (puzzles, sudoku, etc.).	3.24	Moderate
11	When I have math homework, I can't wait to complete it.	4.27	High
12	I feel good when I deal with mathematics.	4.09	High
	Factor (3):Engineering		
13	I like to design things.	4.57	Very High
14	I like drawing a building, car, bridge, plane, etc.	4.25	High
15	I enjoy designing something in my free time.	4.49	High
16	I would like to design a tool/product that will make human life easier.	4.47	High
17	I love to repair something at home.	4.42	High
18	I am interested in designing a product/tool.	4.52	Very High
	Factor (4): Technology		
19	I like to play with technological tools.	4.6	Very High
20	I am interested in the technology.	4.5	Very High
21	I like to deal with technology.	4.25	High
22	I would like to have more courses on technology at school.	4.44	High
23	I want more technology to be used during the course.	4.49	High
24	I closely follow the latest innovations in technology.	4.27	High
	Factor(5):Science-Mathematics- Engineering-Technology		
25	Science, mathematics, engineering and technology are interrelated.	4.06	High
26	Science, mathematics, technology, and engineering skills should be used together when inventing something.	4.49	High
27	Science, mathematics, technology and engineering fields complement each other.	4.25	High
28	Science, mathematics, engineering and technology are very important in our life.	4.41	High
29	If I attend a course that combines science, mathematics, engineering, and technology skills, I may learn about my skills that I am not aware of.	4.31	High
	Factor(6):Career		
30	I would like to choose an engineering profession (machinery, mining, construction, architect etc.) in the future.	3.6	Moderate
31	In the future, I would like to choose a profession related to science (physics, chemistry, biology, medicine, space, etc.).	4	High
32	In the future, I would like to have a job about the technology.	3.98	Moderate
33	I would like to have a profession in the fields of science, mathematics, engineering, and technology so that I can be more beneficial to people.	4.33	High

On the domain, Collaboration and Communication, all items have a high score except, "STEM strategy can be applied in math and science classes at all stages", with moderate mean score.

Research Question 2. What are the attitudes of the secondary school students towards STEM education?

To answer the above research question, the descriptive findings are presented in Table 3.

As depicted in Table 3 secondary school students have a high attitude towards STEM education after being exposed to STEM Lab for a period of over one year.

On the Science domain, 3 items out of four, have got very high mean, "I enjoy participating in science-related activities", "I am interested in science course" and "I am curious about science".

On the Mathematics domain, the highest mean is for, "I am interested in mathematics "and the lowest for. "I do mathematical activities in my spare time".

On the Engineering domain, "I like to design things" and "I am interested in designing a product/ tool" have very high mean and all other items have high mean value.

On the Technology domain, all items have a high mean value, while items like, "I like to play with technological tools", and "I am interested in the technology", have very high mean.

On the Science-Mathematics- Engineering-Technology domain, all items have high mean score and the highest mean score for the item, "Science, mathematics, technology, and engineering skills should be used together when inventing something".

On the Career domain, the highest mean score is obtained for the item and the lowest, "I would like to have a profession in the fields of science, mathematics, engineering, and technology so that I can be more beneficial to people", and the lowest mean score for item, "I would like to choose an engineering profession (machinery, mining, construction, architect etc. in the future".

DISCUSSION AND CONCLUSION

SASTRA University's outreach project on STEM initiative explicitly aims to improve students and teachers' attitude toward STEM and their interest in STEM career pathways, with the ultimate goal of increasing student learning and employability in STEM. A national surge in STEM education programs is taking place. This is partly in response to powerful, societal trends indicating an increase in the need for workers with STEM and 21st century skills. The need for a nationwide effort to transform STEM teaching and learning is clear. Many advances in society will continue to come from the fields of engineering and science. India's job-growth has been accelerating in the STEM fields over the past decade or more, and researchers predict the trend will continue. For these reasons educators, researchers, and policymakers must all work towards improving student attitudes toward STEM and increasing student knowledge and skills in these fields. SASTRA University has always been in the forefront of pioneering social initiatives to uplift the student and teacher community in the remote areas and is contributing remarkably in the field of strengthening STEM education.

Students taught with STEM lab approach developed curiosity about science and in science courses, participated more in classroom discussions and developed more interest in science-related activities. They found mathematical problems and activities engaging and felt good dealing with mathematics. Students after being exposed to STEM lab found designing things, tool/product drawing, repairing enjoyable. These students would like to choose profession in the fields of science, mathematics, engineering, and technology.

Overall students' attitude towards STEM education is very positive after being exposed to SASTRA STEM labs. STEM education has influenced students' attitudes in learning as it is one of active learning. Active learning involves students in activities planned by the teacher. Students are allowed to express their ideas in their STEM activities. Active learning has successfully improved student performance in STEM (Freeman et al., 2014).

The teachers found STEM strategy highly useful in teaching and found this strategy developed the learner's motivation to study. The teachers agreed that STEM labs helped in the development of creative thinking, critical thinking, computational thinking and developed autonomy in thinking among the students. Teachers agreed that STEM labs are interesting and fun for learners, relieved students' anxiety in mathematics and science and increased their self-confidence. The STEM labs fostered collaborative and communicative skills work among students.

Secondary school teachers' attitude towards STEM education is very positive. Teachers with a positive attitude and high awareness towards the importance of STEM education are crucial in this era of educational transformation. Positive attitudes can lead to positive behaviours such as being more motivated and confident in the implementation of STEM education. Teachers' mind set has to be transformed according to the expected learning outcome in STEM. The positive attitude towards STEM among the teachers can expedite the forming of integrated STEM education. The findings provide information that may be useful to relevant stakeholders be taken into consideration in implementing and revising the national curriculum of integrated STEM education. This study only involved teachers from the schools where SASTRA has set up STEM Labs which might be a limitation. Teachers cannot be expected to develop an integration of STEM subjects' knowledge and practice without the support from the government.

Although the sample size is enough to analyse this study, further research may be performed in the schools with STEM centres. Additionally, this study used a self-reported measure to identify students' and teachers' attitudes towards STEM education. Self-reported measures are not enough to comprehend attitudes. It requires in-depth interviews, observations to understand the nature of the attitudes.

In light of the study results, the researchers recommend using STEM approach in teaching students in the schools of India, using modern strategies for teaching mathematics and science courses and conducting studies on students of schools and universities. For these reasons, it is advised that while developing professional development programs for teachers, special emphasis be paid to the specific science subject in which teachers are least confident to teach. Experience in the actual teaching of various science courses must be incorporated into the design of the teacher-education curriculum. The study highly recommends the 'adoption' of a school, or a cluster of schools, by a local STEM industry/ enterprise, University, NGOs and other public, private bodies, to produce the entrepreneurs, thought leaders, and innovators of the future.

ACKNOWLEDGMENT

The author gratefully acknowledges the support extended by SASTRA management for conducting this study. SASTRA University in keeping with its motto of community service has completed phase 2 & phase 3 of setting up STEM labs, as of now.

REFERENCES

Acisli, S., Yalcin, A., & Turgut, U. (2009). The influence of teaching based on 5e learning model on undergraduates' skills of scientific process and their attitudes towards physics laboratory in the electric topics. *Bayburt University Education Faculty Journal, 4*(1-2), 80–92.

Aeschlimann, B., Herzog, W., & Makarova, E. (2016). How to foster students' motivation in mathematics and science classes and promote students' STEM career choice. A study in Swiss high schools. *International Journal of Educational Research, 79*, 31–41. doi:10.1016/j.ijer.2016.06.004

Babbie, E. (2010). *The practice of social research* (12th ed.). Wadsworth.

Başkan, Z., Alev, N., & Kara, I. S. (2010). Physics and mathematics teachers' ideas about topics that could be related or integrated. *Procedia: Social and Behavioral Sciences, 2*(2), 1558–1562. doi:10.1016/j.sbspro.2010.03.235

Büyüköztürk, Ş., Kılıç Çakmak, E., Akgün, Ö. E., Karadeniz, Ş., & Demirel, F. (2013). *Bilimsel araştırma yöntemleri* [Scientific research methods]. Pegem Press.

Bybee, R. (2010). Advancing STEM education: A 2020 vision. *Technology and Engineering Teacher, 70*(1), 30–35.

Bybee, R. W., Taylor, J. A., Gardner, A., Van Scotter, P., Powell, J. C., Westbrook, A., & Landes, N. (2006). The BSCS 5E instructional model: Origins and efec- tiveness. *Colorado Springs Co: BSCS, 5*, 88–98.

Churcher, K. M. A., Downs, E., & Tewksbury, D. (2014). "Friending" Vygotsky: A social constructivist pedagogy of knowledge building through classroom social media use. *The Journal of Effective Teaching, 14*(1), 33–50. Retrieved January 2nd, 2023, from https://uncw.edu/cte/et/articles/Vol14_1/Churcher.pdf

Corlu, S., Capraro, R. M., & Caprar, M. M. (2014). Introducing STEM education: Implications for educating our teachers for the age of innovation. *Education in Science, 39*(171), 74–85.

Cuadra, E., & Moreno, J. M. (2005). Expanding opportunities and building competencies for young people: Anew agenda for secondary education. https://www.researchgate.net/publication/260517903_Introducing_STEM_Education_Implications_for_Educating_Our_Teachers_For_the_Age_of_Innovation

Erkuş, A. (2009). *Davranış bilimleri için bilimsel araştırma süreci* [Scientific research process for behavioral sciences]. Seçkin Press.

Fazelian, P., Ebrahim, A. N., & Soraghi, S. (2010). The effect of 5E instructional design model on learning and retention of sciences for middle class students. *World Conference on Psychology, Counselling and Guidance (WCPCG 2010), 5*, 140143. 10.1016/j.sbspro.2010.07.062

Freemana, S., Eddy, S. L., McDonough, M., Smith, M. K., Okoroafor, N., Hannah Jordt, H., & Wenderoth, M. P. (2014). Active learning increases student performance in science, engineering. *Proceedings of the National Academy of Sciences of the United States of America, 111*(23), 8410–8415. PMID:24821756

Furner, J. M., & Kumar, D. D. (2007). The mathematics and science integration argument: A stand for teacher education. *Eurasia Journal of Mathematics, Science and Technology Education, 3*(3), 185–189. doi:10.12973/ejmste/75397

Gerlach, J. M. (1994). Is this collaboration? In K. Bosworth & S. J. Hamilton (Eds.), Collaborative Learning: Underlying Processes and Effective Techniques, New Directions for Teaching and Learning No. 59. Academic Press.

Lazareva, S. A., & Marchuk, T. L. (2019). STEM technology as a means of forming the engineering thinking of school children. *Perm Pedagogical Journal, 10*, 76–79.

León, J., Domínguez, E. G., Núñez, J. L., Pérez, A., & Martin-Albo, J. (2011). Translation and validation of the Spanish version of the Echelle de Satisfaction des Besoins Psychologiques in academic context. *Anales de Psicología, 27*(2), 405–411 Retrieved from http://digitum.um.es/jspui/handle/10201/26530

Mariani, A., & Ismail, Z. (2013). Pengaruh Kompetensi Guru Matematik Ke Atas Amalan Pengajaran Kreatif. In *2nd International Seminar on Quality and Affordable Education (ISQAE 2013)* (pp. 181-187). Academic Press.

Moe, A., Jansen, P., & Pietsch, S. (2018). *Childhood preference for spatial toys: Gender differences and relationships with mental rotation in STEM and non-STEM students' learning and individual differences*. doi:10.1016/j.lindif.2018.10.003

National Research Council. (2011). *Successful K-12 STEM education: Identifying effective approaches in science, technology, engineering, and mathematics*. Washington, DC: National Academy Press. https://www.researchgate.net/publication/260517903_Introducing_STEM_Education_Implications_for_Educating_Our_Teachers_For_the_Age_of_Innovation

Reeve, E. M. (2015). STEM thinking! *Technology and Engineering Teacher, 74*(4), 8–16.

Sakellariou, C., & Fang, Z. (2021). Self-efficacy and interest in STEM subjects as predictors of the STEM gender gap in the US: The role of unobserved heterogeneity. *International Journal of Educational Research, 109*, 101821. Advance online publication. doi:10.1016/j.ijer.2021.101821

Shaughnessy, J. M. (2013). Mathematics in a STEM Context. *Mathematics Teaching in the Middle School, 18*(6), 324. doi:10.5951/mathteacmiddscho.18.6.0324

Vanykina, G. V., & Sundukova, T. O. (2020). Educational robotics in educational institutions: An empirical study of motivation and STEM skills. *Pedagogika Informatiki, 3*, 1–10.

Willimas, J. (2011). STEM education: Proceed with caution. *Design and Technology Education: An International Journal, 16*(1).

Wiswall, M., Stiefel, L., Schwartz, A. E., & Boccardo, J. (2014). Does attending a STEM high school improve student performance? Evidence from New York City. Academic Press.

Chapter 11
Social Sustainability via Critical Reflexivity:
Strategic Gamification in Higher Education

Catherine Hayes
https://orcid.org/0000-0003-3870-2668
University of Sunderland, UK

ABSTRACT

Whilst individual efforts to achieve social sustainability are positive, the challenges it poses are such that without a collective, proactive, and collective approach, intergenerational success for the most significant critical global issues will become a lost and irretrievable opportunity. Gamification is one pedagogical approach to engaging the collective in issues pertaining to social sustainability, where collective learning and engagement will be central to the creativity and innovation that can ensue. Capacity for critical reflexivity, rather than a reflective stance on the past, is needed to drive the perspective lens of transformation forward for a sustainable and pragmatically implementable future. The process of empowerment and the active gift of agency across educational contexts can be used to frame and contextualise key contemporary issues such as social justice, global poverty, climate, and ecological crisis, within the context of a holistic and humanitarian approach to developmental change and societal progression.

When childhood dies, its corpses are called adults and they enter society, one of the politer names of hell. That is why we dread children, even if we love them, they show us the state of our decay.

Brian Aldiss (1925-2017)

DOI: 10.4018/978-1-6684-9859-0.ch011

INTRODUCTION

Whilst individual efforts to achieve social sustainability are positive, the challenges it poses are such that without a collective, proactive, and collective approach, intergenerational success for the most significant critical global issues will become a lost and irretrievable opportunity. Gamification is one pedagogical approach to engaging the collective in issues pertaining to social sustainability, where collective learning and engagement will be central to the creativity and innovation that can ensue. Capacity for critical reflexivity, rather than a reflective stance on the past, is needed to drive the perspective lens of transformation forward for a sustainable and pragmatically implementable future. The process of empowerment and the active gift of agency across educational contexts can be used to frame and contextualise key contemporary issues such as social justice, global poverty, climate and ecological crisis, within the context of a holistic and humanitarian approach to developmental change and societal progression.

THE NOTION OF LUDIC APPROACHES FOR SUSTAINABLE STRATEGY BUILDING

For the purposes of this chapter, 'ludic' is defined as the capacity to play or integrate games into everyday interaction and dialogue, of which issues surrounding social sustainability are an integral part. Delineating this from the conceptual foundation of 'serious play' is important in terms of the relative purpose both can serve in the context of Higher Education (HE). Whereas serious play combines ideas from constructionism, which has been widely critiqued in relation to the seminal work of Piaget (Sutton-Smith, 1966) and its subsequent progressive development (Harel and Papert's Complex Adaptive System Theory (1991) which was evaluated by Holland (1996) in terms of application to management and organisational infrastructure, ludic approaches are far less formal, and within them, a creative, lateral approach to thinking about complex issues is encouraged as a mechanism of 'thinking outside the box.' For current and topical issues which are often in a state of dynamic flux and require flexible and adaptable approaches to address, a ludic approach to strategic thinking and inferential planning can be an ideal means of ensuring all collaborative partners are afforded individual agency, alongside their collective contribution to informed strategic praxis. The author's own pedagogical contributions to the literature have applied the concept of 'serious play' to learning and teaching in postgraduate level nursing (a contextually and situationally specific environment in United Kingdom (UK) HE). Within this research, the LEGO® SERIOUS PLAY® (LSP®) method was chosen for its specific capacity to engage students in activity that can lead to deeper reflection and collective constructive dialogue (Hayes, 2015). Albeit a contextually different setting, the pedagogical research of Dann (2018) echoes this work, in which he outlines the concept of knowledge co-creation with students. This approach offers important insight into the context of social social sustainability, which this chapter explores further.

FRAMING, CONTEXTUALISING, AND OPERATIONALLY DEFINING 'PLAY'

In common with other contexts of 'adult play', the LSP® method can be operationally defined as 'play' for the purposes of this study since it involves the imaginary, it is limited in time and space and is structured by rules pertaining to the parameters within which the concepts of social sustainability can be explored (James, 2015). Children are progressively educated out of being creative and away from using play to constructively 'think' laterally as they are gradually both collectively and individually shaped and guided away from playfulness to the characteristic format of taught disciplinary signature pedagogies (Robinson and Aronica, 2015). These metaphorical silos of learning often have a negative impact on the potential for emotional freedom, free thinking, and consequently, the emotional self (Resnick and Robinson 2017). In terms of the interprofessional and multi-disciplinary approaches to social sustainability, this is a key mechanism of ensuring that all participants in strategic planning are afforded the agency they need to feel empowered and supported in terms of their active contributions. Indeed, it is possible to harness the agency of several individuals and to ensure that this becomes a valued and collective aspect of professional practice and academic debate.

In contrast to traditional approaches adopted to engage students in critical self-reflection, particularly in nursing and allied health disciplines, the LSP method has become a recognized means of capturing individualism and the acknowledgment of the value of individual thought processes. It is at this stage that whilst we can acknowledge the emotional self as something intrinsically private that it ought also to be acknowledged that thoughts meanings and emotions are rarely, if ever, just held within people, rather they exist between human individualities as part of wider collective aims, values, and experiences at community or societal levels (Kristiansen and Rasmussen, 2014; Schulz and Geithner, 2014). Nowhere is this more apparent than in current culture and context in relation to climate change and the need for pragmatic approaches to the implementation of measures which can address the devastating impact of non-sustainable futures. Whilst the media is full of reports of social activism and protests, it is through dialogue and negotiation that true progress will be made. Kestly's (2014) insight into the interpersonal neurology of play, specifically examined the subcortical motivational systems in the brain. Her findings revealed that playing with blocks powerfully influences the decision-making processes and fundamental behaviours that are regulated in higher brain regions. She also highlighted the notion of 'thinking with your fingers', where the physical processes of creation facilitate reflection, which is such an integral part of the profession/vocation of any practicing healthcare professional, for example.

Within that definitive interrelationship between psychology and neuroscience, cognitive processes such as learning, and memory are strongly influenced by the way the human body is used to interact with the world (Claxton, 2015). The term *emergence* is the hallmark of connection with right mode processing of the brain, which attends to what is happening 'in the present moment/here and now' within the emerging flow of new experience (Schore, 2018). When added to the already active left mode, this is actually the zone of creativity, which is of direct significance to the LSP method. In turn, planning the right mode in the lead is pivotal in allowing what has not yet been thought to emerge via the hands and gives natural integration of the brain for something never before imagined (Corballis, 2017). The process of play is thus temporal and uniquely specific to that moment in time in which it occurs. It also permits the notion of 'flow' where emotionally escapism, excitement, and social connection can flourish unhindered (Seyderhelm, Blackmore and Nesbitt, 2019). It has been established that neural change is supported when humans are in safe connection with others (Wilson, 2010).

INTEGRATING GAMIFICATION INTO STRATEGIC PLANNING

The diverse array of challenges presenting across global, national, and regional HE institutions incorporates a raft of issues that impact on all members of the organizational hierarchy (Hayeter and Cahoy,

2018). The rise of neoliberalism and the consequent commodification of education, in general, has seen changes in the underpinning politics of HE change at a faster rate than at any other time in history (Roberts and Peters, 2019). In the face of the economic austerity accompanying this change, those working in HE have never been more in need of coping mechanisms that accommodate systemic inequity and the need to function effectively on both an individual and collective level (Rodd and Sanders, 2018). At the heart of the capacity to do this is the need to address the emotional self in the workplace (Conley, Durlak and Kirsch, 2015). Awareness of the concept of emotional intelligence has increased radically over the last decade. Intergenerational solidarity and understanding have never been at the fore of positive function in the HE workplace, and correspondingly new approaches to supporting staff in reflecting on experience have been developed and implemented (Parrish, 2015).

Belying any holistic sense of wellbeing, whether organisationally, collectively or individually, is mental health status. As a direct consequence of this, advocating, promoting, and facilitating the positive health and wellbeing of staff has been identified as a key priority for recognition, address and sustained integration into academic practice in HE (Schonert-Reichl and Roeser, 2016).

SITUATIONAL SPECIFICITY AND CULTURE

How ludic approaches, now more commonly termed 'gamification' in the HE arena, can serve to facilitate processes of critical introspection and reflective practice are this highly relevant to the notion of the 'emotional self.' Both are pivotal to making meaning of the experiences which shape the self as an individual and society as a collective, regardless of context. The situational specificity and culture that HE now engenders have been shaped by trajectories of political whim, technological progression, social and historical change (Roos and Victor, 1999). Despite this, the capacity of all people to play is defined by their sense of security and wellbeing, a clear indicator of their mental health status, or the predisposing factors to the development of mental illness. The approaches considered in this chapter are feasible for all members of the academic community, including academic staff, students, and administrative and support staff, all of whom occupy valuable roles in HE.

How ludic approaches can transcend organizational hierarchy, in dispelling power relationships will also be illuminated. This will highlight how the focus of reflective and reflexive praxis can be solution-focused, free from the influence of positional power relationships, and provide a safe arena for all people to contemplate their non-tokenistic engagement in both individual and wider collective agendas.

FOUR KEY OUTCOMES OF PLAY

Exploring each of the four key impacts of play, and its relation to gamification in practice, permits an insight into the potential impact of ludic approaches for people who spend much of their lives in the context of agreeing to social sustainability, they are namely:

- **Social Bonding**

For the last half a century, debate has continued about the role of play in ensuring the complexity and ambiguity that has the capacity to engage humans in critical thinking. Smith (1982) posited that the concept of fantasy was an emergent property in humankind that was assured by the fact they had the capability of symbolic intelligence, language, and consequently the discourse, which served to ensure play was sufficiently complex for the iterative development of skill. Since it is motivation which drives this will, then the delineation between what the aims of play are in terms of intrinsic and extrinsic goal setting are important, since their source impacts not only on the motivational level but the likelihood of success in being able to socially bond with others over play which has shared goals as a central characteristic. The capacity to experience a degree of fulfilment in the expression of creativity is the primary component of intrinsic motivation in the context of the ability to demonstrate any ability of critical expression (Amabile, 1983).

- **Emotional Expression**

It has been posited that characteristic personality traits can also influence the capacity for creativity. Since individuality itself is a multifaceted and inherently complicated notion, this has great relevance to the ability of all people to maximize their potential as adults (Karwowski & Jankowska, 2016). In childhood learning. It is widely accepted that play is an activity that facilitates the progressive development of emotional and consequently, social skills. In the context of adult learning, these often long-established capacities for emotional and social expression can also be greatly enhanced. In relation to LSP, in common with imaginative play, there is a psychomotor activity with the hands that is fundamentally based on making creative associations. This is essentially the basis of divergent or lateral thinking, the capacity for which often provides an opportunity for the management of complexity and ambiguity via non-traditional approaches. Play can thus be a driver of both emotional and social expression, which can be contextually very different (Jankowska and Omelańczuk, 2018).

- **Metacognitive Development**

The capacity to reflect on knowledge acquisition and co-construction takes its formative basis from the development of a cognitive repertoire that extends beyond the immediacy of knowledge available for consideration. This whole capacity for metacognition begins and develops throughout childhood, yet often remains unconsidered in contexts of adult learning such as HE. This is echoed throughout the life trajectory, with gerontological studies of older adults in practice, resonating the same findings, regardless of the presence of organic illness and cognitive disruption (Tse, Kwan and Lee, 2016).

- **Constructive Competition**

Flow, as outlined by Chen (2007), is a key component in the achievement of cognitive 'flow.' As such, the person undertaking the gamification is fully engaged in something without the need for external coercion, and this is an ideal opportunity for the maximizing of learning opportunity, where the person feels a sense of ownership and responsibility for their own contribution. Since this stems from intrinsic motivation, rather than external motivation or the premise of reward, it ensures that optimal learning can continue for as long as the person has the will and the capacity to continue (Fuchs et al, 2014). Fuchs noted that when intrinsic motivation is decreased, there is a corresponding diminished sense of agency, creativity, and self-worth.

At the crux of any academic debate on the emotional self in HE is a wider philosophical consideration that recognizes emotion as an inescapable phenomenon which impacts human interaction, relationality, and the capacity to engage others in collaborative exchanges about the external world, which ultimately remain connected to intrinsic motivations. As such, whether the word competence can ever actually be applied to the concept of human nature or whether its inherently individual variances ought ever to be taxonomized. In a world which has evolved to incorporate technology at every available opportunity, recognition of the emotional self may be no longer a simple process of self-reflection and critical introspection but a systematized and algorithmic binary decision-making approach to categorizing oneself in the world. Interaction in social and emotional environments are key to not only the identification of self but also its place as a temporal entity in personal and professional trajectories through life. Holistic approaches and the integration of perspectives on the personal and the professional are one possible means of advocating individuality in dynamic and ever-changing workplaces.

THE EMOTIONAL SELF AND CREATIVITY IN SOCIAL SUSTAINABILITY PLANNING

As a concept, creativity has become the word of reference to indicate the capacity for problem-solving, critically constructive decision making, higher-order thinking, and the means of dealing with the complex ambiguities that most professionals face each day in the workplace.

Whilst legalistic frameworks and algorithms can irrefutably provide specific frames of reference for dealing with objective processes, the facilitation of the emotional self, necessitates epistemic questioning of often long-held presuppositions and assumptions that can frame and constrain thinking (Freedman, 1983). The central tenet of this recognition lies in the fact that human action often defies logical explanation and where the personal underpins the professional meaning is made by the deconstruction of meaningful experiences so that these can effectively be consolidated with new ones (Finlay and Gough, 2008). It is through this process that challenging and additive conceptual thinking can drive forward the concept of innovation as a bridge to creativity. Where the need for universal creativity really comes to the fore is in dealing with the emotional self and its relationship to others, whether they be colleagues, students, or key stakeholders within the context of HE. Aligning the individual self with collective groups from the same organizational hierarchy is where key challenges arise, where people are often stymied into categories of functionality that limit their inherent capabilities.

SITUATING AUTHENTICITY IN SUSTAINABLE HIGHER EDUCATION

At the heart of authenticity, which can be defined for the purposes of this chapter, as genuine or real, is the concept of individuality. The uniqueness that each person brings to the context of HE ought to ensure the mobilization of a range of skills and capacities for their unique operationalization in academic praxis. One mechanism of ensuring authentic learning experiences for students has been the introduction of simulation – a means of incorporating real-life complexity and ambiguity within the context of safe places. In medical education, this has been transformative in the minimization of risks, as students engage in complex skills and immersion in clinical decision making without fear of harming a patient. The process of debriefing, though, provides a broad scope and platform for critical introspection and reflexivity, which will ultimately serve to ensure that in real-life contexts, many different perspectives and angles of the same scenario can be considered.

EDUCATIONAL ENVIRONMENTS FOR AFFECTIVE IMPACT

Alongside functional skills, affective skills can also be developed in an environment that facilitates reflection on action and a proactive approach to the address of identifiable weakness in attitudes and consequently expressed behaviours in social sustainability practice. Technological advances have ensured an increasing degree of complexity and the capacity for this complexity to be regarded in from a multitude of perspectives, in what can be achieved in immersive learning environments and for a generation of new HE students and academics, the approach provides a degree of currency in interactive learning and critical engagement for social sustainability.

Where authenticity has the greatest impact in educational practice, is in the provision and affordance of learning opportunities that incorporate the expertise of professionals already established in the discipline and in the acknowledgment of how interpersonal dynamics, power relations, and human interaction in practice can shape and define learning potential. It is a strategic pedagogical design of learning opportunity, though, that delineates between how students are afforded immersion and how gamification can be introduced with the specific purpose of increasing intrinsic motivation to learn about social sustainability via exploring their own positional stance regarding its key facets.

PERSPECTIVES ON LUDIC APPROACHES TO SOCIAL SUSTAINABILITY EDUCATION

The ludic approaches that gamification encompasses have been shrouded in criticism since their earliest implementation and no advocating of their implementation ought to ignore this in relation both to their pragmatic execution or their worth to processes of learning and teaching. The mechanisms for the challenge of critical deductive thinking processes usually hinge around the fundamental design principles of being able to benchmark levels of successive improvement or proficiency. However, they can often be regarded as simply driving processes of instructional design rather than informing them. When a negotiated level of choice is integrated, which incorporates individual student preferences for learning in specific ways, then this can also be used to inform processes of data analytics, although their accuracy in relation to gamification is also an area of evident contention. The key issue ought to be posited as to whether ludic approaches serve to build acknowledgment of identified epistemic biases and difference positional stances in the workplace, or whether their novelty value serves simply to entertain.

Another reported key aspect of education and ludic approaches to it is the concept of rationality. Formal learning has often been the very antithesis of fun in transactional, didactic approaches to learning that were launched at the heart of the industrial revolution when education was most highly valued and commodified. Despite progressive changes in both pedagogic practice and technology, learning environments still control the ethos and situatedness of educational experience, and this is often overlooked in relation to acknowledgment of the emotional (Dirkx, 2001). In a world where emotion has been suspended in favor of rationality and the challenge of presupposition, a juxtaposition of what learning is, and ought to be, now prevails.

Whether learning ought also to have a sense of fun, has long been debated in academic scholarship circles. One of the authors who has vocalised an ongoing concern for the implementation of gamification and ludic approaches to educational practice, is Groh (2012), who highlighted the capacity for learning with longevity i.e.) social sustainability might well be lacking in the context of HE.

THE CONCEPT OF EMERGENCE AND EMOTIONAL SELF-REGULATION

The term emergence is the hallmark of connection with right mode processing of the brain, which attends to what is happening 'in this moment' within the emerging flow of new experience. When added to our already active left mode this is the zone of creativity, which is of direct significance to play that involves using the hands and building representations of reality which are otherwise not possible with conventional dialogue and discourse. In turn, planning the right mode in the lead is pivotal in allowing what has not yet been thought to emerge via our hands and gives natural integration of the brain, for something never imagined. The process of play is temporal and most specific to the moment in time in which it occurs. It also permits the notion of 'flow' where emotionally, escapism, excitement, and social connection can flourish unhindered without the need to consider organisational hierarchical positionality. It has been established that neural change is supported when we are in safe connection with others, which emphasises the importance of this context for social connection and escapism from formal processes (Wilson, 2010).

It is the generalised introspection that emotions trigger, which have a temporality worthy of further consideration. Expressed emotions are abstractly representative of belief from epiphanies or even just everyday interactions that nevertheless have a potentially tangible consequence. It is here that the separation of emotion within an event and then emotion of the recollection of an event are two distinct yet related entities, both of dure. Their linkage permits consideration of the difference between perceived and actual reality from a philosophical perspective as well as allowing an insight into the maturation of thinking about experience that is articulated rather than the experience itself. It is the processes of explanation and explication of experiences that the emotional self relies on to produce a semantic report of experience (Gunnarsson, Linell and Nordberg, 2014). Nowhere does this have greater relevance than in HE Institutions that deliver health and medical curricula. It is in these institutions that reflection is not only a pivotal part of being a disciplinary educator, but also an integral taught element in programmes whose students will work in situations and contexts where decision making, in the light of pre-existing experience, and the immediacy of it, can make the difference between life and death.

- The accessing of emotions as the experience occurs / experience 'in situ'
- The reconstruction of experiences from previous incidents of relevance (episodic memory)
- The situated / context specific nature of emotion
- The individualised/personalised belief

All are of significance to the capacity that individuals in HE has to make meaning of their experiences in specific contexts or situations, how key epiphanies or events effect and impact on their future performativity (Kalfa and Taksa, 2017). Since all qualitative expressions hinge only on that specific time, it is impossible to relive either positive or negative experiences, however their recollection can be a new emotive experience (Rubin, Deffler and Umanath, 2019). The universality of human experience is such that as all variables change, then an individual experience of a phenomena at a specific point in time, is unique by virtue of its temporality (Fuchs, 2019). Details of situation and context progressively diminish with the passage of time in terms of our capacity for emotional recall (Rivière et al, 2019). It is this recreation or reconstruction of experience, which may permit a person to reproduce feelings reflective of their temporal emotion. What matters ultimately is the report and recognition of the emotional self in everyday practice and its potential impact on the articulation of approaches to social sustainability.

THE SYMBOLISM OF THE EMOTIONAL SELF

Emotion is in this sense a double-edged sword in that it is a symbolic sign of 'weakness' and of 'power' depending on where it is directed. It is the experience of emotion that serves to weaken but also fuels reflexivity, proactivity and ignites the passion necessary for the intellectual engagement that underpins the emotional self in action.

The difference between knowing in the moment and making inference to a memory of feeling or thought in the past was explored at length in a now seminal article by Ross (1989). He established that the reconstruction of pre-existing feelings is often done in conjunction with consideration of processes of change or stability. This introduces the notion of prospective bias, where previously held feelings can appear more or less negative, depending on current emotional self-status. The psychology of avoidance can be closely linked to this, where people often underestimate their own coping capacity to deal with particular situations and contexts on the premise they believe the outcome will be worse than how it will actually be in reality. Gilbert et al (1998) termed this the concept of 'immune neglect'.

A PRACTICAL APPROACH TO GAMIFICATION IN SOCIAL SUSTAINABILITY PRACTICE

One example of now commonly implemented gamification in practice is the LSP® method. Being able to define a strategic set of aims and objective underpinning the purpose of any ludic approach is what delineates adult from childhood play, but also encourages a state of cognitive flow that otherwise would be more limited and narrower in scope (Gauntlett, 2014; Frick, Tardini and Cantoni, 2013) . A previous research study undertaken by the author, revealed several issues of relevance to the introduction of playfulness in the articulation of the thoughts people had regarding the use of LEGO as a medium in story making for storytelling (Hayes, 2015). The study focused on four key areas with a group of student nurses undertaking their initial undergraduate degrees:

- Whether there was a perceived connection between flow and consciousness in relation to the issues that students explored via introspective reflection on the processes of transformative learning with the LSP® method.
- Whether the perceived impact of the LSP® method by students related to other variables or characteristics such as for example, their capacity for intrinsic motivation.
- Whether students perceived that skills of higher order thinking involved emotional and motivational responses associated with an individual engagement in flow promoting activities.

- Whether there is potentially a role for the LSP® method generally and metaphorical modelling, specifically in the education of future generations of critically reflexive student nurses.

The concept of the 'emotional self' is not a new phenomenon and is now seated at the heart of motivation to equip, predominantly younger generations with a sense of those sustainable behaviour traits, which protect mental health and wellbeing alongside increasing personal resilience in an uncertain world. Being able to harness personal capacity for the adaptation to change by the application of emotional intelligence is fast becoming desirable in relation to the dynamic context of HE which hinges on the pivot of structure, agency and their often very obtuse political drivers.

It is emotion which underpins human motivation, and which distinguishes us from the binary, purely rational thinking of algorithmic artificial intelligence (AI). Whilst this has become a topical futuristic discussion in scientific circles where capacity for decision making is set to become the preserve of future generations of technology, human emotion remains the preserve of the species, which ultimately drives the species agendas, thinking, behaviours and consequential outcomes, whether positive, negative or neutral in terms of societal impact.

One of the commonest misconceptions about HE is that it has ever embedded a sense of awareness as to what it can prepare students for. Just as we can now allude to the rise of AI, our predecessors had no concept of how significant the rise of technological advancement would be in relation to roles that had been occupied for centuries before. Perhaps one of the best examples of this is medicine, where technological advances have revolutionised the assessment, diagnosis and management of patients but ultimately how they are treated remains a pivotal part of any claim of successful care (Forgas and Williams, 2014). At the heart of this is the centrality of human emotion, of compassion and of the capacity to identify people as individuals who matter in the world beyond their physical functionality, and pragmatic usefulness to the world (Bürgi, Jacobs and Roos, 2005).

This outlook does not diminish the inherent value of either conceptual academic disciplines or navigate a steer away from the translation of it into applied practice in real world settings but positions the need for emotional intelligence in the capitalisation of human capability and influence for the future (Finlay and Gough, 2008).

Beyond any acquisition or corpus of academic knowledge, the concept of emotional intelligence as a means of recognising and acknowledging the emotional self is seen as invaluable to HE (Schön, 1983). This is in turn, as seminally posited by Dewey as being reliant on capacity for critical introspection on both pre-existing experience, pre-suppositions and assumptions, alongside the motivation to embed reflexivity as a proactive approach to allowing the past to positively influence the future (Dewey, 1933). It is here that seminal theories of intelligence have most relevance to any academic debate of the concept of the emotional self in action (Thorndike, 1920; Gardner, 1983). It is here that modern counterparts in the literature have added to the body of literature that previously omitted to address the 'self' in the emotive self (Ben-Eliyahu, 2019; Koç, 2019).

Shulman has perhaps progressively identified how professional identities are also founded on the individual academic disciplinarity from which professionals develop and are immersed from the earliest stages of their HE (Shulman, 2005). Alongside the seminal theoretical perspectives of Thorndike, which acknowledge the existence of conceptual or abstract intelligence, his recognition of how signature pedagogies can constrain disciplines at an intellectual level has also been invaluable in considering the emotional self at work (Thorndike, 1920). Resonance has been focused on the engendered binary thinking capacity of empirical scientists versus the ambiguous, nebulous capacity of social scientists, whereas, all are connected by shared human experience of the culture and context within which they work (Clarke and Friese, 2007). It is this delineation between what can be regarded as orthodox versus social intelligence that has been the cause of so much contention and debate about the conceptual limitations of both, when seen in abstraction from each other (Burr, 2015; Augustinos, Walker and Donaghue, 2014).

As reinforced earlier in this chapter, underpinning the concept of the emotional self in the workplace is the notion of emotional competence in being able to prioritise, rationalise and provide a perspective on individuality by members of the HE workforce. The procedural integration of ambiguity, complexity and often indefinable information in practice must be aligned with psychological responsiveness and the willingness and capacity to make sense of it. Once these can be tangibly measured, it is then possible to make informed judgements on the potential individuals must cope with challenging situations and their capacity to elicit positive attitudes, behavioural responses and consequently optimal outcomes.

Primary data collection from postgraduate nursing students revealed themes of direct relevance to the adoption of specific learning and teaching methodologies. As well as having potential implications for pedagogic practice by illuminating the experiences of a specific cohort of students who used the LSP® method. Without exception, students in the cohort concluded that the approach had a positive impact on their potential to increase their hand-mind dynamic. This has implications for the adoption of philosophical approaches to teaching and learning as well as their methodological execution in learning and teaching practice (Cavaliero, 2015). Students could recognise that the approach 'awakened' their capacity to think, engage and trigger deep rather than surface approaches to learning and the inherent value of epistemic cognition to this process (Bürgi, Jacobs and Roos, 2005). It was possible to understand the relationship between meaning making as perceived by students and their identity as learners and students could express acknowledgement of the impact of the LSP® method in practice. Within HE curricula it must be acknowledged that use of the LSP® method necessitates the time and investment in training instructors in order to guide and facilitate a contextual framework conducive to the integration of the facilitation process. This raises important considerations for the integration of the LSP® method within the context of different signature pedagogies and disciplines. Most nursing and allied health practice necessitates critical reflexive practice which ensures learning from experience in a profession where the use of hands in care is never refuted. As such there are direct parallels to accessing epistemic cognition and metaphorically and literally 'grasping' knowledge from practice so that it can be consolidated internally for integration back into practice again. There was, however, no evidence from this specific study to suggest that technical psychomotor skill domain learning could be enhanced using the LSP® method. This contrasts with those signature pedagogies with a high focus on the overlap between cognitive and affective domain learning, which it can be academically debated are inseparable. Therefore, its potential use cannot be underestimated at this stage of exploration of the study. The LSP® method proved a valuable means of transcending language and discourse when students explored their deep understanding and enjoyment of their own learning identities and their individual aspirations. From this perspective the value of social constructivist approaches, of which social constructionism is evident to pedagogic practice. Within group dynamics, the approach successfully contributed to social bonding, constructive competition. Emotional expression and cognate development and from this perspective contributes directly to the body of knowledge dealing with how meaning making attributed to experience can provide an overview of the dynamics of transformative learning.

The development of resilience via critical reflexivity and introspection is of significance to nursing practice and to how this might be formalised in practice. The approaches with this group of students provided as a means of facilitating higher order thinking necessary for students using the LSP® method to recognise their coping mechanisms and indeed the capacity of the brain for building emotional wellbeing when using the LSP® method as they progressed and develop as creative learners. This was achieved by developing a micro community of practice where emotional unwrapping and deconstruction of knowledge through social and collective sharing of experience. Consequently, this communal approach meant that students could engage directly in facilitating the expression of how human individualities contribute to share collective aims, values and experiences in a specific signature pedagogy. It has a direct relevance to how descriptions of what is real are made. The findings of the study from this particular cohort of students are consistent with the work of authors who first acknowledged imagination is linked unconsciously with cultural and linguistic consciousness, where meaning is split into describing, creating and challenging or questioning existing approaches offers a means of moving forward and influencing future pedagogical approaches, particularly in relation to ludic interventions (Papert, and Harel, 1991).

LUDIC APPROACHES TO THE CO-CONSTRUCTION OF KNOWLEDGE

Embracing students as partners alongside other key stakeholders in the co-creation of academic curricula is an integral part of ensuring acknowledgement of the emotional self for educators, stakeholders and most significantly the students themselves. Whilst students as partners has become a recognised and expected norm in the context of HE, ludic approaches to it have not and the co-construction of knowledge through them is a relatively recent phenomenon. Core characteristics of serious play focus on the motivation of participants to engage in reflection on practice in relation to problem-solving, the recognition of key barriers and enablers to learning and the potential societal impact that HE can deliver via processes of transformative learning. All of this occurs amidst an uncertain contextual backdrop of politics, change and dynamic institutional cultures, as individual as the organisations themselves.

Ludic approaches can therefore effectively enhance the concept of student agency by ensuring that authentic learning can be fully embedded across academic curricula. In terms of 'graduateness' this is largely characterised by the transition from superficial to deep learning processes, ensuring that learning can, as far as possible, provide a transferable skill base relevant to the needs of society (Hayter and Cohoy, 2018). It also facilitates the notion that progression to further learning can be advocated throughout programmes, so students embrace the notion of lifelong learning before they enter a specific workforce or individual employment niche. In integrating playful approaches to this transformative learning, there is also the prospect of integrating gamification as a means of building student resilience and a capacity to think laterally as well as introspectively about their overall societal contributions, whether that be in their family, community or wider geographical contexts. The term 'curriculum' can be defined as the overall holistic structure and individual content of a formal programme of learning, but which extends beyond the academic to the social. In this sense, it is indeed a lens through which students' active participation and co-creation (sometimes through gamification) can be illuminated and extended (Bovill, 2014).

When contextualised within individual HE institutions, the co-creation of curricula can be regarded as a fundamentally unique case study that is contextually bound and usually framed within a specific signature pedagogy or academic discipline (Hayes, 2018). These specific case study illustrations are used to further inform evidence based and informed decision-making processes involved in curriculum justification, development and subsequent implementation. This insight essentially provides a means of addressing the need for what is now commonly termed a 'Universal Design for Learning' which is defined by the three core characteristics of engagement, expression and representation. It also a creative tension in defining the culture and commitment of institutions to democratic and fair processes of social development and academically robust progression routes.

ISSUES OF INCLUSION IN THE REGULATION OF THE EMOTIONAL SELF

Any context which contributes a barrier to effective educational integration for either staff or students ought to be regarded as an inclusivity gap warranting immediate address and ongoing monitoring. All members of a HE community are unique and as such have an individualized set of situational issues that frame and define their capacity for learning, personal growth and ultimately self-actualization. If this is not recognised, then the potential to ensure that all students maximise their potential, regardless of academic ability, physical or mental disability, age, gender, sexuality or nationality is functionally diminished. Alongside this is the potential for long term damage of motivation to learn and be part of a learning community. Adopting ludic approaches to support and facilitate these individual trajectories is another means of ensuring that students realize they have an individual voice, their own individual story to make and the means of making sure that they can tell it, to a society which remains democratic enough to ensure it can be heard.

At the heart of finding and facilitating this voice to be heard, is the recognition that students' realities are constructed of multiple variables which provides them with a tangible degree of connection to others or which may highlight how uniquely individual each of us are. This remains the fundamental beauty of education as an academic discipline. A 'one size fits all' curriculum might not exist but where unique individuals can contribute it is in the establishment of an ethos that is truly authentic, collaborative and inclusive – where value for difference is more valuable than the homogenisation of human experience into faceless metrics and correlations. Whilst gamification might be a relatively unconventional approach to ensuring this can happen, it is one approach to ensuring the processes of story making for story telling can be realised in the context of HE. In this sense, fundamental and characteristic natures of all academic curricula are shaped by formality and informality – what is bracketed under each can be startlingly different and can be engaged with by students as part of formal or informal learning that ultimately contributes to an holistic and societally driven education system.

As the concept of 'learning gain' or learning beyond the constraints of a purely academic curricula, develops an increasingly elevated profile across all HE institutions, as academics we need to identify the need for Universal Design for Learning as one potential mechanism for incorporating our students as the ultimate resource in democratic change agency. In conjunction with this, if ludic approaches are to become an integral part in shaping change, we also need mechanisms of assuring the quality and robustness of a HE system facing the greatest generational changes we have ever encountered.

MEANING MAKING, TRANSFORMATION AND EPISTEMIC COGNITION

For the purposes of this chapter, we define 'meaning making' to be the interpretation of phenomena or events that can be deconstructed, interpreted and reconstructed as students make sense of the information, experience and values they are facilitated in developing. The potential to use the LSP® method as a vehicle for this provides students with the opportunity to embed an ethos of informed creativity in the development of co-created curricula. Since co-creation has historically been one of the critical yet tokenistic elements of integrating and extending democratic values into, HE then this ludic approach has resonance in how this can be facilitated. Power relations are also dispelled using the process of building meaning through objects such as the individual bricks, since they become mediators of meaning through which the students can articulate and express themselves.

DEEP LEARNING AND INTRINSIC MOTIVATION

There is much published evidence to suggest that the significance of epistemic cognition to learning and teaching is grossly underestimated, particularly in relation to the concept of deep learning and how best students can be motivated to engage at this cognitive level (Pick and Saltzman, 2014). Theoretically, reflection and self-awareness contribute to students' capacity to function at a metacognitive level and as part of the process undergo truly authentic transformative learning (Mezirow, 1991). On an epistemological level this is directly dependent on students' capacity not only to acknowledge their pre-existing knowledge base but also their acknowledgement of how knowledge is socially constructed. How knowledge is interpreted during the process of transformative learning hinges on their capacity for both epistemic cognition and self-awareness (Moshman, 2014). Academic learning at postgraduate level (doctoral level in particular) hinges on the capacity of students to develop a pragmatic and working knowledge of what acknowledging their epistemic cognition entails (Sandoval, 2014; Sinatra, Kienhues and Hofer, 2014).

CONCLUSION

Facilitating ludic approaches to critical introspection and processes of reflection and reflexivity for HE communities has the potential to afford participants a raft of opportunity to engage in critical thinking in a democratic, inclusive and power-free environment with due regard for the conceptual basis of social sustainability in practice. It adopts the principles of Universal Design for Learning, which are characterised by expression, engagement and representation and contributes directly to the individual 'reach' and confidence of students in constructing stories for storytelling in a safe learning environment. Where this can be framed in an opportunity to take risk, develop confidence, and bridge evident inclusivity gaps, it is hard to refute that the process of co-creating academic curricula can be beneficial in building the social learning relationships that define educational provision in the context of HE. Increasing agency and the capacity for democracy in HE is just one part of this wider picture for social sustainability. It can be debated that focus on engaging students in not just what they learn but how, has the potential to dilute academic rigour and that whilst closure of the inclusivity gap ought to be an aim of all educationalists, there is also a role for pragmatism and the recognition of the need for quality assurance in the process. This raises the question of whether any pedagogic approach can ever truly be a 'one size fits all' but gamification as a driver for social inclusion and intrinsic motivation undoubtedly offers a different approach to traditional methods of affording students a recognisable voice that they otherwise might never have the opportunity of recognising and using. At the heart of the approach though, and perhaps best of all is the chance to recognise the individual nature of humanity through a collective approach which embraces the peculiarity and brilliant uniqueness of personhood in the context of HE. It is these aspects of peculiarity and brilliant uniqueness that offer the opportunity to be harnessed and used to provide empowerment and agency in the context of social sustainability.

REFERENCES

Amabile, T. M. (1983). *The Social Psychology of Creativity*. doi:10.1007/978-1-4612-5533-8

Augoustinos, M., Walker, I., & Donaghue, N. (2014). Social cognition: An integrated introduction. *Sage (Atlanta, Ga.)*.

Ben-Eliyahu, A. (2019). Academic Emotional Learning: A critical component of self-regulated learning in the emotional learning cycle. *Educational Psychologist, 54*(2), 84–105. doi:10.1080/00461520.2019.1582345

Bovill, C. (2014) An investigation of co-created curricula within HE in the UK, Ireland and the USA. *Innovations in Education and Teaching International, 51*(1), 15-25.

Bürgi, P. T., Jacobs, C. D., & Roos, J. (2005). From metaphor to practice in the crafting of strategy. *Journal of Management Inquiry, 14*(1), 78–94. doi:10.1177/1056492604270802

Burr, V. (2015). *Social constructionism*. Routledge. doi:10.4324/9781315715421

Cavaliero, T. (2015). 'Creative blocs': Action research study on the implementation of Lego as a tool for reflective practice with social care practitioners. *Journal of Further and HE*, 1-10.

Chen, J. (2007). Flow in games (and everything else). *Communications of the ACM, 50*(4), 31–34. doi:10.1145/1232743.1232769

Clarke, A. E., & Friese, C. (2007). Grounded theorizing using situational analysis. The Sage handbook of grounded theory, 363-397.

Claxton, G. (2015). *Intelligence in the Flesh: Why Your Mind Needs Your Body Much More Than It Thinks*. Yale University Press.

Conley, C. S., Durlak, J. A., & Kirsch, A. C. (2015). A meta-analysis of universal mental health prevention programs for HE students. *Prevention Science, 16*(4), 487–507. doi:10.1007/s11121-015-0543-1 PMID:25744536

Corballis, M. C. (2017). The evolution of lateralized brain circuits. *Frontiers in Psychology, 8*, 1021. doi:10.3389/fpsyg.2017.01021 PMID:28670296

Dann, S. (2018). Facilitating co-creation experience in the classroom with Lego Serious Play. *Australasian Marketing Journal, 26*(2), 121–131. doi:10.1016/j.ausmj.2018.05.013

Dewey, J. (1933). *How We Think: A Restatement of the Relation of Reflective Thinking to the Educative Process*. DC Health.

Dirkx, J. M. (2001). The power of feelings: Emotion, imagination, and the construction of meaning in adult learning. *New Directions for Adult and Continuing Education, 2001*(89), 63–72. doi:10.1002/ace.9

Finlay, L., & Gough, B. (Eds.). (2008). *Reflexivity: A practical guide for researchers in health and social sciences*. John Wiley & Sons.

Forgas, J. P., & Williams, K. D. (2014). *The social self: Cognitive, interpersonal and intergroup perspectives*. Psychology Press. doi:10.4324/9781315800516

Freedman, N. (1983). On psychoanalytic listening: The construction, paralysis, and reconstruction of meaning. *Psychoanalysis and Contemporary Thought*.

Frick, E., Tardini, S., & Cantoni, L. (2013). *LEGO® SERIOUS PLAY®*. Academic Press.

Fuchs, M., Fizek, S., Ruffino, P., & Schrape, N. (2014). *Rethinking gamification*. Meson Press.

Fuchs, T. (2019). The Experience of Time and its Disorders. In *The Oxford Handbook of Phenomenological Psychopathology* (p. 431). Oxford University Press.

Gardner, H. (1983). *Frames of mind*. Basic Books.

Gauntlett, D. (2014). *The LEGO System as a tool for thinking, creativity, and changing the world*. Academic Press.

Gilbert, D. T., Pinel, E. C., Wilson, T. D., Blumberg, S. J., & Wheatley, T. P. (1998). Immune neglect: A source of durability bias in affective forecasting. *Journal of Personality and Social Psychology*, *75*(3), 617–638. doi:10.1037/0022-3514.75.3.617 PMID:9781405

Groh, F. (2012). Gamification: State of the art definition and utilization. *Institute of Media Informatics Ulm University*, *39*, 31.

Gunnarsson, B. L., Linell, P., & Nordberg, B. (Eds.). (2014). *The construction of professional discourse*. Routledge. doi:10.4324/9781315842820

Hayes, C. (2015). Meaning-making through transformative learning for HCA education. *British Journal of Healthcare Assistants*, *9*(7), 354–361. doi:10.12968/bjha.2015.9.7.354

Hayes, C. (2018). Case Study Research. In Methodologies for Practice Research: Approaches for Professional Doctorates. SAGE Publications Limited.

Hayter, C. S., & Cahoy, D. R. (2018). Toward a strategic view of HE social responsibilities: A dynamic capabilities approach. *Strategic Organization*, *16*(1), 12–34. doi:10.1177/1476127016680564

Holland, J. H. (2006). Studying complex adaptive systems. *Journal of Systems Science and Complexity*, *19*(1), 1–8. doi:10.1007/s11424-006-0001-z

James, A. (2015). Learning in Three Dimensions: Using Lego Serious Play for Creative and Critical Reflection Across Time and Space. In Global Innovation of Teaching and Learning in HE (pp. 275-294). Springer International Publishing.

Jankowska, D. M., & Omelańczuk, I. (2018). Potential Mechanisms Underlying the Impact of Imaginative Play on Socio-Emotional Development in Childhood. *Creativity. Theories–Research-Applications*, *5*(1), 84–103. doi:10.1515/ctra-2018-0006

Kalfa, S., & Taksa, L. (2017). Employability, managerialism, and performativity in HE: A relational perspective. *HE, 74*(4), 687-699.

Karwowski, M., & Jankowska, D. M. (2016). Four faces of creativity at school. *Nurturing creativity in the classroom*, 337-354.

Kestly, T. A. (2014). *The Interpersonal Neurobiology of Play: Brain-Building Interventions for Emotional Well-Being (Norton Series on Interpersonal Neurobiology).* WW Norton & Company.

Koç, S. E. (2019). The relationship between emotional intelligence, self-directed learning readiness and achievement. *International Online Journal of Education & Teaching*, *6*(3), 672–688.

Kristiansen, P., & Rasmussen, R. (2014). *Building a Better Business Using the Lego Serious Play Method.* John Wiley & Sons.

Mezirow, J. (1991). Transformative dimensions of adult learning. Jossey-Bass.

Moshman, D. (2014). *Epistemic cognition and development: The psychology of justification and truth.* Psychology Press. doi:10.4324/9781315884684

Papert, S., & Harel, I. (1991). Situating constructionism. *Constructionism*, *36*, 1–11.

Parrish, D. R. (2015). The relevance of emotional intelligence for leadership in a HE context. *Studies in HE*, *40*(5), 821–837.

Pick, H. L. Jr, & Saltzman, E. (2014). *Modes of perceiving and processing information.* Psychology Press. doi:10.4324/9781315802565

Resnick, M., & Robinson, K. (2017). *Lifelong kindergarten: Cultivating creativity through projects, passion, peers, and play.* MIT press. doi:10.7551/mitpress/11017.001.0001

Rivière, E., Jaffrelot, M., Jouquan, J., & Chiniara, G. (2019). Debriefing for the Transfer of Learning: The Importance of Context. *Academic Medicine*, *94*(6), 796–803. doi:10.1097/ACM.0000000000002612 PMID:30681450

Roberts, P., & Peters, M. A. (2019). Neoliberalism, HE and Research. Brill Sense.

Robinson, K., & Aronica, L. (2015). *Creative schools: Revolutionizing education from the ground up.* Penguin UK.

Rodd, P., & Sanders, K. (2018). The imperative of critical pedagogy in times of cultural austerity: A case study of the capacity to reimagine education as a tool for emancipation. *New Zealand Sociology*, *33*(3), 33.

Roos, J., & Victor, B. (1999). Towards a new model of strategy-making as serious play. *European Management Journal*, *17*(4), 348–355. doi:10.1016/S0263-2373(99)00015-8

Ross, M. (1989). The relation of implicit theories to the construction of personal histories. *Psychological Review*, *96*(2), 341–357. doi:10.1037/0033-295X.96.2.341

Rubin, D. C., Deffler, S. A., & Umanath, S. (2019). Scenes enable a sense of reliving: Implications for autobiographical memory. *Cognition*, *183*, 44–56. doi:10.1016/j.cognition.2018.10.024 PMID:30412854

Sandoval, W. (2014). Science education's need for a theory of epistemological development. *Science Education*, *98*(3), 383–387. doi:10.1002/sce.21107

Schön, D. A. (1983). *The reflective practitioner: How professionals think in action* (Vol. 5126). Basic books.

Schonert-Reichl, K. A., & Roeser, R. W. (2016). Mindfulness in education: Introduction and overview of the handbook. In *Handbook of mindfulness in education* (pp. 3–16). Springer. doi:10.1007/978-1-4939-3506-2_1

Schore, A. N. (2018). The right brain implicit self: A central mechanism of the psychotherapy change process. In *Unrepressed unconscious, implicit memory, and clinical work* (pp. 73–98). Routledge. doi:10.4324/9780429484629-4

Schulz, K. P., & Geithner, S. (2014). *Creative tools for collective creativity–the serious play method using LEGO bricks. In Learning and Collective Creativity: Activity-Theoretical and Sociocultural Studies*. Routledge.

Seyderhelm, A. J., Blackmore, K. L., & Nesbitt, K. (2019, November). Towards Cognitive Adaptive Serious Games: A Conceptual Framework. In *Joint International Conference on Entertainment Computing and Serious Games* (pp. 331-338). Springer. 10.1007/978-3-030-34644-7_27

Shulman, L. S. (2005). Signature pedagogies in the professions. *Daedalus, 134*(3), 52–59. doi:10.1162/0011526054622015

Sinatra, G. M., Kienhues, D., & Hofer, B. K. (2014). Addressing challenges to public understanding of science: Epistemic cognition, motivated reasoning, and conceptual change. *Educational Psychologist, 49*(2), 123–138. doi:10.1080/00461520.2014.916216

Smith, P. K. (1982). Does play matter? Functional and evolutionary aspects of animal and human play. *Behavioral and Brain Sciences, 5*(1), 139–155. doi:10.1017/S0140525X0001092X

Sutton-Smith, B. (1966). Piaget on play: A critique. *Psychological Review, 73*(1), 104–110. doi:10.1037/h0022601 PMID:5906212

Thorndike, E. L. (1920). Intelligence and its uses. *Harper's Magazine.*

Tse, M. M. Y., Kwan, T. S., & Lee, P. H. (2016). The development and psychometric evaluation of the Perception of Play Questionnaire for older adults. *Educational Gerontology, 42*(2), 79–88. doi:10.1080/03601277.2015.1071600

Wilson, F. R. (2010). *The hand: How its use shapes the brain, language, and human culture.* Vintage.

KEY TERMS AND DEFINITIONS

Creativity: Is the use of collective or individual imagination or the capacity for fundamentally original ideas to create something; can often be used as a bridge to innovation in terms of the process of invention.

Emergence: Is the process of revealing something which has previously remained unseen or unrecognized; the outcome of emergence may be to challenge long held presuppositions or assumptions of knowledge.

Epistemic Cognition: Is can be most straightforwardly be defined as knowledge of pre-existing knowledge, particularly those which belie capacity for objective, subjective and tacit rationality; epistemic cognition is a central concept in relation to the notion of truth and verisimilitude.

Gamification: The transference and application of the characteristics of game playing to a formal context, such as Higher Education, with the purpose of enhancing capacity for creative thinking and the harnessing of internal motivation.

LEGO Serious Play®: An innovative and creative methodology for the formalization of gamification in a range of academic and business contexts; a means of articulating meaning making in learning environments.

Meaning Making: Is the process of how people can construe, understand, or make sense of life events, relationships, and the self; linked to the establishment of individuality, positional stance and interpretation of external and/or internal lifeworld(s).

Metacognition: Part of the awareness and understanding of one's own thought processes; an integral component of the capacity for reflection and critical reflexivity.

Signature Pedagogy: Refers to the characteristic forms or styles of teaching and instruction that are common to specific professional disciplines, academic subjects or aspects of professional identity.

Social Constructionism: A general sociological term, originating from the Chicago School which is often applied to theories that emphasize the socially created nature of social life. The original theory is traditionally traced back at least to the work of Thomas and the Chicago sociologists, as well as the phenomenological sociologists and philosophers such as Schutz. Such approaches emphasize the idea that society is actively and creatively produced by human beings.

Chapter 12
The Impact of VR-Based Learning on Student Engagement and Learning Outcomes in Higher Education

Chai Lee Goi
https://orcid.org/0000-0003-0131-2818
Curtin University, Malaysia

ABSTRACT

This research investigates virtual reality (VR) in higher education, covering types of VR, implementation in academic settings, effects on student engagement and learning outcomes, and exploring potential benefits for enhancing learning and developing graduate capabilities. A comprehensive review covered VR types like simulations, virtual labs, 360-degree experiences, and virtual classrooms, enabling experiential learning. The implementation process involves goal setting, content creation, hardware/software selection, infrastructure setup, curriculum integration, assessment, and evolution. Effective VR content and prepared instructors positively impact engagement, influencing comprehension, retention, and academic performance. AR/VR tech refines cognitive skills, practical knowledge, and personalised learning, fostering teamwork and critical thinking for graduates' career readiness.

INTRODUCTION

The rapid advancements in Virtual Reality (VR) technology have sparked a transformative shift in the way education is delivered and experienced. VR, with its immersive and interactive capabilities, holds great potential for revolutionising higher education. Traditional approaches to education often struggle to keep pace with the evolving needs and preferences of today's digital-native students. The introduction of VR-based learning offers a promising solution to bridge this gap, providing a more engaging and experiential mode of learning. By simulating real-world scenarios, virtual environments allow students to actively participate in their learning, leading to increased student engagement and motivation.

DOI: 10.4018/978-1-6684-9859-0.ch012

VR technology enables students to step into virtual worlds where they can manipulate objects, conduct experiments, and explore complex concepts with unprecedented depth. This level of interactivity fosters a sense of ownership over the learning process, promoting higher levels of intrinsic motivation. As a result, students are more likely to retain knowledge and develop critical thinking skills. Moreover, the multisensory nature of VR-based learning enhances information retention and recall. By incorporating sight, sound, and touch, VR experiences tap into multiple modalities, catering to various learning styles. This inclusive approach to education can benefit diverse student populations, including those with learning disabilities, by offering personalised and adaptable learning experiences.

VR technology is seen as having great potential for incorporation into the educational system, as it can foster motivation to learn and develop student skills in simulating learning models, making learning efficient and effective (Kustandi et al., 2020). The potential impact of VR-based learning extends beyond individual students. Collaborative virtual environments enable seamless group interactions, promoting teamwork and communication skills. Students can collaborate on projects, conduct group experiments, or engage in role-playing simulations, fostering a cooperative learning environment that mirrors real-world professional settings.

VR technologies offer immersive, three-dimensional experiences that allow students to interact with virtual objects and environments. Consequently, it is important to explore the different types of VR technologies and applications used in higher education. The most common type of VR technology is the head-mounted display (HMD). HMDs are worn on the head and provide an immersive, stereoscopic view of the virtual environment. They are typically used in combination with a handheld controller to allow the user to interact with the environment. Additionally, some HMDs incorporate a tracking system that allows the user to move around the environment. Another type of VR technology is the Cave Automatic Virtual Environment (CAVE). A CAVE is a room consisting of multiple walls that project images of the virtual environment. This type of technology is often used in collaborative settings, such as group projects or virtual meetings. The last type of VR technology is the desktop virtual reality (DVR) system. This type of system is more affordable than the other VR technologies and is typically used for basic tasks, such as product design or 3D modeling (Radianti et al., 2020).

This research aims to investigate various facets of VR within higher education. It encompasses a thorough examination of the types of VR utilised in higher education, an analysis of the implementation of VR technologies in this academic setting, an evaluation of the effects of VR use on student engagement, an exploration of the impact of VR on learning outcomes, and a review of the potential benefits of leveraging VR to enhance learning outcomes and cultivate graduate capabilities.

METHODOLOGY

In order to complete this research, we reviewed relevant literature on the topic. To achieve this objective, a rigorous and systematic methodology was employed to identify, select, and analyse the most relevant articles. The methodology for conducting this review involved the following steps:

1. Database selection: The first step in the methodology was to select appropriate databases for conducting the literature search and the selection of databases was a strategic decision informed by the unique strengths of each platform. Google Scholar, known for its broad coverage, was chosen to capture a wide range of sources. Scopus, recognised for its multidisciplinary focus, aimed to ensure

inclusivity across diverse academic fields. Web of Science, with its emphasis on citation indexing, provided an additional layer of validation and completeness. This triangulation of databases aimed at minimising the risk of overlooking relevant literature in the exploration of VR-based learning in higher education.

2. Search strategy: Crafting the search strategy involved a meticulous process of refining keywords and phrases. The initial identification of relevant terms was followed by iterative testing and adjustment, ensuring that the search queries were finely tuned. Boolean operators (e.g., AND, OR) were applied judiciously to create nuanced queries that balanced comprehensiveness with specificity. This process sought not only to cast a wide net but also to capture the nuances of VR-based learning's impact on student engagement and learning outcomes.
3. Inclusion and exclusion criteria: The establishment of inclusion and exclusion criteria was a thoughtful exercise in defining the boundaries of the research. The decision to include only English-language articles was made to facilitate a deep understanding of the content. Criteria excluding duplicates and non-peer-reviewed publications were not arbitrary but aimed at upholding the quality and reliability of the literature under consideration. This step was critical for ensuring a robust and focused review.
4. Article screening: Screening articles based on titles and abstracts was not merely a preliminary sorting process but a strategic categorisation. Rapid exclusion of irrelevant articles ensured that subsequent stages of the review focused on a refined set of literature. The screening process involved not only assessing relevance but also identifying preliminary patterns and themes, guiding the direction of the review.
5. Full-text review: The in-depth review of full texts went beyond a surface-level assessment. It involved a critical examination of the methodologies employed in the selected articles. This phase was not just about confirming relevance but about scrutinising the research design, data collection methods, and overall rigor. The aim was to ensure that each selected article contributed substantively and methodologically sound insights to the overarching research objectives.
6. Data extraction and synthesis: Systematic data extraction was a nuanced process that went beyond assembling basic information. It involved categorising data elements, identifying recurrent themes, and extracting contextually rich insights. The synthesis of findings was not a mechanical aggregation but a purposeful construction of a narrative that captured the complexity and diversity of the literature. This phase aimed at creating a cohesive and nuanced storyline that reflected the multifaceted impact of VR-based learning.
7. Quality assessment: This phase aimed at ensuring that the included literature met rigorous standards of credibility and reliability. It was a discerning examination to filter out potential biases and methodological shortcomings.
8. Data analysis: Thematic and content analysis techniques were applied with a keen eye for subtleties. Beyond identifying common themes, this phase involved exploring divergent perspectives, trends, and outliers. The analysis sought not only to summarise findings but to distill meaning, generating insights that contributed to a nuanced understanding of the impact of VR-based learning on student engagement and learning outcomes in higher education. This was not just an analytical process but a sense-making exercise that enriched the overall research narrative.

TYPES OF VR USED IN HIGHER EDUCATION

VR provides an array of immersive and interactive learning experiences, enabling students to develop essential skills, deepen their knowledge, and prepare for real-world challenges in various disciplines.

Table 1. Categories of VR technology

Category	Description	Example	Description of Example
Simulations (Freina and Ott, 2015)	Simulations in higher education offer virtual environments that mimic real-life scenarios and situations. These simulations are particularly valuable in fields like medicine, where medical students and healthcare professionals can practice surgeries, patient interactions, and emergency situations in a risk-free setting. Business management simulations allow students to run virtual companies, analyse market conditions, and make strategic decisions with real-time consequences, enhancing their understanding of business principles. Engineering prototyping simulations enable students to design and test prototypes, providing insights into complex engineering processes before physical manufacturing.	Medical Simulations	Medical simulations offer medical students and healthcare professionals the opportunity to practice surgeries, patient care, and emergency procedures in a controlled and safe virtual environment. This enhances their clinical skills and decision-making abilities, preparing them for real-world medical challenges.
		Business Management Simulations	Business management simulations immerse students in virtual business environments, allowing them to make strategic decisions, analyse market conditions, and run virtual companies. Through these simulations, students gain practical experience in business management and strategic planning, developing essential skills for their future careers in the business world.
		Engineering Prototyping Simulations	Engineering prototyping simulations enable engineering students to design and test virtual prototypes, providing valuable insights into the performance and functionality of their designs. Students can iterate and optimise their designs before physical manufacturing, saving time and resources while enhancing their creative problem-solving abilities in engineering disciplines.
Virtual Labs (Potkonjak et al., 2016)	Virtual labs provide students with simulated laboratory experiences, overcoming the limitations of physical space and equipment. In disciplines like physics and chemistry, students can manipulate virtual equipment, perform experiments, and collect data, deepening their understanding of scientific principles. Engineering and technology virtual labs allow students to experiment with virtual components and systems, enhancing their practical skills and knowledge in engineering disciplines. In biology and life sciences, virtual labs offer interactive environments for dissecting specimens, studying cellular structures, and exploring biological phenomena.	Physics and Chemistry Virtual Labs	Physics and chemistry virtual labs enable students to conduct experiments, manipulate virtual equipment, and collect data to explore scientific principles. These labs provide a safe and cost-effective way to practice experiments and develop scientific skills, enhancing students' understanding of the fundamental concepts of physics and chemistry.
		Engineering and Technology Virtual Labs	Engineering and technology virtual labs allow students to experiment with virtual equipment and components, enabling them to troubleshoot, build circuits, and simulate technological systems. These labs provide hands-on learning experiences, helping students gain practical skills and insights into various engineering concepts and technologies.
		Biology and Life Sciences Virtual Labs	Biology and life sciences virtual labs provide interactive environments for students to dissect virtual specimens, study cellular structures, and explore biological phenomena. These labs enhance students' understanding of living organisms and ecosystems, fostering scientific inquiry and deepening their knowledge of biological concepts.
****360-Degree** (Shadiev et al., 2022)	360-degree experiences in higher education offer immersive journeys to historical events, cultural landmarks, and natural wonders. Students can explore ancient civilisations, visit museums virtually, and witness significant historical moments, enhancing their historical and cultural understanding. Cultural immersions provide virtual visits to different cultures and traditions, fostering cultural awareness and empathy. Virtual field trips enable students to study geological formations, ecological systems, and geographical phenomena from anywhere in the world, enriching their field-based learning and environmental awareness.	Historical Tours and Reconstructions	Historical tours and reconstructions offer immersive experiences of historical events and landmarks. Students can explore ancient civilisations, visit museums virtually, and witness significant historical moments, deepening their historical understanding and providing a sense of presence and context to the past.

continued on following page

Table 1. Continued

<table>
<tr><th>Category</th><th>Description</th><th>Example</th><th>Description of Example</th></tr>
<tr><td rowspan="2">Experiences**
(Hodgson et al., 2019)</td><td></td><td>Cultural Immersions</td><td>Cultural immersions provide virtual visits to different cultures, customs, and traditions. Students can engage with diverse perspectives, fostering cultural awareness and empathy, and developing global competency and intercultural communication skills. These experiences promote understanding and respect for different cultures and traditions, enhancing students' intercultural competencies in a diverse world.</td></tr>
<tr><td></td><td>Virtual Field Trips</td><td>Virtual field trips take students on immersive journeys to remote locations and natural wonders. Students can study geological formations, ecological systems, and geographical phenomena from anywhere in the world, transcending physical constraints and providing educational experiences beyond traditional classroom settings. These trips deepen students' environmental awareness and knowledge of different ecosystems and geographical features.</td></tr>
<tr><td rowspan="3">Virtual Classrooms
(Bower et al., 2017)</td><td>Virtual classrooms facilitate remote learning and collaboration among students and instructors. Collaborative virtual environments create spaces for group projects, discussions, and presentations, enhancing teamwork and communication skills. Online lectures and presentations in VR enable students to attend classes from remote locations, engaging with interactive content and accommodating diverse learning needs. Interactive group discussions in VR promote active learning, encouraging students to participate in debates, brainstorming sessions, and problem-solving exercises, fostering deeper engagement and critical thinking among learners.</td><td>Collaborative Virtual Environments</td><td>Collaborative virtual environments create virtual meeting spaces for students and instructors to interact and collaborate. These environments facilitate group projects, discussions, and presentations, enhancing teamwork and communication skills. Students can work together on projects, exchange ideas, and develop skills in collaborative problem-solving, preparing them for collaborative work environments in their future careers.</td></tr>
<tr><td rowspan="2"></td><td>Online Lectures and Presentations in VR</td><td>Online lectures and presentations in VR enable instructors to deliver content in a virtual setting. Students can attend classes from remote locations, engage with interactive content, and benefit from flexible learning opportunities. These virtual lectures accommodate diverse learning needs, making education more accessible to students with varying schedules or geographical constraints.</td></tr>
<tr><td>Interactive Group Discussions in VR</td><td>Interactive group discussions in VR promote active learning and engagement. Students can participate in debates, brainstorming sessions, and problem-solving exercises in an immersive and dynamic environment. These discussions foster critical thinking and collaboration skills, encouraging students to share ideas, listen to different perspectives, and develop well-rounded insights into complex topics and issues.</td></tr>
</table>

Moreover, the flexibility and accessibility of VR technologies contribute to making education more inclusive and engaging, catering to the diverse needs and learning styles of students in higher education.

The application of VR in learning can be broadly categorised into four main types: simulations, virtual labs, 360-degree experiences, and virtual classrooms. VR technologies used in higher education can be classified into four categories:

- Simulations: Simulations offer virtual environments that replicate real-life scenarios, allowing students to practice and apply theoretical knowledge in a safe and controlled setting. This category includes medical simulations, business management simulations, and engineering prototyping simulations.

Table 2. A general overview of VR components

Component	Description
Hardware components	
VR Headset (Lenses)	The primary device that displays the virtual environment to the user.
Head Strap	Provides secure and comfortable wear for extended VR sessions.
Sensors (Tracking)	Various sensors track the user's head movements for an immersive experience.
Speakers	Deliver audio feedback, enhancing the overall immersive experience.
Hand Controllers (Optional)	Controllers for interacting with the virtual environment, if supported.
Software and Apps	
Virtual Learning Platforms	Platforms that host educational content, lectures, and collaborative activities.
Simulation Software	Software that simulates real-world scenarios for practical learning experiences.
Educational VR Applications	Applications designed specifically for educational purposes, covering various subjects.
Content Creation Tools	Tools that enable educators and students to create their own VR content.
Collaboration Tools	Tools that facilitate group activities and discussions within the virtual environment.
Educational Use Cases	
Virtual Labs	Simulated laboratory environments for scientific experiments and learning.
Field Trips	Virtual tours to distant places, museums, or historical sites for experiential learning.
Medical Simulations	VR applications for medical training and practicing surgical procedures.
Language Learning	Immersive environments for language practice and cultural immersion.
Arts and Design	Platforms for creating and experiencing 3D artworks, architecture, and design.
Learning Process	
Educator Preparation	Training for instructors on how to use VR technology effectively in teaching.
Student Training	Orienting students with VR equipment, controls, and etiquette for optimal usage.
VR Content Delivery	Educators deliver lessons, presentations, and content using VR applications.
Interactive Engagement	Students engage with interactive simulations, discussions, and collaborative activities.
Assessment and Feedback	Evaluating student progress through VR-based assignments, quizzes, and assessments.

- Virtual Labs: Virtual labs provide students with access to realistic laboratory experiences without the need for physical equipment or materials. They enable learners to conduct experiments and collect data in various disciplines, such as physics, chemistry, engineering, and life sciences.
- 360-Degree Experiences: This category involves immersive experiences that allow students to explore and interact with 360-degree content. It includes historical tours and reconstructions, cultural immersions, and virtual field trips, offering opportunities to explore historical sites, cultural landmarks, and natural wonders.
- Virtual Classrooms: Virtual classrooms create virtual meeting spaces for students and instructors to interact and engage in educational activities. This category includes collaborative virtual environments, online lectures and presentations in VR, and interactive group discussions, facilitating remote learning and fostering a sense of community among learners.

A general overview of components, tools, apps, and educational use cases associated with VR in Higher Education can be referred to Table 2.

Sources: Radianti et al. (2020), Hamilton et al. (2021), Luo et al. (2021), Rauf et al. (2021), Li et al. (2022), Marks and Thomas (2022), Dengel et al. (2022), Ding and Li, (2022), Bermejo et al. (2023), and McGrath et al. (2023)

THE IMPLEMENTATION OF VR TECHNOLOGIES IN HIGHER EDUCATION

The implementation of VR technologies into Higher Education represents a meticulously planned and executed process, characterised by a series of detailed steps aimed at ensuring not only the successful adoption of VR but also its optimal impact on the learning environment. Each step is underpinned by research and informed decision-making:

1. Identifying learning objectives: The process of identifying learning objectives involves a comprehensive examination of the educational landscape. Researchers like Bailenson (2018) emphasise the need for a nuanced understanding of disciplines that can benefit from VR immersion. This includes not only acknowledging the potential benefits but also recognising the specific educational goals that can be achieved through VR-enhanced learning experiences.
2. Content development: Collaborative efforts between educators and content developers extend beyond mere alignment with the curriculum. Freina and Ott's research (2015) stresses the importance of a pedagogical approach in VR content development. This involves tailoring scenarios not only to curriculum requirements but also ensuring that the content aligns with best practices in educational theory, enhancing its overall effectiveness.
3. Selecting VR hardware and software: Decision-making regarding VR hardware and software involves a delicate balance of technical specifications, budget considerations, and understanding user needs. Radianti et al. (2020) provide a detailed framework for these decisions, emphasising the importance of considering not only the current technological landscape, but also potential future developments to ensure a sustainable and adaptable VR infrastructure.
4. Infrastructure setup: The establishment of dedicated VR labs or spaces on campus goes beyond physical logistics. Dalgarno and Lee (2010) highlight the pedagogical implications of the physical environment. Creating a space conducive to collaborative and immersive learning is paramount, requiring thoughtful consideration of spatial design and ergonomic elements to optimise the learning experience.
5. Instructor training: Faculty training, as advocated by Beckem and Watkins (2012), extends beyond technical competence. Instructors need to develop a deep understanding of how to seamlessly integrate VR into their pedagogical approach. This involves not only mastering the technical aspects but also refining instructional techniques to leverage the unique affordances of VR for enhanced student engagement and learning outcomes.
6. Student access and training: Providing students with access to VR equipment is only the first step. Merchant et al. (2014) emphasise the need for a structured training program. This includes orientation sessions or workshops designed to familiarise students not only with the operation of VR tools but also with the potential educational applications and benefits, fostering a proactive and informed user base.

7. Incorporating VR into the curriculum: The strategic integration of VR experiences into the curriculum, as highlighted by Parong and Mayer (2018), involves more than selecting relevant content. Educators need to carefully design how VR experiences align with specific learning objectives, ensuring that they enhance and complement traditional teaching methods rather than being perceived as supplementary or extraneous.
8. Monitoring and assessment: The ongoing monitoring of students' progress and engagement is a dynamic process. Makransky and Lilleholt's research (2018) emphasises the importance of not only assessing outcomes but also understanding the learning processes facilitated by VR. This involves deploying a mix of qualitative and quantitative assessment tools to capture the nuanced impact on student cognition and understanding.
9. Feedback and improvement: Feedback loops involving students and educators, as outlined by Allcoat and von Mühlenen (2018), should be ingrained in the educational process. Beyond identifying technical glitches, these mechanisms should actively seek insights into the educational efficacy of VR content. This iterative feedback process is crucial for refining content and ensuring its continual relevance and effectiveness.
10. Scale-up and expansion: Scaling up VR initiatives requires a strategic approach. Jensen and Konradsen's work (2018) emphasises the need for institutions to carefully evaluate the scalability of their VR programs. This involves considerations of infrastructure, faculty readiness, and ongoing support structures, ensuring that growth occurs organically without compromising educational quality.
11. Continuous support and maintenance: Continuous support and maintenance are critical components in the successful integration of VR technologies into Higher Education. VR technologies, with their ability to create immersive and interactive learning environments, offer substantial educational benefits. However, their integration is not without challenges. Continuous support is necessary to ensure that both educators and students can effectively use these tools. This includes regular training sessions, the availability of technical support, and the provision of up-to-date resources (Bailenson, 2018). Maintenance is equally important, as VR technologies can quickly become outdated or malfunction. Regular updates and repairs are essential to maintain the efficacy of these tools in an educational setting (Freina and Ott, 2015). Furthermore, as VR technology continues to evolve, educational institutions must stay abreast of new developments to ensure that their VR resources remain relevant and effective (Radianti et al., 2020).
12. Research and evaluation: The research and evaluation phase, as advocated by Radianti et al. (2020), is not a one-time event but a continuous process. Institutions should engage in ongoing studies to understand the evolving impact of VR on student learning outcomes, employing sophisticated methodologies that capture both short-term and long-term effects. This research-driven approach ensures that VR initiatives remain at the forefront of educational innovation.

Impact on Student Engagement

The Impact of VR-Based Learning on Student Engagement

Virtual Reality (VR) technology is increasingly being used to improve student engagement in higher education. Research has identified six intervention strategies with positive effects on student engagement. These interventions include the use of two situated learning environments (animation studio, theatre with animation artifacts), designed 3D environments in Virtual Reality (VR) to prime the effect within VR and the use of eye-tracking metrics to measure cognitive engagement. While the impact of VR-based learning on student engagement can be positive, it depends on the teacher's familiarity with the technology and courses being adapted for AR/VR usage. Studies have shown that when courses are well-designed and teachers are qualified to use AR/VR, student engagement increases. On the other hand, when teachers are not adequately prepared, a decrease in performance can be observed (Nesenbergs et al., 2021). VR-based learning can increase cognitive and emotional engagement processes in higher education. This is because it allows participants to move, sense, touch and act upon simulated computer graphics, creating a digital psychological sense of immersion in a synthetic, computer-generated virtual world. This increases learners' affective arousal and boosts their cognitive processing. Moreover, eye-tracking metrics of fixations and saccades, and blink rate were used to capture cognitive engagement (Dubovi, 2022). In addition, user experience (UX) and affective surveys related to the immersive VR experience were positive, although no significant difference was observed between the Prime and No Prime conditions (Hawes and Arya, 2022).

The Impact of VR-Based Learning on Student Motivation and Collaboration

VR and AR technologies are increasingly used in higher education to develop the skills of students and to provide an engaging and realistic learning experience, which can lead to deep comprehensive learning and the "learning-by-doing effect". However, there is limited information regarding the impact of these technologies on student motivation and collaboration in higher education. XR technologies include VR, AR, and MR, and provide opportunities for experimentation and simulations. Students prefer to use new technologies in education such as XR, as they require motivation and enthusiasm. The adoption of these technologies in higher education can lead to the development of new methodologies for training and teaching, which can meet millennial expectations and the technological revolution. In addition, the use of modern mobile technologies, interactions with AI, ubiquitous computing and technology, real-time communication with students, and continuous interactions between universities and companies can enhance the use of XR technologies in education (Bucea-Manea-Ţoniş et al., 2020). This can help to foster student motivation and collaboration in higher education and contribute to the development of more effective learning opportunities.

The Use of VR-Based Learning to Enhance Student Engagement

The advances in Augmented and Virtual Reality (AR/VR) technology have opened up an opportunity for its use in Higher Education (HE). AR/VR based learning, also known as 'immersive learning' has been touted as a revolutionary way to engage students, improve learning outcomes and broaden access to education (Nesenbergs et al., 2021). However, its success has been limited due to the cost and complexity of the required hardware such as HMDs, controllers, high-capacity computers, etc. (Hajirasouli et al., 2023). This has largely restricted its use to research and development rather than mainstream education. Despite this, a new blended online pedagogy based on the metaverse has been developed with the aim of enhancing student engagement in Higher Education. The efficacy of this blended teaching model was evaluated for its ability to improve learners' confidence in self-directed learning, communication skills, collaborative learning, problem analysing and solving, innovation, critical reflection and proficiency in clinical skills (Chen et al., 2022). Furthermore, a novel BIM-enabled VR-based pedagogical framework was developed and tested to assess its use for design studio teaching in architectural courses. The results showed that this framework could successfully be implemented into design studios and could be used to enhance student engagement in higher education. It was also seen that the use of VR-based learning could bridge the gap between technological advancements in the ACE industry and higher education teaching and learning methods and outcomes (Hajirasouli et al., 2023). However, it was also observed that whenever VR-based learning was tried as a complete replacement, student grades suffered. This indicates that while it can be used to improve student engagement, it cannot completely replace on-site studies. To take full advantage of its potential, further experiments could be conducted to determine the best way to use VR-based learning to enhance student engagement in Higher Education, and also to identify the novelty requirements to keep the engagement and performance of students until the end of the study year (Nesenbergs et al., 2021).

Impact on Learning Outcomes

The Impact of VR-Based Learning on Student Learning Outcomes

Virtual Reality (VR) based learning has the potential to drastically improve student learning outcomes in higher education. It can provide students with a live experience that is closer to reality, thereby increasing the motivation and level of interest in the learning process (Radianti et al., 2020). VR can also maximise the delivery of complex material in the classroom and is compatible with students' current learning styles. Moreover, VR can enhance comprehension and generate interest in science subjects at all levels of education (Kuncoro et al., 2023). The study investigated the effectiveness of a newly created VR game-based English mobile learning application on student learning outcomes in a higher education setting. It was found that the interaction feature of the VR application and the challenges of game-based design enable students to enter the state of flow easily, thus enhancing their motivation to learn. Student self-regulation in the learning environment was found to be at a moderate level, and it was found that self-efficacy and self-regulation affect each other. Immersion, flow and presence were found to enhance self-efficacy while absorption and immersion enhance self-regulation. Additionally, VR features were found to have a causal path to usability, presence, motivation, control and active learning. The study also investigated how VR enhances the learning outcomes, and it was found that usability has a causal path to motivation, cognitive benefits, control and active learning, and reflective thinking. Furthermore, the determinants and their relationships for effective desktop VR-based learning in a learning environment that supports constructivism model were examined (Lee et al., 2010).

The Impact of VR-Based Learning on Knowledge Retention and Academic Performance

The use of Virtual Reality (VR) technologies in higher education has been the subject of much research. VR technologies often involve the user in an immersive experience, stimulating the senses and creating a sense of presence, which is associated with increased enjoyment of the media. AR technologies, on the other hand, are less likely to elicit feelings of presence, which could be a factor in the effectiveness of learning with VR. Studies have shown that VR technology-based instruction may improve students' learning outcomes in various subjects and contexts (Huang et al., 2019). Furthermore, a study conducted to verify the learning effectiveness of a desktop VR-based learning environment (Lee and Wong, 2014) found that students in the VR condition allocated more attention to the mediated environment, resulting in more psychological and cognitive responses to the media. This was reflected in the increased knowledge retention score in the VR condition, as well as increased levels of enjoyment (Huang et al., 2019). The Cognitive Theory of Multimedia Learning (Hamilton et al., 2021) postulates that high-fidelity graphics and animations in I-VR could significantly increase cognitive load, detracting from the learning task at hand. However, studies have shown that application level and structural accuracy after learning were positively correlated, and the interplay of pre-accuracy and prompt had a significant impact on structural accuracy after learning. Furthermore, the use of elaboration prompts before entering the VRLE on learning robotics was found to increase knowledge level. In addition, the use of VR-based learning reduce extraneous cognitive load and increases germane cognitive load, resulting in better performance. The study also found that desktop VR-based learning environment benefits more to low spatial ability students and the use of VR-based learning improves learning outcomes for students. Thus, it can be concluded that the use of VR-based learning improves knowledge retention and academic performance (Lee and Wong, 2014).

Strategies to Improve Learning Outcomes With VR-Based Learning

More recently, the development of VR-HMDs has allowed for immersive sensory experiences and direct interaction with learning material. Additionally, the use of motion tracking, which allows for six degrees of freedom, improves the experience of immersion, leading to increased engagement and time spent on the task. VR-based learning can allow learners to experience situations that are difficult or dangerous to replicate in reality and can also be used to improve procedural, practical and declarative knowledge in higher education. Moreover, VR-HMDs can be used to improve cognitive skills such as spatial and visual knowledge, visual scanning, observational skills, and psychomotor skills. Furthermore, low-end VR-HMDs suit 360° videos but may have limited computing power (Albus et al., 2021), while high-end VR-HMDs are capable of more advanced functions. It follows that careful consideration of the VR device to be used is crucial for effectively utilising the capabilities of the VR-HMD in supporting the desired learning outcomes. In this regard, the instructional design model proposed by the text can guide the design of VR-based learning environments, as it outlines five new principles that were generated to enhance the robustness of the instructional design model through formative research. However, to further assess the effectiveness of these strategies, experimental studies should be conducted to test the efficacy of the newly derived hypothesised principles (Chen and Teh, 2013). Lastly, VR-based learning can be used in addition to conventional learning methods such as textbooks, to provide learners with interactive learning experiences that respond dynamically to their movements and behaviour. Moreover, signals in VR can be used as orientation aids to reduce difficulties in spatial orientation and facilitate the formation of mental models. However, it is important to be aware that the effects of various design variants in VR are less straightforward than in classical multimedia learning material. Thus, further investigation is needed to determine whether signals can assist in deeper processing of learning content, and the use of multimedia principles that have robust effects in real learning environments should be considered (Albus et al., 2021) in order to optimally utilise working memory, prevent cognitive overload and promote successful learning.

LEVERAGING VR FOR ENHANCED LEARNING OUTCOMES AND CULTIVATED GRADUATE CAPABILITIES

Virtual Reality (VR) has emerged as a transformative tool in higher education, offering innovative ways to achieve learning outcomes and develop graduate capabilities. The integration of VR technology brings about a range of benefits that positively impact both students' academic growth and their preparation for the professional world (Refer to figure 1).

- Learning outcomes enhanced by VR:
    1. Active learning and practical application: VR creates an immersive learning environment conducive to active engagement. Merchant et al. (2014) posit that VR simulations provide hands-on experiences, significantly contributing to a deeper understanding and retention of concepts. This is impactful in fields, such as medicine and engineering, where practical skills are paramount (Kyaw et al., 2019). Learners are immersed in realistic scenarios, allowing them to apply theoretical knowledge in practical settings, thereby enhancing their learning outcomes.
    2. Personalised learning journey: The adaptability of VR to individual learning styles and paces contributes to a personalised learning journey. Mikropoulos and Natsis (2011) emphasise that VR environments can be tailored to address specific educational needs, making learning more effective and engaging. The personalised nature of VR learning ensures that students receive content at a pace suitable for their comprehension, thereby optimising the educational experience.
    3. Collaborative experiences: VR's capacity for immersive experiences extends to collaborative learning, even in remote settings. Jensen and Konradsen (2018) found that VR supports interactive group learning, fostering teamwork and communication skills. Learners can engage in shared experiences, working together on complex tasks, and thereby cultivating essential interpersonal skills crucial for various professional contexts.
    4. Critical thinking and problem-solving: VR scenarios often necessitate the application of critical thinking and problem-solving skills. Radianti et al. (2020) highlight how immersive VR environments challenge students to think critically and make decisions in complex situations. By placing learners in scenarios that require thoughtful analysis and decision-making, VR contributes significantly to the development of critical thinking skills, a key outcome in higher education.
- Graduate capabilities nurtured by VR:
    1. Practical application of knowledge: VR facilitates the practical application of theoretical knowledge, a vital aspect for professional readiness (Kyaw et al., 2019). Graduates equipped with hands-on experience gained through VR simulations are better prepared to apply their knowledge in real-world professional settings. This aligns with the growing demand for practical skills in various industries.
    2. Critical thinking and analytical skills: The interactive nature of VR enhances critical thinking and analytical skills. Freina and Ott (2015) assert that VR simulations present scenarios requiring analytical thinking, which is essential for graduates entering the modern workforce. The ability to analyse complex situations and make informed decisions is a valuable skill that is honed through VR-based learning experiences.

Figure 1. Elevating learning outcomes and graduate capabilities through VR

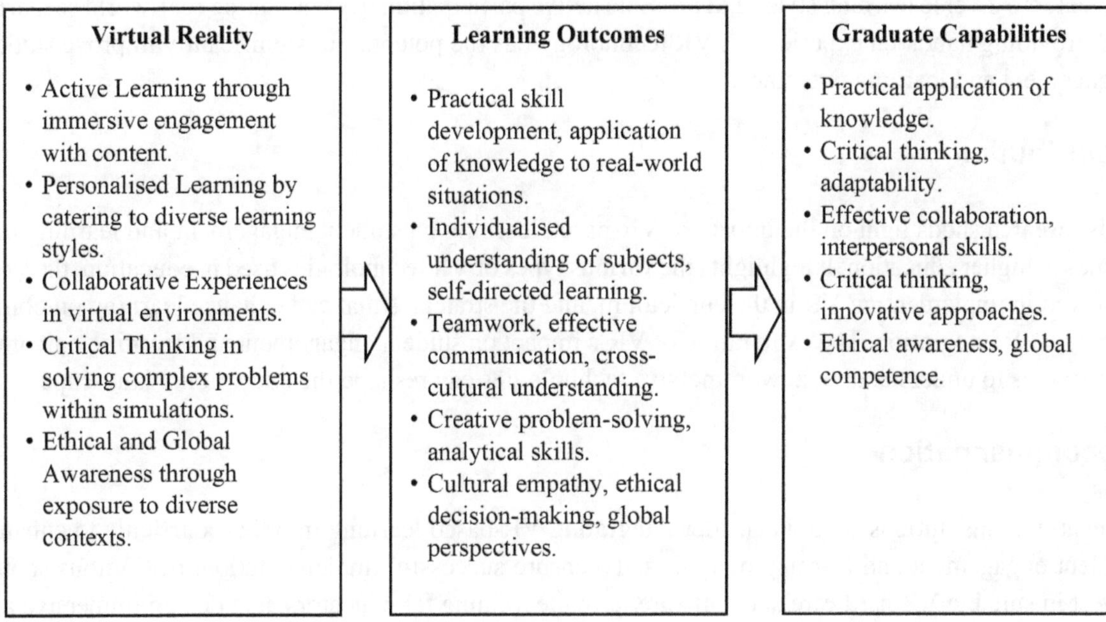

3. Adaptability and flexibility: VR's diverse and dynamic environments train students to be adaptable and flexible, preparing them for the rapidly changing landscapes of professional work. Huang et al. (2010) suggest that VR experiences expose students to varied scenarios, fostering adaptability and resilience. Graduates thus emerge with a capacity to navigate and thrive in dynamic work environments.
4. Effective collaboration and communication: Collaboration in VR settings enhances communication skills, a critical aspect of graduate capabilities. Allcoat and von Mühlenen (2020) emphasise the importance of communication in VR learning environments, promoting teamwork and interpersonal skills. Graduates who have engaged in collaborative VR experiences are better equipped for effective communication in diverse professional contexts.
5. Ethical and global awareness: VR's ability to simulate global and ethical scenarios contributes to the cultivation of broader understanding and awareness. Herrera et al. (2018) demonstrate how VR can be employed to cultivate empathy and ethical thinking by immersing learners in scenarios that foster awareness of diverse perspectives and ethical considerations. This aspect enriches the graduate's capacity for global citizenship and ethical decision-making in professional settings.

CONCLUSION

Virtual Reality (VR) has emerged as a powerful tool in higher education, revolutionising the way students engage with learning material and enhancing their overall learning outcomes. The immersive and interactive nature of VR-based learning offers a unique opportunity to create engaging educational experiences that cater to diverse learning styles and promote active participation. By simulating real-world scenarios and providing hands-on experiences, VR technology has the potential to significantly improve student engagement and learning outcomes.

Contribution

This research sheds light on the impact of VR-based learning on student engagement and learning outcomes in higher education. It highlights the various types of VR technologies used in education, the steps involved in implementing VR in the curriculum, and the strategies that can enhance learning outcomes through VR integration. The exploration of VR's impact on student engagement and learning outcomes contributes to understanding how immersive technologies can reshape the educational landscape.

Recommendation

Educational institutions should consider integrating VR-based learning into their curricula to enhance student engagement and learning outcomes. To ensure successful implementation, institutions should invest in suitable VR hardware and software, provide training for educators, and design immersive and interactive VR experiences aligned with learning objectives. Collaboration between educators, content developers, and technology experts is crucial to create effective VR-based learning materials.

Limitation

Despite the promising potential of VR-based learning, there are limitations that need to be addressed. The cost of VR equipment, software development, and technical support can pose financial challenges for educational institutions. Additionally, not all subjects and topics may be suitable for VR-based learning, and the effectiveness of VR may vary depending on individual student preferences and aptitudes. Ensuring equal access and inclusivity for all students, including those with disabilities, remains a challenge.

Future Research

Future research should delve deeper into the long-term effects of VR-based learning on student retention of knowledge and skills. Comparative studies could be conducted to assess the effectiveness of VR-based learning against traditional methods and other technology-enhanced approaches. Exploring the optimal design principles for creating effective VR-based learning experiences and understanding how VR impacts different disciplines and student demographics would also contribute to the field. Furthermore, investigating the scalability and sustainability of VR-based learning in various educational settings and cultures is an avenue for future research.

REFERENCES

Albus, P., Vogt, A., & Seufert, T. (2021). Signaling in virtual reality influences learning outcome and cognitive load. *Computers & Education*, *166*, 104154. doi:10.1016/j.compedu.2021.104154

Allcoat, D., & von Mühlenen, A. (2018). Learning in virtual reality: Effects on performance, emotion and engagement. *Research in Learning Technology*, *26*(0), 1–13. doi:10.25304/rlt.v26.2140

Bailenson, J. (2018). *Experience on demand: What virtual reality is, how it works, and what it can do.* W. W. Norton & Company.

Beckem, J. M., & Watkins, M. (2012). Bringing life to learning: Immersive experiential learning simulations for online and blended courses. *Journal of Asynchronous Learning Networks*, *16*(5), 61–70.

Bermejo, B., Juiz, C., Cortes, D., Oskam, J., Moilanen, T., Loijas, J., Govender, P., Hussey, J., Schmidt, A. L., Burbach, R., King, D., O'Connor, C., & Dunlea, D. (2023). AR/VR Teaching-Learning Experiences in Higher Education Institutions (HEI): A Systematic Literature Review. *Informatics (MDPI)*, *10*(2), 1–16. doi:10.3390/informatics10020045

Bower, M., Lee, M. J. W., & Dalgarno, B. (2018). Collaborative learning across physical and virtual worlds: Factors supporting and constraining learners in a blended reality environment. *British Journal of Educational Technology*, *48*(2), 407–430. doi:10.1111/bjet.12435

Bucea-Manea-Ţoniş, R., Bucea-Manea-Ţoniş, R., Simion, V. E., Ilic, D., Braicu, C., & Manea, N. (2020). Sustainability in Higher Education: The Relationship between Work-Life Balance and XR E-Learning Facilities. *Sustainability (Basel)*, *12*(14), 1–19. doi:10.3390/su12145872

Chen, C. J., & Teh, C. S. (2013). Enhancing an instructional design model for virtual reality-based learning. *Australasian Journal of Educational Technology*, *29*(5), 699–716. doi:10.14742/ajet.247

ChenY.ChenY.LinW.ZhengY.XueT.ChenC.ChenG. (2022). Application of Active Learning Strategies in Metaverse to Improve Student Engagement: An Immersive Blended Pedagogy Bridging Patient Care and Scientific Inquiry in Pandemic. https://ssrn.com/abstract=4098179 doi:10.2139/ssrn.4098179

Dalgarno, B., & Lee, M. J. (2010). What are the learning affordances of 3-D virtual environments? *British Journal of Educational Technology*, *41*(1), 10–32. doi:10.1111/j.1467-8535.2009.01038.x

Dengel, A., Auer, A., Urlbauer, P., & Läufer, T. (2022). Game-Based Teaching of Basic Hardware Components With an Educational Virtual Reality at Different Levels of Immersion. ITiCSE 2022: Innovation and Technology in Computer Science Education, Dublin, Ireland.

Ding, X., & Li, Z. (2022). A review of the application of virtual reality technology in higher education based on Web of Science literature data as an example. *Frontiers in Education*, *7*, 1048816. doi:10.3389/feduc.2022.1048816

Dubovi, I. (2022). Cognitive and emotional engagement while learning with VR: The perspective of multimodal methodology. *Computers & Education*, *183*, 104495. doi:10.1016/j.compedu.2022.104495

Freina, L., & Ott, M. (2015). A literature review on immersive virtual reality in education: State of the art and perspectives. *Proceedings of the 11th International Scientific Conference eLearning and Software for Education.*

Hajirasouli, A., Banihashemi, S., Sanders, P., & Rahimian, F. (2023). BIM-enabled virtual reality (VR)-based pedagogical framework in architectural design studios. Smart and Sustainable Built Environment, ahead-of-print.

Hamilton, D., McKechnie, J., Edgerton, E., & Wilson, C. (2021). Immersive virtual reality as a pedagogical tool in education: A systematic literature review of quantitative learning outcomes and experimental design. *Journal of Computers in Education*, *8*(1), 1–32. doi:10.1007/s40692-020-00169-2

Hawes, D., & Arya, A. (2022). *VR-Based Context Priming to Increase Student Engagement and Academic Performance.* 8th International Conference of the Immersive Learning Research Network (iLRN), Vienna, Austria.

Herrera, F., Bailenson, J., Weisz, E., Ogle, E., & Zaki, J. (2018). Building long-term empathy: A large-scale comparison of traditional and virtual reality perspective-taking. *PLoS One*, *13*(10), e0204494. doi:10.1371/journal.pone.0204494 PMID:30332407

Hodgson, P., Lee, V. W. Y., Chan, J. C. S., Fong, A., Tang, C. S. Y., Chan, L., & Wong, C. (2019). Immersive Virtual Reality (IVR) in Higher Education: Development and Implementation. In Augmented Reality and Virtual Reality, Progress in IS. Springer.

Huang, H. M., Rauch, U., & Liaw, S. S. (2010). Investigating learners' attitudes toward virtual reality learning environments: Based on a constructivist approach. *Computers & Education*, *55*(3), 1171–1182. doi:10.1016/j.compedu.2010.05.014

Huang, K. T., Ball, C., Francis, J., Ratan, R., Boumis, J., & Fordham, J. (2019). Augmented Versus Virtual Reality in Education: An Exploratory Study Examining Science Knowledge Retention When Using Augmented Reality/Virtual Reality Mobile Applications. *Cyberpsychology, Behavior, and Social Networking*, *22*(2), 105–110. doi:10.1089/cyber.2018.0150 PMID:30657334

Jensen, L., & Konradsen, F. (2018). A review of the use of virtual reality head-mounted displays in education and training. *Education and Information Technologies*, *23*(4), 1515–1529. doi:10.1007/s10639-017-9676-0

Kuncoro, T., Ichwanto, M. A., & Muhammad, D. F. (2023). VR-Based Learning Media of Earthquake-Resistant Construction for Civil Engineering Students. *Sustainability (Basel)*, *15*(5), 1–12. doi:10.3390/su15054282

Kustandi, C., Fadhillah, D., Situmorang, R., Prawiladilaga, D., & Hartati, S. (2020). VR Use in Online Learning for Higher Education in Indonesia. *International Journal of Interactive Mobile Technologies*, *14*(1), 31–47. doi:10.3991/ijim.v14i01.11337

Kyaw, B. M., Saxena, N., Posadzki, P., Vseteckova, J., Nikolaou, C. K., George, P. P., Divakar, U., Masiello, I., Kononowicz, A. A., Zary, N., & Car, L. T. (2019). Virtual reality for health professions education: Systematic review and meta-analysis by the digital health education collaboration. *Journal of Medical Internet Research*, *21*(1), 1–13. doi:10.2196/12959 PMID:30668519

Lee, E. A. L., & Wong, K. W. (2014). Learning with desktop virtual reality: Low spatial ability learners are more positively affected. *Computers & Education*, *79*, 49–58. doi:10.1016/j.compedu.2014.07.010

Lee, E.A.L., Wong, K.W., & Fung, C.C. (2010). How does desktop virtual reality enhance learning outcomes? A structural equation modeling approach. *Computers & Education, 55*(4), 1424-1442.

Li, P., Fang, Z., & Jiang, T. (2022). Research Into improved Distance Learning Using VR Technology. *Frontiers in Education*, *7*, 757874. doi:10.3389/feduc.2022.757874

Luo, H., Li, G., Feng, Q., Yang, Y., & Zuo, M. (2021). Virtual reality in K-12 and higher education: A systematic review of the literature from 2000 to 2019. *Journal of Computer Assisted Learning*, *37*(3), 887–901. doi:10.1111/jcal.12538

Makransky, G., & Lilleholt, L. (2018). A structural equation modeling investigation of the emotional value of immersive virtual reality in education. *Educational Technology Research and Development*, *66*(5), 1141–1164. doi:10.1007/s11423-018-9581-2

Marks, B., & Thomas, J. (2022). Adoption of virtual reality technology in higher education: An evaluation of five teaching semesters in a purpose-designed laboratory. *Education and Information Technologies*, *27*(1), 1287–1305. doi:10.1007/s10639-021-10653-6 PMID:34257511

McGrath, O., Hoffman, C., & Dark, S. (2023). Future Prospects and Considerations for AR and VR in Higher Education Academic Technology, Emerging Technologies and Trends. *EDUCAUSE Review*. https://er.educause.edu/articles/2023/4/future-prospects-and-considerations-for-ar-and-vr-in-higher-education-academic-technology

Merchant, Z., Goetz, E. T., Cifuentes, L., Keeney-Kennicutt, W., & Davis, T. J. (2014). Effectiveness of virtual reality-based instruction on students' learning outcomes in K-12 and higher education: A meta-analysis. *Computers & Education*, *70*, 29–40. doi:10.1016/j.compedu.2013.07.033

Mikropoulos, T. A., & Natsis, A. (2011). Educational virtual environments: A ten-year review of empirical research (1999-2009). *Computers & Education*, *56*(3), 769–780. doi:10.1016/j.compedu.2010.10.020

Nesenbergs, K., Abolins, V., Ormanis, J., & Mednis, A. (2021). Use of Augmented and Virtual Reality in Remote Higher Education: A Systematic Umbrella Review. *Education Sciences*, *11*(1), 1–12.

Parong, J., & Mayer, R. E. (2018). Learning science in immersive virtual reality. *Journal of Educational Psychology*, *110*(6), 785–797. doi:10.1037/edu0000241

Potkonjak, V., Gardner, M., Callaghan, V., Mattila, P., Guetl, C., Petrović, V. M., & Jovanović, K. (2016). Virtual laboratories for education in science, technology, and engineering: A review. *Computers & Education*, *95*, 309–327. doi:10.1016/j.compedu.2016.02.002

Radianti, J., Majchrzak, T. A., Fromm, J., & Wohlgenannt, I. (2020). A systematic review of immersive virtual reality applications for higher education: Design elements, lessons learned, and research agenda. *Computers & Education*, *147*, 103778. doi:10.1016/j.compedu.2019.103778

Rauf, H. L., Shareef, S. S., & Othman, N. N. (2021). Innovation in Architecture Education: Collaborative Learning Method Through Virtual Reality. *Journal of Higher Education Theory and Practice*, *21*(16), 33–40.

Shadiev, R., Yang, L., & Huang, Y. M. (2022). A review of research on 360-degree video and its applications to education. *Journal of Research on Technology in Education*, *54*(5), 784–799. doi:10.1080/15391523.2021.1928572

Chapter 13
Listen to What Anatolia Is Saying for a Sustainable Future for All

Pınar Ayyildiz
Ankara Medipol University, Turkey

Adem Yilmaz
Kastamonu University, Turkey

ABSTRACT

This chapter intends to illustrate and get inspired by rituals and beliefs belonging to several societies and cultures rooted in Anatolia to facilitate students' understanding of what climate and climate change mean for local and global communities. In fact, the protection and the transfer of socio-cultural identities of local people and groups via assuring an intergenerational means of knowledge dissemination through oral communication/dialogues, observations, and participation will be both a meaningful and a purposeful act. With the help of activities like readings, on-site visits, storytelling, and portfolio work that can be integrated into the current program, the researchers attempt to welcome learners studying at high school as the key agents of the process to help develop pro-environmental values and related behaviors, as well as discussing the collective consciousness of the past and present civilizations of Anatolia, whilst embracing a non-Western outlook.

Anatolia has always been an eminent center of civilizations and each of these civilizations is known to have their unique ways of celebrating seasons and acknowledging nature. One of these ceremonies is Nevruz, which is a holiday celebrated almost everywhere in today's modern Türkiye, namely, Turkish Republic and we know that the awakening of nature on March the 21st is embraced through numerous activities and rituals (Kildiroglu, 2020). Another popular ceremony is entitled Hıdırellez and is portrayed in the line of literature as amongst the initial summer feasts celebrated in Anatolia, whose origins date back to ancient times and as one that is kept alive together with other traditions and customs that are linked to solidarity (Uca, 2007). It has also been pinpointed that "although the myths, rituals, play, and

DOI: 10.4018/978-1-6684-9859-0.ch013

performances of Hıdrellez have become constrained due to urbanization and modernization, Hıdrellez is still a significant event celebrated by a large number of people." And what makes Hıdrellez a sign of collective thought and understanding is hinted at these words: "the conviviality and sociality of Hıdrellez contribute to social inclusion. The celebrations engage various groups, particularly socially excluded groups, women, elderly, and children, into rituals, prayer, plays, and performances. In urban areas, a socially disadvantaged community, namely the Roma community, come to the fore due to its lively performances during celebration (Gursoy, 2021: 594, 603).

With that being said, as is the case with the rest of the globe, it has become relatively more difficult to mark the beginnings and endings of seasons and it appears to citizens that seasons have already lost their inherent features that are distinct. As a matter of fact, the very situation is getting severe and now has reached a point which has been italicized in a fair number of recent studies on climate change, on agriculture and others indicating "the increased temperature would lead to an increase in evapotranspiration demand, and thereby to more severe drought events due to unsatisfied demand by precipitation" (Arslantas & Yesilirmak, 2020: 14).

Taking into account all the aforementioned discussions, this paper intends to illustrate and get inspired by rituals and beliefs belonging to several societies and cultures (ancient mythologies, minorities, and others) rooted in Anatolia, Türkiye, to facilitate students' understanding of what climate and climate change mean for local and global communities. In fact, the protection and the transfer of socio-cultural identities of local people and groups via assuring an intergenerational means of knowledge dissemination through oral communication/dialogues, observations, and participation will be both a meaningful and a purposeful act (UNESCO, 2017).

With the help of activities like readings, on-site visits, storytelling, and portfolio work that can be integrated into the current program, the researchers attempt to welcome learners studying at high school as the key agents of the process to help develop pro-environmental values and related behaviors, as well as discussing the collective consciousness of the past and present civilizations of Anatolia, whilst embracing a non-Western outlook on the subject. In so doing, the said students being both the subjects (as data collectors) and the objects (learners) will be doing their share pertaining to the research. To exemplify, an agenda notebook of each student will be of assistance in this regard. Such notebooks would include visual elements and instructions, suggestions to conduct outdoor observations, recommendations for visits, notes with traditional sayings, descriptions of weekly activities or the ceremonies of peoples that formerly inhabited the lands of Anatolia, alongside those of the indigenous groups living there at present. These students will accompany the researchers during the face-to-face interviews with the target groups such as locals. Suggested modifications to the current syllabi, in the form of an alternative track, will be made possible again with the help of the students in question, for instance, via some hyperlinks to be created through augmented reality, i.e., the videos of the elderly speaking about e.g., the past winters, when "it actually snowed heavily."

To this end, the following will constitute the key questions to propose a novel model; an innovative trajectory of education in a collaborative fashion with learners taking over active roles, which still respect the value systems of the past: a) What are the signs associated with the arrival of Spring, and what celebrations are performed to welcome Spring and Summer in Anatolia by diverse communities? b) What are the names and specific dates, for example, of the seasonal storms observed in various documents like historical records? c) How can this invaluable collection of inclusive experiences belonging to different societies and the refined knowledge of the past be implemented in the secondary school curricula of Türkiye to raise student awareness about sustainability, post-anthropocentrism, and ecocentrism on the road to a hopeful future for all?

In this direction, with a view to addressing the first research question stated above, the researchers and the participating students will visit local groups and communities as well as nursing homes so as to be able to obtain the relevant information by carrying out interviews and focus group discussions to be recorded and transcribed later on following ethical codes. For the second research question, the students and the researchers get together to conduct a thorough review of the related documentation in a cooperative manner adhering to the conventions of the document analysis method. To seek answers to the third question, the students and the researchers work together and also invite field experts (curriculum makers, teachers, authorities, education researchers, teacher candidates, academics) alongside other student groups to help construct an alternative syllabus to be integrated into the available/existing curricula in light of the data obtained through the varying steps of the present study.

INTRODUCTION

Anatolian lands have a deep history. As a matter of fact, it has hosted a fair number of civilisations along with states from the past to the present. These civilisations that Anatolian lands have hosted have also led to the emergence of a variety of cultures and beliefs over time (Esigul, 2021). When approached from this point of view, culture can be expressed as the total of several kinds of beliefs, feelings, thoughts, language and living elements that have emerged in a society and have turned into hence become traditions (Turan, 1990). Culture includes traditions, customs, skills and habits that human beings learn, produce and transfer to future generations as well. For this very reason, in order for a society to be sustainable, it is also deemed necessary to transfer its culture and ensure that all the members of that society is sensitive to this issue in an equal fashion. In this direction, amongst the most important things pertaining to what civilisations should be doing is to keep their language and culture alive to assure they still exist in the future. For sure, societies have their unique perspectives on life, beliefs, clothing and festivals. Many of these elements are also of great importance for societies to gain a reputation as a state and ensure its continuity. Apart from that, societies are in an effort to protect, preserve and carry their traditions to the future within their official or unofficial plans (Kapagan, 2019).

People living in Anatolian lands have gained experiences by interacting with their environment and nature. These invaluable experiences accumulated and refined throughout ages are now cultural elements and unwritten rules that can be passed on to future generations. Further, civilisations living in Anatolian geography have affected not only themselves but also many regions around them. As a result, common values and shared forms of traditions and customs began to form between and among civilisations. One of these celebrations is "Nevruz", a holiday celebrated almost everywhere in modern Turkey, the Republic of Turkey, and the awakening of nature on 21 March is embraced with a number of joyful activities and rituals (Kıldıroğlu, 2020).

Nevruz is celebrated on 21-22 March as one of the oldest holidays in Anatolia. It indeed is a holiday celebrated with great festivities in distinct parts of the globe, in particular in the regions where Turks live (Özcan, 2015). Nevruz, which is celebrated with great enthusiasm every year by Turks, has an origin of thousands of years. It symbolises the end of the winter months and the arrival of abundance and fertility. It points out to the warming of air, water, and soil. In other words, this is the resurrection of nature and the harbinger of 'good days' to come. Yet another popular celebration such as Nevruz is the Hıdırellez holiday.

Hıdrellez is a holiday whose origin dates back to ancient times. It is also a symbol of unity, solidarity and solidarity (Uca, 2007). The myths, rituals, games and shows of Hıdrellez have been restricted today due to urbanisation and modernisation. Howbeit, it is known that Hıdrellez is still acknowledged and thus celebrated by many. It is a symbol of collective thought and understanding too (Turkish Republic Ministry of Culture and Tourism [MoCT], 2019). This is because Hıdrellez socialises the community and contributes to social recovery of all by their participating in a series of celebrations, prayers, games and performances, especially for marginalised groups, women, the elderly and children. Within the scope of these, the Roman community, which can be considered a socially and demographically disadvantaged community, plays an important role in the cohesion of the society through coming to the fore with their live(ly) performances during the said celebrations (Gursoy, 2021: 594, 603). In addition to these features, Nevruz and Hıdrellez holidays basically serve as a harbinger of the arrival of the spring season and climatic changes.

Today, it would be fair to state that the climate has started to change in a way that humans are not very much used to, and seasons have already begun to be felt and experienced rather strangely owing to a bunch of reasons involving anthropecenic acts. The advancement of science, technology and industrialisation at an unimaginable speed has brought with it the problems of global temperature and climate change. Resultatntly, it has become relatively difficult to mark the beginning and/or end of the seasons. In other words, the arrival of spring and seasonal transitions can no longer be determined; sensed and embraced as easily as before. Global warming is on the stage and the activities carried out in agriculture and other areas are getting more difficult. This results in a decrease in precipitation, and it becomes hard to meet the needs of that many people living in the world. It is predicted that much more severe droughts may occur in the coming years due to rises in the global temperature (Arslantas & Yesilırmak, 2020: 14). From this point of view, it is of utmost importance to fully comprehend what climate and climatic change are.

In order for the natural environment we live in to be more sustainable and to be left to future generations preserving its desired form, environmental awareness of individuals should be developed. With this research, it is intended that some practicum will be planned and carried out by and for high school students to help them gain environmental awareness and also assist in their getting to know about the places they live in. Firstly, rituals and beliefs of societies and cultures (ancient mythologies, minorities, and others) with their roots in Anatolia and Türkiye are scrutinized. In this regard, the issue of climate change and sustainability are taken as a basis, and what they actually mean for local and social communities is delved into taking into account that the ability to transfer knowledge from generation to generation, to protect the sociocultural identities of local people and groups, and to create a sustainable climate today recognised as both meaningful and purposeful (UNESCO, 2017).

The 21st century has brought innumerable innovations, technological applications, and opportunities of novel sort to make our lives easier. That said, these changes and transformation have led to the emergence of several new skills and new practices. Now, everything can change very quickly (Hoeckel, 2008). This undoubtedly consumes our environment and its natural resources. On the other hand, if we are intensively engaged in this, we have to be extremely cautious. In light of these, one can comfortably put forth that if environmentally conscious individuals can be raised particularly at the secondary education level, it will be easier to talk about a sustainable living environment. At this point, concepts like climate change, the Anthropocene era and sustainability gain more importance.

The notion of the Anthropocene can be expressed as the human age. This age becomes meaningful with the fact that humans started to interact with nature, and over time, the direct and indirect effects of humans have begun to affect and change nature more (Erginsoy, 2022). The impact of the Anthropocene era accelerated much more with the transition of humans to the Industrial Revolution. With the Industrial Revolution, humans started using, influencing, and changing nature in biological, chemical, physical, and geological means. Nowadays this is true at a level that can interfere with the functioning of nature causing global warming, natural disasters and alike. In the 21st century, accelerated population growth, boosted consumption of natural resources, excessive use of air, soil and water resources, the decrease in biodiversity, increase in radioactive wastes and environmental problems are merely some of the effects of humans in question. These actions bring with them the problem of climate change, a severe issue. As a result of the global warming increase, the melting of glaciers has accelerated, and carbon and other biogeochemical cycles have changed. This has led to a decline in biodiversity. These bring another matter viz. sustainability, to the agenda. As has been mentioned earlier, it is becoming increasingly difficult for the world to remain sustainable and leave a healthy future for the next generations. These clearly shed light to that humanity has reached a critical point in its relations with its environment. People oftentimes think that they are the sole drivers of global change. However, earthquakes, tsunamis, cyclones, and other natural events show that this power is not solely in the hands of humans but also in the hands of nature. For this interaction between human beings and the world to proceed in a positive direction, arrangements need to be made specifically through education taking place in secondary schools, which constitute a transition stage (Ruddiman, 2007).

In Türkiye, in 9th, 10th, 11th, and 12th-grade curricula improvements have been made in many areas, especially in the areas such as information technologies, industrial automation and renewable energy technologies (http://mufredat.meb.gov.tr/Programlar.aspx). With that being said, areas such as nature conservation, environmental awareness, sustainability, climate change, Anthropocene age, design technologies, environmental education, recycling and waste management entail more attention. In addition, it has been determined that teaching materials focus on basic courses (mathematics, science, Turkish) and are at times insufficient in subjects involving human-nature interaction. To work on environmental awareness in the Anthropocene era, the General Directorate of Lifelong Learning in Türkiye opens courses annually. As part of these, courses such as cultivation of open-seeded plants, disaster awareness, conscious tree cutting, waste management, ecology and ecological literacy and sustainable agricultural practices are observed. It is worth noting that, these courses are entirely voluntary. Thence, it is vital that these courses are made widespread, that they are taught in a way that they turn into behaviours gained during school years and that the awareness is raised in students at an early age.

The current study is believed to be filling a gap in the relevant line of literature alongside those in the implementations. Firstly, high school students are the main actors in the research. In the research process, they served both as learners and as researchers in collecting data. To put differently, the researchers conducted the present study in cooperation with high school students. To this end, a random sample was determined, and the study was carried out with 15 students studying in a state high school. Within the scope of the study, activities such as readings, on-site visits, storytelling, and portfolio work that can be integrated into the available programme were carried out. It was attempted to back up students so that they can develop pro-environmental approaches. Moreover, the collective consciousness of the past and present civilisations living in Anatolia was dwelled upon.

During the research process, each researcher and student was provided with an agenda notebook. These notebooks contained elements such as visual elements and instructions, suggestions for outdoor observation, recommendations for visits, notes containing traditional sayings, descriptions of weekly activities or ceremonies of the peoples who used to live in Anatolian lands, as well as the indigenous groups living there. Students and researchers completed face-to-face interviews with the target groups in the research. At the end, recommendations were prepared as an alternative to the existing curriculum. In brief, this research sought answers to the below questions so as to propose a new model to accompany the existing curriculum:

a) What are the signs associated with the arrival of Spring, and what celebrations are performed to welcome Spring and Summer in Anatolia by diverse communities? What are Nevruz and Hıdrellez celebrations in Anatolia in the literature?
b) What are the names and specific dates, for example, of the seasonal storms observed in various documents like historical records?
c) How can this invaluable collection of inclusive experiences belonging to different societies and the refined knowledge of the past be implemented in the secondary school curricula of Türkiye to raise student awareness about sustainability, post-anthropocentrism, and ecocentrism on the road to a hopeful future for all?

There are three different problem situations in the research process. Thereupon, a dissimilar approach was taken for each problem situation. For the first one, the researchers and students acted together and took part in the interviews and focus group studies. With a view to doing so, local groups, communities, and nursing homes were visited. The related interviews were recorded and transcribed adhering to ethical rules. For the second problem situation, the students and researchers came together to review the bulk of literature and the selected documents were analysed through a comprehensive document analysis. In this process, attention was also paid to developing high school students' research, access to information and scientific process skills. For the third problem situation, the students and researchers worked together, and new stakeholders were added to the group. These stakeholders consisted of field experts (curriculum makers, teachers, authorities, educational researchers, teacher candidates, and academics) and experienced individuals in the field. The research findings were subjected to a comprehensive analysis, and suggestions and additions to the existing curriculum were then determined. These implementation steps are presented in Figure 1.

SEASONAL TRANSITIONS IN ANATOLIA AND THE CELEBRATION OF CHANGES BY DIFFERENT SOCIAL GROUPS

Under this heading, the first problem situation of the research, *"What are the signs associated with the arrival of Spring, and what celebrations are performed to welcome Spring and Summer in Anatolia by diverse communities?"*, is discussed. During the research process, the researcher got together with the students in the high school. Afterward, the interviews were conducted with the local and different social groups. What is more, the interviews were conducted by visiting nursing homes. Direct quotations were resorted to from time to time to elevate the leve of authenticity. The names of the participants were covered and coded as P1, P2, P3... Within the scope of the research, first, the question *"What do you think are the signs for the arrival of spring?"* was asked. Some of the answers received from the participants are given below:

Figure 1. Application steps

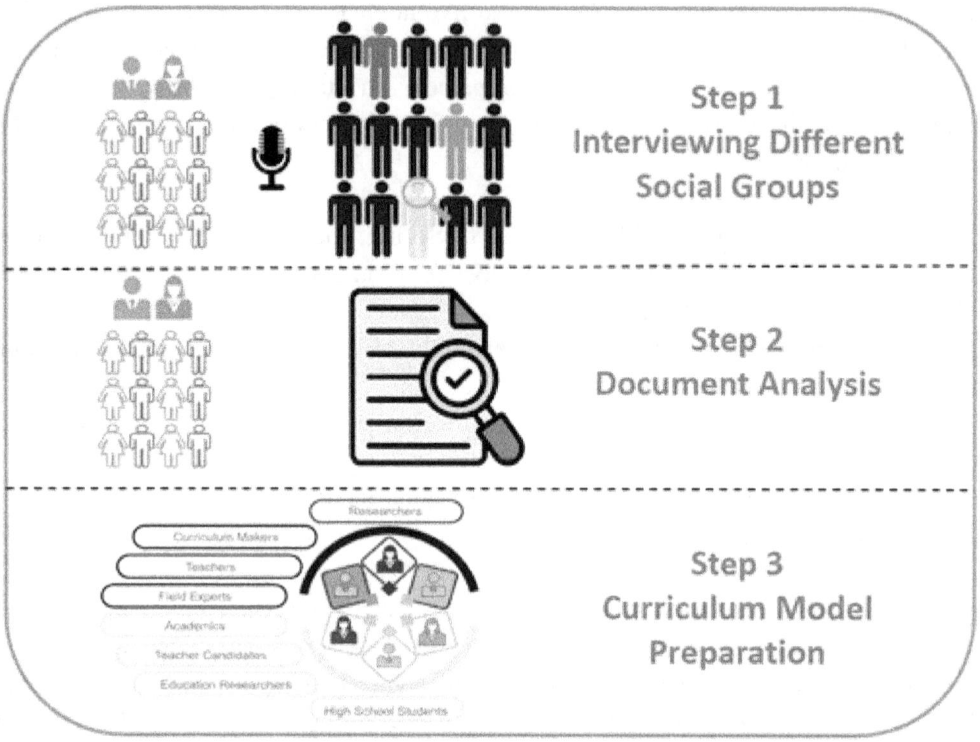

P1: We know that spring has arrived by our lambs and animals. One month after our lambs give birth to their young, the land turns green, the snow starts to melt and the first flowers bloom.

P3: When I was little, my mother always showed us snowdrops and crocus flowers. These flowers are the first flowers to sprout with the melting of the snow. Whenever I see these flowers starting to bloom, I knew that spring was about to arrive.

P5: We are farmers. So, we have water wells in our houses and storages where we store our seeds. I first realise that spring has come from the water wells. The water has a certain height. When it reaches that height, I can understand that the first snow melts and signals spring. Also, when insects and pests start to breed in our seed stores, I know that the weather has warmed up and the soil is ready to be planted.

P6: I follow the arrival of spring with "Cemres". Cemre first falls in the air, then in the water and then in the soil. The fall of Cemre means the warming of air, water, and soil. First off, I mean, towards the end of February, sun gets warmer. Then we understand that Cemre has fallen into the air. Then, water and soil get their shares in one-week intervals. But in recent years, these calendars have got confusing. The first Cemre, which used to fall in February, is now sensed later, let's say, at the end of March.

What the participants hint at that people who are involved in animal husbandry, agriculture, nature, and soil have different checkpoints for the arrival of spring. This stems from the culture of these different social groups. A nomadic group expresses the arrival of spring with the following sentences:

P9: We are a nomadic family. We have a big family. We migrate during a year. In winter, we sleep with our clothes made of felt and woollen quilts. When spring comes, first of all, we change our clothes that give us discomfort. This is the first signal of spring. Then our animals feel that spring is coming. They smell the earth. When they start to move but do not stand still, we know that the soil will begin to give its first sprouts.

P10: We are a large and crowded nomad family. As nomads, we are trying to survive in difficult conditions and we are always in an intense struggle with the nature. Sometimes we have to walk for hours to graze our animals. If our surroundings become green in the area where our animals are and if our animals are fully fed, then we realise that spring has come.

These social groups living a nomadic life get to know the arrival of spring in so many ways thanks to their living conditions and the animals they feed. The arrival of spring and the changes occurring in nature are witnessed more in groups that make a living with nature as they are intertwined with nature. Another question directed to groups within the scope of the research is *"What kind of activities do you celebrate the arrival of spring?"*. Some of the answers are below:

P12: The arrival of spring is celebrated in our village with communal events. First of all, a gathering is organised in the village square on the Fridays of February and March in the afternoon and food is prepared and served. Families with more economic power give more support. Leading elders of the village and older people make a short speech addressing the crowd. Then the food is eaten, and it is time to reconcile those who are offended. The coming of spring means brotherhood for us. No matter how many resentful people there are, they are reconciled. At the end of the day, we pray, and everyone goes back home.

P11: When March comes, spring comes for us. In order for spring to start in March, we tie wishing strings called "yazgara" on our wrists and pray. If we remove these bracelets, we believe that winter will come again. When the streams start flowing at the end of March, then we think our prayers have a response.

P14: When Spring comes, we decorate our doors and windows to bring abundance and prosperity to our house. If our streams are flowing, we take the water, ferment dough and make pastries and distribute them.

P17: Spring is abundance and fertility. Thusly, we also combine the arrival of spring with our religious worship. People who have troubles, families who cannot have children, elder people who are ill come together in the evenings and pray.

P20: When Spring comes, we do not forget our ancestors. We visit them in their graves and clean around. We water the trees and flowers. We pray for them and ask forgiveness from the Creator for the health of our family.

P22: Spring is not only important for humans. It is also important for animals and plants. In this sense, when spring comes, we take care of our animals and plants and make sure their needs are met. For instance, we clean the harmful weeds and spray the plants. We clean and ventilate the areas our animals are and take care of them. Usually we carry out these activities in a co-operative manner.

The views shared cast light to that the arrival of spring is celebrated both individually and as a community. In these celebrations, apparently the sense of unity, solidarity and brotherhood are italicized, and social solidarity activities take place.

Nevruz and Hıdrellez ceremonies are generally seen with festivities. In this respect, activities are organised in varying regions of the world:

- In Altai, Siberia and Tatars, milk raki, sacrificed meat and boiled eggs are served. There are also horse races, wrestling, javelin games, singing, and dancing.
- As another different community, thc Anatolian Alevis and Bektashis, go round all the houses and collect festive clothes. These collected things are then distributed to poor families and orphans. During Nevruz, houses are cleaned, and clean outfits are preferred.
- Central Asian Turks have celebrations that last for a week from 21 March to the 28th. Fires are lit and people jump over it, they ride horses, get engaged in collective entertainment activities and folk dances are performed.
- Turkmens in Anatolia go to the highlands on March 22 and 23 to organise festivities. Cemeteries are visited. Coffees are cooked with the nomads.
- In Southern Anatolia, events called "Sultan Narrız" are organised. Women frequently wear "halhals" on their feet and play games.
- In the Western Thrace area, tables are set up in large spaces, meals are cooked, and all mingle together.

TRADITIONAL NAMES OF WEATHER EVENTS IN ANATOLIA

This section explores the second problem statement of the research: *"What are the names and specific dates, for example, of the seasonal storms observed in various documents like historical records?"*. Whilst looking into this subject, a literature review was conducted with the participating high school students as co-reseurachers. For this, salient information about the weather events in Anatolia and the months when these events occur are shared through Table 1 (https://evvelzamansoylencelerim.com/2012/02/21/ne-varsa-eskilerde-var-dedeler-hesabi-mevsimler/; https://www.kalyeta.com/?pnum=52&pt=F):

It appears that a variety of names are entitled to seasons locally. One of the reasons why these are called in different ways is that cultures meet in Anatolian lands. It is visible both through the cultural documents and the literature that seasonal transitions and the events that are common during the transition processes are calendarised. Table 2 shows the changes and activities that occur on certain dates (https://www.marasgundem.com.tr/makale/halk-takviminde-mevsim-kis-yaz-16299).

It is clear from the table above that the events progress within a certain calendar range and until recent years people used to carry out their seasonal work according to this calendar. One can infer that there exist regressions and alterations now on account of climate change and global warming. Another

Table 1. Traditional names for climate and weather phenomena, moons, and storms

Months	Traditional Names Given to the Months	Storm Names
January	*Don Ayı, Çileler Ayı, Zemheri*	*Zemheri, Ayandon, Karancalos*
February	*Gücük, Gücük Ay*	*Hamsi*
March	*Döl Dökümü, Kuzu Ayı, Yazbaşı*	*Kocakarı, Çaylak, Hüsun*
April	*Yağmur Ayı, Yağar Ay, Abrul*	*Kırlangıç, Abrıl Beşi, Kuğu*
May	*Çiçek Ayı, Dut Ayı*	*Mevsimsiz Soğuklar, Çiçek, Filizkıran, Kokulya, Bevarih*
June	*Yayla Ayı, Kiraz Ayı*	*Ülker Doğumu, Gün Doğumu, Kızılcık*
July	*Kotan Ayı, Orak Ayı*	*Çark Döndü, Kızıl Erik, Yaprak*
August	*Biçim Ayı, Çürük Ay*	*Mihrican*
September	*Harman Ayı, Böğrüm Ay, İlk Güz*	*Çaylak, Turna Geçimi, Kestane Karası*
October	*Şarap Ayı, Değirmen Ayı, Sultan Ay, Orta Güz*	*Koç Katımı*
November	*Koç Ayı, Koç Katımı, Son Güz*	*Kuş Geçimi, Lodos*
December	*Nahır-Kovan, Kara-kış*	*Karakış*

Table 2. Seasonal events

Months	Seasonal Changes
January	---
February	---
5 March 8 March 9 March	Water goes up to trees Vineyard pruning time Pencil vaccination time
2 April 14 April	Time for the flowers to bloom Tulips bloom
18 May 20 May	The blossoming of roses It may snow
23 June	Leaf vaccination time
16 July	Increasing temperatures
15 August 26 August	Fruit ripening Yellowing of leaves
3 September 9 September 19 September 28 September	Collection of August berries Walnut picking time Vintage time Decreasing water supply to trees
20 October 30 October	Tree planting time Tree pruning time
29 November	Drainage of water from trees
9 December	End of the leaf fall

remarkable result of the literature review is that all the four seasons are fully experienced in Anatolia in certain periods. Since this is valid throughout the year the procedures for each month proceed systematically. Bearing in mind the deviations and drastic changes in the usual climate, informing the society of the issues such as climate change, global temperature, sustainability and emphasizing environmental literacy and environmental awareness from an early age have become prominent actions to take place.

A NEW MODEL: CULTURAL ELEMENTS, SUSTAINABILITY, CLIMATE CHANGE, POST-ANTROPOCENE ERA, AND ENVIRONMENTAL AWARENESS ARE INTEGRATED IN SECONDARY EDUCATION

This part investigates the third problem statement of the research i.e., *"How can this invaluable collection of inclusive experiences belonging to different societies and the refined knowledge of the past be implemented in the secondary school curricula of Türkiye to raise student awareness about sustainability, post-anthropocentrism, and ecocentrism on the road to a hopeful future for all?"*. Again, the students and researchers worked together, and new stakeholders were invited to the group. These stakeholders are comprised of field experts (curriculum makers, teachers, authorities, educational researchers, teacher candidates, academics) and people with a certain level of experience in the field. The main purpose of this study, which is for the most part a model proposal, is to add units and applied activities to the present courses in the curricula. First off, the curricula were analysed in this framework. Accordingly, the courses available for the 9th, 10th, 11th, and 12th grades, which are directly or indirectly related to cultural elements, climate change, sustainability, post-anthropocene era and environmental awareness, were classified through Table 3.

The table implies that there are courses of various sort in the secondary education curricula. At this juncture, it would be more advantageous to propose additional models for directly related courses instead of the indirectly related ones. Within this frame of reference, the courses considered directly related were taken into consideration. Prior to making some additions to these courses as planned, the literature was analysed thoroughly again, and the prominent stand points of the curricula were inspected. Republic of Türkiye Ministry of National Education General Directorate of Secondary Education made all the curricula and the applications accessible via online platforms and in relation to these, many materials seem to have been developed (https://ogmmateryal.eba.gov.tr/). These materials cover interactive textbooks,

Table 3. The courses

Directly Related Courses	Indirectly Related Courses
Chemistry	Contemporary Turkish and World History
Biology	Philosophy
Social Activity	Physics
Geography of the Turkish World	History of Science
History	Culture and Civilisation
Turkish Culture and Civilisation	Turkish Literature
Geography	Turkish Language and Literature
Democracy and Human Rights	Art History

Figure 2. The model

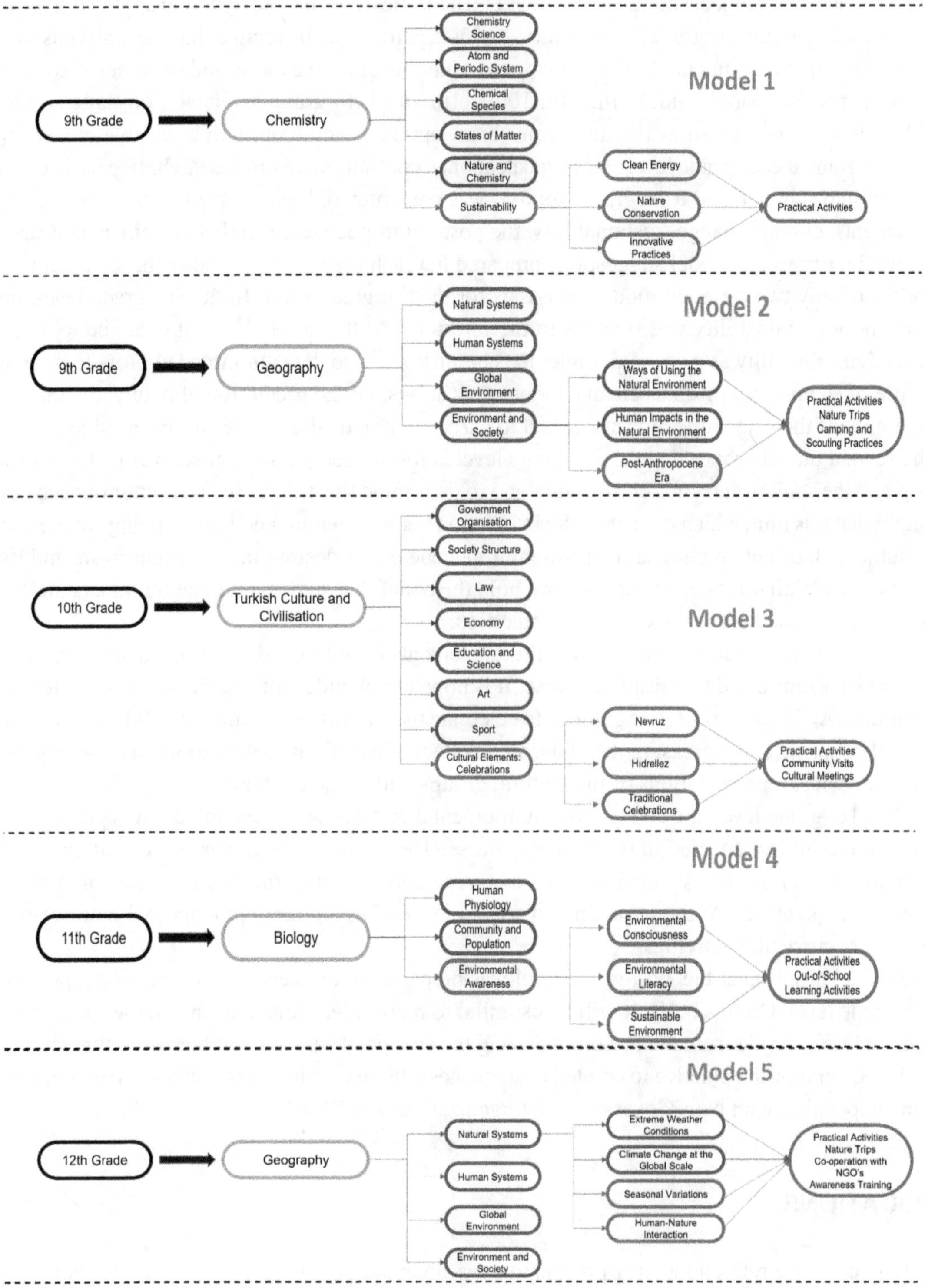

skill-based books, acquisition and comprehension activities, concept teaching materials, a question bank, lecture presentations, subject summaries, games and activities, mobile and dynamic applications, virtual museums, and an administrator/teacher library. In this regard, it can be argued that the materials offer an enriched learning environment. After receiving the opinions of field experts and the other shareholders, concentrating upon a single unit at the 9th, 10th, 11th and 12th-grade levels and on three sub-topics related to these units seemed as the most convenient option to come up with a suitable model for the existing program at each grade level and to speak about a continuum. In our research, five basic concepts were taken into account, and a program proposal was constructed. It encompasses the concepts of cultural elements, climate change, sustainability, the post-anthropocene era and environmental awareness.

Figure 2 portrays the model suggestions prepared for each grade level. Among these model proposals, there are only two different model proposals for the 9th grade level. In the 9th-grade level model, the concept of sustainability was brought to the forefront. At this grade level, it is aimed to teach the concept of sustainability as a new unit under the chemistry course. It is attempted to provide training on the nature of chemistry as well as clean energy applications, on the protection of nature and innovative applications. At the same time, it is suggested to carry out practical activities for each subject.

The second model proposal at the 9th-grade level is for the geography course. It is planned to add a sub-topic to the society and environment unit and to determine this title as the post-anthropocene era. It is thought that this unit, which describes the interaction between nature and human beings in the context of the subject, does not emphasise the post-anthropocene era as desired in its current form, and its not in-depth either. With this subject addition, we hold the belief that students will better understand nature through nature trips, camping and scouting practices.

At the 10th-grade level, the concept of cultural elements is underlined. We think about adding a unit to the Turkish Culture and Civilisation course. It is hoped to include cultural elements and celebrations with this unit. At the same time, it is planned to celebrate the Nevruz and Hıdrellez holidays, symbolising the arrival of spring and increasing the widespread effect of traditional celebrations, as a large group of teachers, students, representations of the cultural groups within the country.

At the 11th-grade level, the concept of environmental awareness was examined. It is planned to add a unit onto the subject covered in the biology course. The environmental awareness unit aims to help gain environmental awareness, environmental literacy and to make students be a part of sustainable environmental practices. At the same time, it strives for reinforcing these practices with out-of-school, namely, extra curricular activities.

The model proposal at the 12th-grade level was prepared as an alternative to the model proposal at the 9th-grade level. On this basis, it will be essential to reconsider climate change issues at the 9th and 12th grade levels. Aside from that, re-emphasizing the change of seasons and human-nature interaction would be an appropriate practice to create consciousness. In this setting, it is planned to organize nature trips in cooperation with Non-Governmental Organizations (NGOs).

IMPLICATIONS

This study can have reflections over prospective research and policy making. For the former, it might be articulated that cross cultural and interdisciplinary studies can be of help to escalate the quality and quantity of the perspectives that come into play while designing and implementing research. Similarly, longitudinal studies can be useful to monitor the awareness levels of students as they proceed in their academic life. For the latter, decision making bodies need to cooperate and collaborate with all the stakeholders such as students, teacher candidates, faculty, marginal groups and many more to target a collective network of understanding when planning these aspects of the curricula for all levels, which we hope can synthesize ecowisdom.

REFERENCES

Arslantas, E. E., & Yesilirmak, E. (2020). Changes in the climatic growing season in western Anatolia, Turkey. *Meteorological Applications.*

Erginsoy, G. (2022). *Antroposen çağ.* https://ansiklopedi.tubitak.gov.tr/ansiklopedi/antro posen_cag

Esigul, T. (2021). Innovative approaches to the ancient timbre of Anatolian kabak kemane. *Online Journal of Music Sciences*, *6*(2), 195–221. doi:10.31811/ojomus.1026560

Gursoy, I. T. (2021). Spring celebration, Hıdrellez: Myth and play. *Leisure Sciences*, *43*(6), 593–605. doi:10.1080/01490400.2020.1870593

Hoeckel, K. (2008). *Costs and benefits in vocational education and training*. OECD Publishing. https://www.oecd.org/education/innovation-education/41538706.pdf

Kapagan, E. (2019). Gelenekten geleceğe "Nevruz". *Karabük Türkoloji Dergisi*, *1*, 46–60.

Kildiroglu, M. (2020). Geçmişten günümüze Türklerde nevruz. *Kafkas Üniversitesi Sosyal Bilimler Enstitüsü Dergisi*, *25*, 315–338. doi:10.9775/kausbed.2020.018

Özcan, S. (2015). Climate, seasons, weather events and public calender in Turkish proverbs. *The Journal of International Social Research*, *8*(36), 179–187.

Ruddiman, W. F. (2007). The early anthropogenic hypothesis: Challenges and responses. *Reviews of Geophysics*, *45*(4), 1–37. doi:10.1029/2006RG000207

Turan, S. (1990). *Türk kültür tarihi*. Bilgi Yayınevi.

Turkish Republic Ministry of Culture and Tourism. (2019). *Hıdrellez traditions.* https://www.ktb.gov.tr/EN-98558/Hıdrellez-traditions.html

Uca, A. (2007). Türk toplumunda hidirellez. *İ. A. Ü. Türkiyat Araştırmaları Enstitüsü Dergisi*, *34*, 113–138.

UNESCO. (2017). *Spring celebration, Hıdrellez.* https://ich.unesco.org/en/RL/spring-celebrationhdrellez-01284

Chapter 14
Exploring Sustainability Education in the Business Schools of Australian Branch Campuses in Malaysia:
Now and the Way Forward

Chorng Yuan Fung
https://orcid.org/0000-0002-2007-6286
Swinburne University of Technology, Sarawak, Malaysia

Kiat Sing Heng
Swinburne University of Technology, Sarawak, Malaysia

Chia Hua Sim
Swinburne University of Technology, Sarawak, Malaysia

ABSTRACT

Over the past decade, sustainability education has been an essential subject that many universities have infused into their curricula. However, no research has analysed the infusion of sustainability education in the curriculum. Such knowledge is essential for educators and policy-makers to develop relevant and contemporary curricula to produce future-ready graduates. This chapter is closing this gap by tracing sustainability education in the business schools of two Australian university branch campuses in Malaysia. An exploratory case study approach was used. The unit course details have been analysed thematically according to sustainability education literature. It was discovered that most keywords were found in the core and capstone units across a three-year bachelor's degree but with a higher emphasis on year-one units. This finding suggests that building the fundamentals of sustainability education from year one is essential.

DOI: 10.4018/978-1-6684-9859-0.ch014

INTRODUCTION

Sustainability in education refers to integrating the concepts and practices of sustainable development into educational settings. In this regard, studies have shown that education substantially impacts how young adults see sustainability issues like the environment, as personal values are observed to develop early in life and are more challenging to alter as people age (DeSombre, 2018). Specifically, Triana (2022) asserts that values are developed early on and at transitional years in a person's life (e.g. during college) and become more difficult to change with time. Consequently, scholars strongly argue for integrating sustainability issues inside educational settings, such as corporate social responsibility (CSR) and ethics.

In this regard, higher education institutions (HEIs) play a crucial role in cultivating a culture of sustainability for a sustainable future by incorporating sustainability into their curricula and assessments and educating future professionals (Kim, 2023). The drive to have sustainability embedded in higher education has gained momentum in recent years, given initiatives such as the United Nations' Sustainable Development Goals (SDGs). Many HEIs have committed to making sustainability in education a pressing concern. Specifically, HEIs in Australia are at the forefront of the movement to promote sustainability education. In some cases, higher education offers various courses centred around sustainability to help produce responsible citizens who can contribute towards developing a more sustainable future.

In the Asian setting, sustainability in education has been gaining momentum in higher education, especially at the foreign branch campuses of Australian universities. Drawing on the experiences of their counterparts in Australia, these branch campuses are uniquely positioned to combine the best practices from both countries to create an approach that is more inclusive of the concept of sustainability. In the Malaysian context, the Australian branch campuses are actively working to address sustainability challenges through collaborations with local communities, governments, and businesses. By embedding sustainability in education, these branch campuses are cultivating a generation of graduates with the knowledge and skills to address pressing sustainable issues and understand the local context to help manage regional challenges. Importantly, these branch campuses have the potential to make significant contributions towards developing a more sustainable future in Malaysia and in the Southeast Asian region as a whole.

Concerning business education, Charroin et al. (2022) highlighted that global business schools must transform to ensure their long-term viability and contribute positively to the world. In the same context, Bryant et al. (2022) pointed out that Business Schools are encountering significant challenges due to evolving market demands and increased expectations from external stakeholders. Specifically, Business schools and universities have started to realise the necessity for sustainability, CSR, and ethics in business education (Weybrecht, 2017). Students at tertiary education institutions who will become future managers rely on Business schools to educate them on the value of sustainability. In the United States context, researchers thoroughly evaluated the literature and highlighted the need for sustainability and social impact curricula to be integrated into American undergraduate Business education (Triana, 2022).

Meanwhile, extant literature indicates that there are many studies (e.g., Anastasiadis et al., 2021; Hil & Wang, 2018; Jun & Moon, 2021) that have affirmed the importance of incorporating sustainability into higher education and even provided frameworks and tools to integrate sustainability into university culture and curricula. In the area of management, the Principles for Responsible Management Education (PRME) were developed in 2007. PRME comprises six principles to help universities better align with the United Nations Global Compact and prepare future business leaders to strike a fair and sustainable balance between economic, social and environmental concerns (PRME, 2021). The literature, however, noted numerous challenges associated with integrating sustainability into management education at the higher education level (Galleli et al., 2022). Meanwhile, many Business schools are also facing challenges in integrating CSR into their curriculum despite students' high level of interest and the increasing pressure from accreditation agencies (Deer & Zarestky, 2017). In essence, it was noted that many HEIs have yet to adequately incorporate sustainability principles into the curricula of their undergraduate courses (Etse & Ingley, 2016). While the problems involve numerous aspects of the educational process, one of the initial indications of HEIs moving towards sustainability lies in integrating this theme into the academic curriculum (Brunnquell et al., 2015). Based on these considerations, it is logical to infer that any shortcomings or inadequacies in this initial step could impede the pursuit of the HEI in its goal towards sustainable education integration.

In this regard, there is a dearth of recent research examining the extent of sustainability integration in higher education, particularly in emerging economies, including Southeast Asia. There is a lack of conclusive evidence indicating that HEIs in this region have effectively incorporated sustainability into their curriculum despite increasing emphasis on its significance. The present study will bridge the gap by examining how sustainability is being integrated into undergraduate courses at the Business Schools of two Australian branch campuses in Malaysia. Findings from this study should provide further insights into the degree of sustainability education incorporated in the curriculum of the Business Schools of these Australian branch campuses in Malaysia. The outcome would aid HEIs in formulating and implementing appropriate strategies to expand sustainability education in their institutions.

LITERATURE REVIEW

Despite employers and students globally demanding business schools to incorporate sustainability into their curricula, there has not been much progress made by business schools in fulfilling these demands (Barber et al., 2014). As such, few initiatives have been identified for promoting sustainability-focused education within business schools. For example, a multi-institutional program is created on the United Nations Global Compact principle as the core of the management education curriculum to train students to become responsible leaders in their future employment. This program takes place outside the business school by extending faculty resources through institutional partnerships and using the Internet for course delivery (Dickson et al., 2013).

Meanwhile, the study by Barber et al. (2014) demonstrated the implementation of sustainable development education through interdisciplinary and cooperative relationships with internal and external stakeholders. Moreover, an interdisciplinarity-based sustainability management framework in business education was developed upon discussions with experts from the United Nations Principles for Responsible Management Education (UN-PRME) (Martins et al., 2022). Besides that, the empirical study conducted by Anastasiadis et al. (2021) outlines how teachers who view sustainability as a complex problem included sustainability in their lessons using WikiRate, a cutting-edge teaching resource. In contrast, teachers in a Brazilian business school used an active learning strategy in teaching sustainability-oriented skills (SOCs), which has increased student participation and engagement in the curriculum and improved grades (Claro & Esteves, 2021).

Apart from that, among the Association to Advance Collegiate Schools of Business (AACSB) accredited business schools in Lebanon (Chedrawi et a.l., 2019), there is a goal to illuminate a new function for corporate social responsibility (CSR) in the accreditation process following this accreditation.

The goal is to adhere to the proposition of the AACSB that emphasises the value of ethics, CSR, and sustainability education in business schools. Secondly, in Budapest, a business school in the context of AACSB accreditation candidacy co-created an innovative educational tool based on the UN Sustainable Development Goals and the concepts of sustainable development to promote sustainability practises at higher education institutions (Andrási et al., 2022).

Undeniably, there was an upward trend among business schools in South Korea to integrate sustainability-related issues into their curriculum; however, the course offerings are fragmented and typically concentrate on a limited understanding of ethics rather than creating a comprehensive curriculum that integrates sustainability throughout functional majors (Jun & Moon, 2021). Similarly, this issue happened to a business school in New Zealand where some courses are considered standalone sustainability courses, and others incorporate a technical focus on accounting as required by the professional accounting bodies in Australia and New Zealand (Sharma & Stewart, 2022).

Researchers like Gottardello and Pàmies (2019), and Shanika et al. (2023) have presented recommendations addressing the fragmented integration of sustainability components. For instance, Gottardello and Pàmies (2019) emphasise that sustainability should be promoted and encouraged within every course to nurture future sustainability leaders. Shanika et al. (2023) propose integrating fundamental sustainability topics into mandatory course modules.

Since 2022, some universities have implemented various innovative ideas in their courses to tackle sustainability issues. For instance, in early childhood education programs, Gal (2023) introduced a transformative teaching methodology involving reflective analysis, visual analysis through drawings, and the analysis of course summaries. This approach aims to reshape students' understanding, perspectives, emotions, and practical capacity to address the climate crisis. Additionally, researchers such as Affolderbach (2022) demonstrated how a problem- and project-based group learning exercise designed to emphasise the power of geographic thinking can teach green economy concepts; and Kim (2023) showcased how a problem-based service-learning (PBSL) initiative can be systematically integrated into the curriculum of a Product Innovation and Management course.

Meanwhile, researchers like Tijsma et al. (2023) have emphasised the significance of innovative educational programs based on transdisciplinary principles. These programs effectively prepare students for both interdisciplinary and transdisciplinary work. Furthermore, Kasch et al. (2023) have highlighted how a Challenge-Based Learning (CBL) initiative can support interdisciplinary collaboration when addressing sustainability issues.

In Malaysia, Michael et al. (2020) surveyed the students' awareness, attitudes, and actions regarding sustainable development in a human resource development course at a Malaysian public university. They discovered that there was a variation in the sustainability levels among students. Final-year students exhibited the highest degree of sustainability awareness compared to first-year and second-year students. Therefore, there is a need to address how sustainability has been embedded into the curriculum across various levels of units offered by the universities.

Meanwhile, Saleem et al. (2023) investigated the practices and perceptions of education for sustainable development (ESD) among students and staff across four public universities. Their findings revealed that Malaysian universities commonly employ holistic, pluralistic, and action-oriented ESD teaching and learning approaches. Significant associations were identified between these ESD approaches and sustainability consciousness (SC). Since their focus was primarily on two stakeholders' perspectives and did not include an analysis of curriculum design, it is crucial to examine the integration of sustainability into the curriculum. Since these were perception studies, there is a need to examine how ESD has been incorporated into the curriculum of the universities.

Research Objectives

The existing body of literature on sustainable education demonstrates notable progress but also highlights areas that require additional research, particularly in the business curricula of higher learning institutions.

In the context of Swedish business schools, a comparative study conducted by Kanwal et al. (2023) offers valuable insights regarding integrating academic and operational engagement to foster sustainability in business education. While the study emphasises the significance of employing a dual-engagement approach, it focused on Swedish institutions, which may restrict its applicability to other contexts. Another European study by Obrecht et al. (2022) presents a framework to analyse the incorporation of environmental sustainability into academic courses. While this research demonstrates a rigorous methodological approach, its application is centred on the Slovenian education system.

Also, Zguir et al. (2021) studied the curriculum models in New Zealand and Singapore and compared how sustainability values are taught in those two countries. The study focuses on establishing a framework for developing sustainability competencies. However, it does not include a comprehensive analysis of other regions when it is crucial to consider that the socio-political environment in different regions may impact the integration of sustainability into business education.

In the Canadian context, Driscoll et al. (2017) examine the possibility of greenwashing practices in incorporating sustainability into business education. A discrepancy was identified between the business school's mission and the MBA programme's policies and practices. In a more recent study, Gil-Doménech et al. (2021) introduced a Sustainability Competencies Map (SCM) designed specifically for the field of Business Administration and Management (BAM). This valuable resource serves as a tool to enhance sustainability education within the discipline.

In the context of Australia, Hay and Eagle (2020) reveal that curriculum modifications can improve business students' understanding of and adherence to sustainable practices. In China's context, Winter et al. (2022) highlight the practical challenges Chinese faculty members encounter when integrating sustainability into their teaching. The difficulties noted suggest the possibility of comparable problems in other Asian institutions; however, direct comparisons are difficult to make without detailed local investigation.

In sum, existing literature provides a fundamental comprehension of sustainability in the context of business education, yet it also highlights the necessity for additional research. In particular, Southeast Asia's diverse economic, cultural, and political circumstances suggest that incorporating sustainability concepts into business courses may have a special character unique to this region. Conducting localised research that can inform tailored strategies for embedding sustainability into business education is essential. Such research will ultimately contribute to cultivating future business leaders with the necessary skills and knowledge to effectively address the pressing sustainability challenges in the Southeast Asian region.

Overall, the literature review on sustainable education reveals a fragmented picture, highlighting the gradual and incomplete implementation of sustainability concepts in different geographical areas. For example, Jose (2016) highlights significant obstacles in Indian business schools, including inadequate institutional capacity and a lack of suitable teaching materials. These challenges impede the integration of sustainability into the fundamental curriculum.

Similarly, Barber et al. (2014) emphasised the difficulties encountered by a business school in the United States, highlighting the need for innovative teaching methods and partnerships with stakeholders to integrate sustainability in business education. In the case of South Korea, Jun and Moon (2021) observe a noteworthy rise in the incorporation of sustainability-focused courses in business education. However, they critique the lack of consistency and depth in the courses, as they tend to narrowly concentrate on ethics while neglecting the broader range of sustainability education.

In a broader European context, Painter-Morland et al. (2016) call for a systemic institutional approach to examine the incorporation of sustainability into the core aspects of educational institutions. Further, Figueiró and Raufflet (2015) argue that there is a lack of consistency and depth in management education regarding the integration of sustainability, revealing the need for a more cohesive and thorough approach to address this issue. The call for systemic institutional integration made by Painter-Morland et al. (2016), combined with the findings of Figueiró and Raufflet (2015) on inconsistency in sustainable education, suggests further research regarding the consistent integration of sustainability across various disciplines of business education.

On the other hand, Mburayi and Wall (2018) emphasised the gap in sustainability integration within accounting and finance due to stringent professional standards, recommending further research into how sustainability competencies are being introduced into business curricula. Similarly, Kemper et al. (2019) argue for comprehending the educational methodologies that form the foundation of sustainability educators within the marketing field. These studies highlight the incomplete and fragmented investigation of sustainability integration in business education.

The existing literature indicates the need for additional study into sustainable education in business courses. Further study is essential to ensure that future business leaders possess the necessary skills to tackle complex sustainability issues. In particular, sustainable education in business curricula is essential for foreign branch campuses in Southeast Asia. However, there is a noticeable lack of research in this area.

Based on the literature review, the current body of research primarily focuses on sustainable education in Western contexts and the challenges encountered by traditional institutions while overlooking the distinctive circumstances of foreign branch campuses in Southeast Asia. These campuses frequently face unique cultural, regulatory, and resource challenges that may impact their capacity to integrate sustainability into their business courses effectively. While the benefits of sustainability education are widely acknowledged, the extent to which it can be adapted and implemented within diverse cultural and economic settings, such as those found in Southeast Asia, remains underexplored.

Hence, the objectives of this study are:

(i) To identify how sustainability education is incorporated into the curriculum of business schools of Australian branch campuses in Malaysia, **an emerging market in the Southeast Asia region;**
(ii) To analyse the degree of incorporation of sustainability education in the curriculum **of the Australian branch campuses in Malaysia;** and
(iii) To propose future research in sustainability education in business schools.

BACKGROUND OF HIGHER EDUCATION IN MALAYSIA

Malaysia's higher education sector is characterised by its diversity, comprising public and private universities, foreign branch campuses, and vocational institutions. The development of Malaysia's higher education system into a binary structure began in 1963, and it was accompanied by the enactment of various legislations from 1971 to 2007. These laws were established to regulate higher education institutions, oversee the delivery of higher education, and ensure the quality and recognition of both local and foreign institutions in Malaysia (Morshidi & Wan, 2022). However, the most notable advancement during this timeframe was implementing the Private Higher Education Institutions Act (PHEIA) in 1996, which was crucial in acknowledging private universities and colleges as integral components of the Malaysian higher education system (Symaco & Wan, 2017).

International higher education has emerged as Australia's third-largest export sector (Baer, 2022). Implementing the Malaysia–Australia Free Trade Agreement (MAFTA) allows Australian service providers to maintain full ownership of privately funded higher education institutions in Malaysia. This agreement has fostered the export of Australian academic degrees offered through collaborative ventures with Malaysian institutions and Australian university branch campuses (Wan & Abdullah, 2021). It also signifies that upon introducing the 2030 Agenda for Sustainable Development, it became evident that higher education institutions, including Australian branch campuses, must take measures to pursue the global sustainability goals in Malaysia (Wan & Abdullah, 2021).

METHODOLOGY

This study adopted an exploratory and intrinsic case study to examine sustainability education in the undergraduate curriculum of business schools of Australian branch campuses in Malaysia. Stake (2005) categorises case studies into intrinsic and instrumental categories, which guides the purposive sampling strategy. In an intrinsic case study, the case is inherently exciting and purposively preselected, with the researcher exploring specific characteristics to achieve a deep understanding without emphasising representativeness or generalisation. On the other hand, in the instrumental case study design, purposive sampling selects cases that provide insights into the investigated phenomenon and uses the cases to understand a research problem better. Furthermore, this study used deductive and inductive approaches to provide rich information about sustainability education and its degree of incorporation.

Samples

The preselected intrinsic case studies blend well with purposive sampling (Stake, 2005), especially involving the deliberate selection of cases that optimise the efficient utilisation of limited research resources (Palinkas et al., 2015). The main criterion in the sample section was the availability of course information on their websites. The business schools of two branch campuses of Australian universities in Malaysia, namely Alpha University Malaysia and Beta University Malaysia (keeping the original names of the universities anonymous), were chosen as the sample of this study. Most universities in Australia have incorporated sustainability education into their curriculum, and these branch campuses would provide a good picture of this curriculum in a local context.

Data Collection Procedures

The online course information about the undergraduate degrees in these business schools has been obtained from the official websites of these universities. Figure 1 below shows the search criteria for the inclusion of data for analysis.

Figure 1. Search criteria for the data to be included for analysis

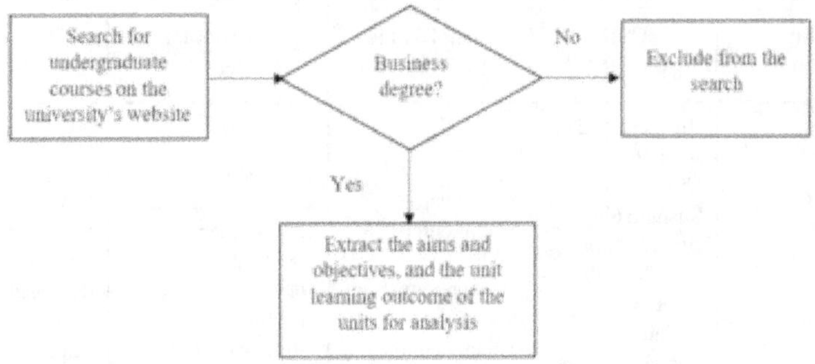

Data regarding the business degree units of study, which covers the unit aims and objectives, unit learning outcomes, contents, and brief descriptions, were obtained from the official websites of these universities from June to August 2023. This data extracted was organised into a spreadsheet to facilitate an analysis of sustainability keyword coverage. A matrix table was created to streamline the data collection process to show the frequency recorded for each sustainability keyword category across the two institutions' different business degree core units.

The matrix table in Excel was set up using sustainability keyword codes identified by Jun and Moon (2021). These keyword codes are grouped under specific categories, offering insights into the sustainability elements incorporated into business education at tertiary institutions. To ensure consistency and reliability, the three authors cross-checked the codes to the aims and objectives and the unit learning outcomes. Any discrepancies were then discussed and resolved. To facilitate analysis and results reporting, a keyword representation is assigned to each sustainability theme using the following abbreviated format: Ethics = eth; CSR = CSR; Social Value = SV; Environment = env; Finance = fin; Sustainable Business = SB.

Data Analysis Procedures

This study used a two-stage data analysis approach (Saldana, 2013) and carried out manually. The first stage was a deductive approach. Keyword codes and categories were developed based on Jun and Moon (2021). These keyword codes were used when analysing the content of the aims and objectives, unit learning outcome, content and brief description of the units. In the second stage of analysis, an inductive approach was used. Additional codes were developed inductively based on the contents analysed. Table 1 shows the keyword codes and categories used in the analysis.

Firstly, the keyword category of Ethics "eth" emphasises the moral principles that influence corporate decisions, operations, and management. The central concept focuses on ethical concerns in business and management. According to Jun and Moon (2021), the keyword codes for this category cover Ethics, Business ethics, Ethical management, and Management ethics. Meanwhile, the keyword category of CSR "CSR" is concerned with Corporate Social Responsibility and its implications. This category encompasses the firm's duty to impact society and the environment positively. As identified by Jun and Moon (2021), this category's keyword codes include business and society, CSR, Sustainable supply chain, CSV/Shared value, Philanthropy, Social responsibility, and Development cooperation.

Table 1. Sustainability keyword codes and category

Sustainability Keyword Category	Ethics "eth"	CSR "CSR"	Social Value "SV"	Environment "env"	Finance "fin"	Sustainable Business "SB"
Stage One Codes according to Jun & Moon (2021)	Ethics (Or ethical) Business ethics Ethical management Management ethics	Business and society CSR Sustainable supply chain CSV/Shared value Philanthropy Social responsibility Dev. Cooperation	Social value Social enterprise Social entrepreneurship Social venture Social economy Sharing economy	Env. management Green management Environmental sustainability (or environmentally sustainable) Climate change Eco-friendly Env. Value	ESG Social finance Socially responsible investing (SRI) Impact investing	Sustainability Sustainable business Sustainable management
Stage Two Codes from the actual data	Cultural, academic and professional integrity Ethical thinking Ethical issues Ethical and just Ethical theories Ethical theories and approaches		Social impact Societal needs Societal expectations Creating value for wider community Broader society Social agenda Cultural contexts Social welfare Consumer protection Cultural diversity value creation Social responsibilities Social accounting Human right Global citizenship	Environmental contingencies Environmental pressures Responsible management Environmental protection	Financial literacy Performance assessment Business planning Risk management	Responsible business practices Business Strategy Framework Strategic management Market efficiency and welfare Welfare effects of international trade Market efficiency Sustainable Development Goals (SDGs)

Further, the keyword category of Social Value "SV" centred around businesses making a societal impact or creating value that benefits society. Hence, Jun and Moon's (2021) keyword codes for this category cover Social value, Social enterprise, Social entrepreneurship, Social venture, Social economy, and Sharing economy. Meanwhile, the Environment "env" keyword category emphasises the importance of eco-friendly business practices and revolves around environmental concerns in business. Hence, Jun and Moon (2021) identified its keyword codes to encompass Environmental management, Green Management, Environmental Sustainability, Climate change, Eco-friendly, and Environmental value.

Next, the Finance "fin" keyword category focuses on investment and finance strategies that prioritise sound social consequences over returns. According to Jun and Moon (2021), the keyword codes cover ESG (Environmental, Social, and Governance), Social finance, Socially responsible investing (SRI), and Impact investing. Finally, the Sustainable Business "SB" keyword category emphasises a business approach that seeks long-term viability, integrating economic profitability with societal and environmental responsibility. Hence, Jun and Moon (2021) identified that the keyword codes cover Sustainability, Sustainable Business, and Sustainable Management.

In short, Jun and Moon (2021) offer an insightful overview of sustainability keywords in Sustainable Education at tertiary institutions. Categorising the keywords codes under specific categories provides a structured understanding of the various facets of sustainability in the context of Business Education at the tertiary level. For this study, related keywords were added to each sustainability theme by reviewing relevant articles, extant literature, and course information. Table 1 displays the updated sustainability keyword codes used to examine the unit aims and objectives, learning outcomes and contents of business courses. A score of '1' was assigned if a particular category's keyword codes were covered. Otherwise, the result for this Sustainability keyword category is a '0'. Lastly, the number of core units under each Sustainability keyword category was totalled up, and the coverage (in percentage) was computed. Subsequently, the outcome was analysed by courses, year of study, weightage in the syllabus, assessments used, and the depth and breadth of the topics covered in the curriculum, where the needed information is available.

FINDINGS AND DISCUSSION

The findings depend on the availability of the information from each university. Although it varies, there was a trend across the three-year degree courses and the two universities. Each finding is discussed below.

Alpha University Malaysia

This foreign branch campus offers nine Bachelor of Business (BBus) majors: Accounting, Accounting and Finance, Finance, International Business, Human Resource Management, Management, Marketing, an untagged discipline, and Management and Digital Media (M&DM). The first eight BBus courses consist of a common set of eight core (or common) units, indicating that the eight courses have a common fundamental structure in Business education, with differences emerging only in the specialised or disciplinary majors. Meanwhile, the Bachelor of Business (M&DM) course distinguishes itself by incorporating four additional core units while excluding a core unit in other BBus courses. The additional core units in this specialisation show a higher emphasis placed on management, especially within digital media. Except for the M&DM specialisation, all BBus courses comprise 24 units. While they share the same core units, each course is differentiated by some electives or specialised units. Compared to other BBus courses, the M&DM major has a more focused curriculum with fewer elective options. Besides the 11 Business core units, this specialisation consists of 12 Design core units and 1 unit that carries double credit.

Table 2. BBus (M&DM) course – Sustainability keywords by category and by year of study

Sustainability Keyword Category	Ethics "eth"	CSR "CSR"	Social Value "SV"	Environment "env"	Finance "fin"	Sustainable Business "SB"	Total
Year 1 Units	2	1	2	1	2	-	8
Year 2 Units	1	1	2	1	-	1	6
Year 3 units	2	1	2	1	-	2	8
Total	5	3	6	3	2	3	

Table 3. Other BBus courses – Sustainability keywords by category and by year of study

Sustainability Keyword Category	Ethics "eth"	CSR "CSR"	Social Value "SV"	Environment "env"	Finance "fin"	Sustainable Business "SB"	Total
Year 1 Units	2	1	2	1	2	-	8
Year 2 Units	-	-	-	-	-	-	0
Year 3 units	2	-	-	-	-	2	4
Total	4	1	2	1	2	2	

Tables 2 and 3 provide an overview of BBus integration of sustainability elements by keyword categories.

Based on Table 2 and 3, it is obvious that double major degree from two different discipline would cover more

- *The Ethics "eth" and the CSR "CSR" Keyword Categories*

Based on BBus (M&DM) data, "eth" keyword codes have been integrated into 5 out of its 11 core units, accounting for a 45.5% coverage. For eight other BBus courses, the "eth" keyword codes have been covered in 4 out of the 8 core units, resulting in a keyword coverage rate of 50%. On the other hand, the "CSR" keyword codes are present in 3 out of the 11 core units of the BBus (M&DM) course, yielding a coverage of 27.3%. As for the rest of the BBus courses, the "CSR" keyword category has only been covered in 1 out of the 8 core units, which translates to a coverage of 12.5%. Notably, both "ethics" and "CSR" themes are covered in 3 core units of M&DM specialisation. For the other eight BBus courses, one core unit thoroughly covers ethical and CSR-related themes. Across all 9 BBus courses, the focus on the "eth" theme is significantly higher than "CSR".

- *The Social Value "SV" and Environment "env" Keyword Categories*

The M&DM course data shows its noteworthy integration of the Social Value "SV" keyword category, which was incorporated into 6 out of the 11 core units (or a coverage of 54.5%). The presence of the "SV" theme in more than half of the course's core units signifies the important role of businesses in effecting societal changes. As for the other eight BBus courses, only 2 out of the 8 core units covered keyword codes from the Social Value (SV) category, resulting in a coverage rate of 25%. Meanwhile, the Environment "env" keyword code appears in 3 of the 11 core units of the M&DM course, representing 27.3% of integration. Incorporating the "env" theme in this Business discipline indicates a growing recognition of environmental considerations that extends beyond conventional environmental science curricula. For the rest of the BBus courses, the "env" keyword codes have only been covered once in the 8 core units, resulting in a coverage of 12.5%. Notably, three core units of the M&DM course have integrated both the "SV" and "env" keyword codes into the curricula. One Management core unit in the other eight BBus courses stands out as it covers both "SV" and "env" themes.

- *The Finance "fin" and Sustainable Business "SB" Keyword Categories*

Figure 2. Keyword category combination – Core units coverage [BBus (M&DM)]

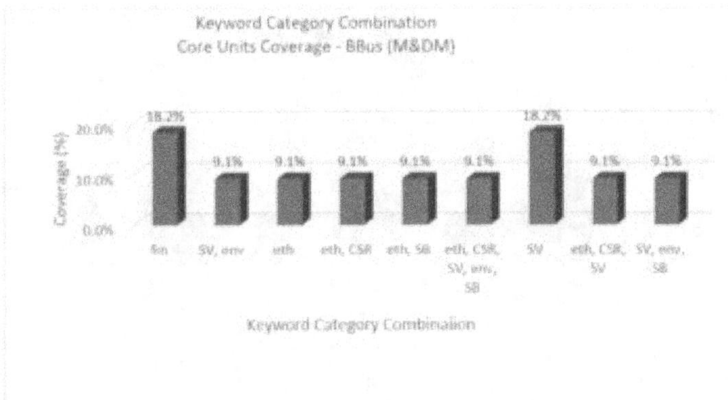

This Finance "fin" keyword has been included in 2 out of the 11 core units of the M&DM course, translating to an integration rate of 18.2%. Likewise, 2 out of the 8 core units (25%) of the rest of the BBus courses have integrated the keyword codes from the "fin" category. From M&DM course information, keyword codes under the "SB" theme have been integrated into 3 of the 11 core units, resulting in a coverage of 27.3%. The "SB" theme coverage indicates that strategic sustainability concepts are increasingly incorporated into Business education. For the rest of the BBus courses, 2 out of the 8 core units (25%) have incorporated keyword codes from the "SB" category.

In sum, the sustainability coverage of M&DM is highest in the "SV" theme (inclusion by 54.5% of the core units), followed by "eth" (covered by 45.5% of its core units). The "CSR", "env", and "SB" themes share the same mid-level prominence with a 27% integration into the core units, respectively. Notably, a 100% coverage of the six Sustainability keyword themes across the core units of M&DM highlights the increasing importance of sustainability. Even though each core unit may have a different focus, all of them integrate sustainability to a certain extent. The result is comparable to Leal Filho et al. (2021), who reported that sustainability elements are being incorporated in more than 80% of the sampled universities in Latin American institutions.

As for the other eight BBus courses, the most prominent sustainability keyword category appears to be "eth" (integration into 50% of the core units), followed by "SV", "fin", and "SB" themes (incorporated by 25% of the core units, respectively). Notably, all six Sustainability keyword themes have been incorporated into the core units of these eight BBus courses, resulting in a 100% coverage rate. By providing students with a multidimensional perspective, the course contributes to developing future leaders with a greater awareness of sustainability. Overall, the findings offer valuable insights into the dynamic nature of Business education at this branch campus, where sustainability is not merely a supplement but a fundamental or core aspect of the curriculum. The multidimensional approach to sustainability inclusion in its courses helps ensure that students are better prepared to face the sustainability challenges of the business world.

Next, Figure 2 shows the keyword category combinations for BBus (M&DM). Specifically, the theme "fin" surfaced as a significant theme, covered by 18.2% of the core units. It is matched in prominence by the theme "SV" which likewise stands alone and integrates into 18.2% of the core units.. For this course, the keyword category combinations that emerged were "SV, env", "eth, CSR", "eth, SB", "eth, CSR, SV, env, SB", "eth, CSR, SV", and "SV, env, SB", which have an equal core unit coverage of 9.1%.

Figure 3. Keyword category combination – Core units coverage (other BBus courses)

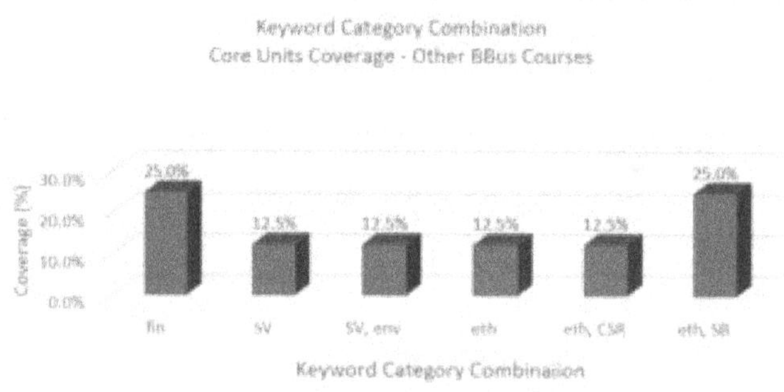

These keyword category combinations signify a wide-ranging integration of sustainability themes into the common units, spanning from finance or SV standalone theme to keyword category combinations encompassing ethics, CSR, social value, environment and sustainable business.

Figure 3 shows the keyword category combinations for other BBus courses. Again, the theme "fin" was noted to be a standalone theme, accounting for a relatively higher 25.0% integration by the core units. Meanwhile, combined keyword categories that surfaced include "eth, SB" which was integrated into 25.0% of the core units. Other combinations of sustainability themes that emerged were "SV, env" and "eth, CSR" with core units coverage of 12.5% . This data signifies the significance of the intersection of ethics and sustainable business ("eth, SB"), as well as ethics and CSR ("eth, CSR").

Meanwhile, Tables 2 and 3 also provide structured insights into the presence of sustainability elements by level of study. For the first eight BBus Courses, two Year 1 core units have each integrated the keyword themes "eth", "SV", and "fin", recognising the significant roles of ethical challenges, finance considerations, and the social value of sustainable business practises. Meanwhile, no Year 2 units have been designated as core units for these courses, whilst two Year 3 units have been identified as core units. Specifically, the Sustainability themes "eth" and "SB" were incorporated into these two core units, making these themes the focus of sustainability.

As for the M&DM course, the integration of sustainability at first-year common units ranges from finance-related theme "fin" to ethics "eth", CSR "CSR", social value "SV", and environment "env", highlighting the multifaceted nature of sustainability in this course. There were two units, each integrating the "eth", "SV," and "fin" keyword themes, making them the popular sustainability theme in Year 1. At level 2 of the study, all sustainability themes, except "fin", have been incorporated into at least one of the core units, recognising the significant role sustainable business practices play. For Year 2's core units, the "SV" category stands out as sustainability's focal point, given its incorporation into two core units. At the third-year level, there were two units each that included the "eth", "SV", and "SB" categories, making them the popular sustainability themes. In essence, incorporating five distinct Sustainability themes across all levels of study indicates the Business School's commitment to providing students with a broader grasp of sustainability in today's business context. Research on value formation has demonstrated that individuals tend to establish their value system early in life and during significant transitional phases, such as the college years. Once these values are set, they tend to grow more resistant to change as time progresses. Hence, scholars advocate for incorporating sustainability principles within

Table 4. Bachelor of business and commerce courses' core and capstone units – Sustainability keyword coverage (%)

Sustainability Keyword Code	Ethics "eth" (%)	CSR "CSR" (%)	Social Value "SV" (%)	Environment "env" (%)	Finance "fin" (%)	Sustainable Business "SB" (%)
	15.4	23.1	53.8	15.4	15.4	30.8

Business Education (Sidiropoulos, 2014; Triana, 2022). In this regard, integrating sustainability across different levels of BBus course study is noteworthy and essential. The above discussion demonstrates the current state of sustainability integration into the business education at this branch campus, which acknowledges its importance in nurturing a resilient, well-informed, and socially conscious generation.

Beta University Malaysia

Sustainability Keywords Category Analysis

In terms of the analysis of sustainability keyword categories, 6 out of 7 core units contain at least one keyword category. As for the capstone units, students are required to take only one capstone unit from a choice of six. Based on the keyword category analysis, 4 out of 6 capstone units contain at least one keyword category.

As presented in Table 4 (below), the keyword 'Social Value' (SV) accounts for the highest percentage of coverage at 53.8%, with 7 out of 13 units integrating this keyword. Meanwhile, 'Sustainable Business' (SB) is covered in 4 units, resulting in a keyword coverage rate of 30.8%. Lastly, the keyword 'Corporate Social Responsibility' (CSR) is present in 3 units, generating a coverage of 23.1%. From the data gathered, the keyword 'ethics' has been integrated into 2 out of the 13 units, including core and capstone units, accounting for a 15.4% coverage. Similarly, this percentage holds for the 'Environment' (env) and 'Finance' (fin) keywords.

Furthermore, Figure 4 below provides a detailed exploration of the sustainability keyword combinations within these 10 units. Notably, one of the capstone units, 'Ethics and Sustainability in a Business Environment,' includes five keyword categories: 'eth, CSR, SV, env, SB,' (7.7%) in its curriculum.

Figure 4. Sustainability keywords combined for Beta University

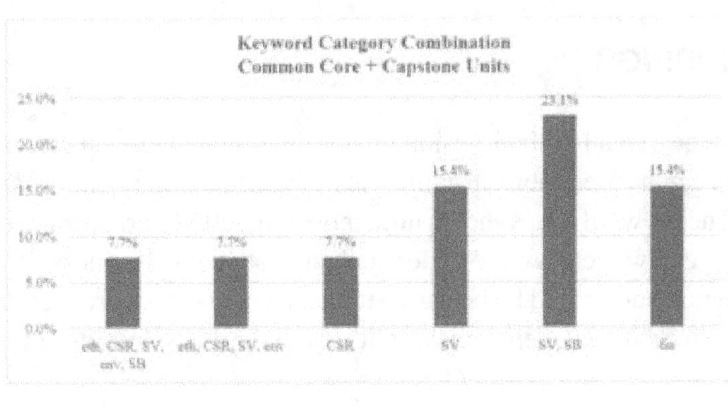

Table 5. Sustainability keywords by category and by year of study

Sustainability Keyword Category	Ethics "eth"	CSR "CSR"	Social Value "SV"	Environment "env"	Finance "fin"	Sustainable Business "SB"	Total
Year 1 Units	1	2	3	1	2	1	10
Year 2 Units							NA – Level 2 units are discipline-based units
Year 3 units	1	1	4	1	-	3	10
Total	**2**	**3**	**7**	**2**	**2**	**4**	

Meanwhile, one of the core units, 'Introduction to Management,' integrates four keyword categories: 'eth, CSR, SV, env,' (7.7%) into its curriculum. Core units in Accounting and Finance incorporate the keyword category 'fin' into their respective curricula, accounting for 15.4% of the course curriculum. Moreover, the Business Law core unit and the Global Issues in Accounting capstone unit integrate the keyword category 'sv' (15.4%). Additionally, three units (one core: Microeconomics and two capstone units: Experiential Learning Project and International Study Program in Business) feature a combination of two keyword categories, 'SV, SB,' in their curriculum. This combination of keywords constitutes the highest percentage (23.1%) within the course curriculum, encompassing both core and capstone units. Lastly, the Marketing core unit integrates the keyword category 'CSR' in its curriculum, which counts for 7.7%.

Apart from that, Table 5 (below) demonstrates insights into the presence of sustainability elements across different levels of study. In the BBus/BCom Courses, one Year 1 core unit and one Year 3 capstone unit have each integrated the keyword themes "eth" and "env," recognizing the significant roles of ethical challenges and environmental considerations. Moreover, three Year 1 core units and four Year 3 capstone units have embedded components related to societal values, denoted by the keyword "SV." Additionally, one Year 1 and three Year 3 units have integrated the " SB " component for sustainable business. However, for the corporate social responsibility element, 'CSR' is embedded in two Year 1 core units and one Year 3 capstone unit. In contrast, the finance element, 'fin,' is only embedded in two Year 1 units. The analysis highlights the importance of emphasising societal values across Year 1 and Year 3 units. Lastly, none of the Year 2 units have been designated as core or capstone units for these courses as they fall under the discipline major units.

SUMMARY OF FINDINGS

Overall, there seems to be a trend in the infusion of sustainability education across a three-year degree. These two universities emphasised more sustainability education in their first year and the final year as presented in Table 6. The keyword codes shown in year one suggested a broader coverage of sustainability education in the entry year, while keyword codes such as sustainable business are being emphasised in the later year. This phenomenon could be because students would have a broader understanding of business practices and the environment, hence making it more relevant to introduce topics on sustainability

Table 6. Overview of keyword codes and units of study over a three years degree of the two business schools

Alpha:	Year One	Year Two	Year Three
Units with sustainability in the aims and objectives or content/outline	6 units	2 units	4 units
Keywords	fin, SV, env, eth, CSR	eth, CSR, SV, env, SB	eth, SB, CSR, SV, env
Beta:	**Year One**	**Year Two**	**Year Three**
Units with sustainability in the aims and objectives or outline	6 units	None	4 units
Keywords	fin, SV, env, SB, eth, CSR,	None	eth, SB, CSR, SV, env

that have more overall coverage. Interestingly, the keywords found in years one and three across the two universities are substantially the same. Such similarity could be due to their curriculum design and development from the same country (Mori Junior, et al, 2019). As seen in this study, the movement to incorporate sustainability in higher education in Australia can be seen as an asserted effort nationwide, harvesting promising outcomes.

CONCLUSION

This chapter traced sustainability education in the business schools of two Australian university branch campuses in Malaysia. The unit course details have been analysed thematically according to sustainability education literature. The Bachelor of Business and Commerce courses at these two Australian branch campus universities have incorporated sustainability into their curriculum at various levels of study but with a higher emphasis on year-one units. The keywords were identified from the aims and objectives, unit learning outcomes, unit contents, and brief descriptions of the units. It was found that there were additional Sustainability Keyword Codes and Categories beyond those identified by Jun and Moon (2021). Such findings could be attributed to the quality assurance frameworks these universities must adhere to: the Australian Qualifications Framework (AQF) and the Malaysian Qualifications Framework (MQF). On the one hand, this could mean more stringent compliance effort is needed for the Australian branch campuses in Malaysia. However, from the stakeholder's perspective, such a requirement could highlight that these institutions have a higher role in promoting sustainability education in Malaysia. Therefore, the curriculum of these universities could be a model for other universities to follow suit, fulfilling the call for good corporate citizenship in society.

This unprecedented study provided an understanding of the degree of sustainability education incorporated into Malaysia's Business Schools of Australian Branch Campuses curriculum. The outcome would help policy-makers formulate and implement suitable policies to expand sustainability education in their institutions. The result could also be helpful to researchers in furthering comparative studies on this topic with other universities in the same country or other countries. In addition, the outcome could help educationists further enhance the curriculum to produce graduates with a higher knowledge and positive attitudes toward sustainability. Furthermore, the outcomes of this study could be used as a benchmark for some universities in the infant stage of implementing sustainability in their curriculum. Academics and curriculum developers could infuse sustainability education in more year-one units to build the foundations for advanced application in the later years.

The common pitfall of any case study is the generalisability of its results. These two universities offer various disciplines in their Bachelor of Business and/or Commerce courses. However, for consistency, the comparison and review were limited to the common core units (consisting of capstone units) that all students must complete for these courses. Universities could infuse sustainable education into many units of study and spread across multiple disciplines. Future research could expand the scope of the study by increasing the number of universities and courses of study to provide a broader understanding of the depth and breadth of sustainability education in higher education institutions. The analysis of this study is only confined to the keywords identified from the aims and objectives, unit learning outcomes, unit contents, and brief descriptions of the units. Future research could incorporate an impact study on the students using surveys and interviews, expanding the work of Figueiró and Raufflet (2015), Michael et al. (2020) and Saleem et al. (2023). Such a study would be able to determine the effectiveness of sustainability education and provide feedback on the suitability of the curriculum design. Societal context is crucial in sustainability awareness and plays a significant role in sustainability education. Countries with a strong knowledge of sustainability would incorporate more elements of sustainability education into their degree. Future studies could explore the curriculum of those in those countries at the forefront of sustainability education to provide valuable benchmarks and targets to work towards.

REFERENCES

Affolderbach, J. (2022). Translating green economy concepts into practice: Ideas pitches as learning tools for sustainability education. *Journal of Geography in Higher Education*, *46*(1), 43–60. doi:10.1080/03098265.2020.1849063

Anastasiadis, S., Perkiss, S., Dean, B. A., Bayerlein, L., Gonzalez-Perez, M. A., Wersun, A., Acosta, P., Jun, H., & Gibbons, B. (2021). Teaching sustainability: Complexity and compromises. *Journal of Applied Research in Higher Education*, *13*(1), 272–286. doi:10.1108/JARHE-02-2020-0029

Andrási, G., Körtvési, D., & Szegedi, K. (2022). Developing Ethical Future Managers Embracing Sustainability: Case Study of Budapest Business School. *International Journal of Web-Based Learning and Teaching Technologies*, *17*(8), 1–19. doi:10.4018/IJWLTT.300786

Baer, H. A. (2022). How Environmentally Sustainable Is the Internationalisation of Higher Education? A View from Australia. In K. Bjørkdahl & A. S. Franco Duharte (Eds.), *Academic Flying and the Means of Communication* (pp. 103–132). Palgrave Macmillan. doi:10.1007/978-981-16-4911-0_5

Barber, N. A., Wilson, F., Venkatachalam, V., Cleaves, S. M., & Garnham, J. (2014). Integrating sustainability into business curricula: University of New Hampshire case study. *International Journal of Sustainability in Higher Education*, *15*(4), 473–493. doi:10.1108/IJSHE-06-2013-0068

Brunnquell, C., Brunstein, J., & Jaime, P. (2015). Education for sustainability, critical reflection and transformative learning: Professors' experiences in Brazilian administration courses. *International Journal of Innovation and Sustainable Development*, *9*(3-4), 321–342. doi:10.1504/IJISD.2015.071858

Bryant, S. M., Cullen, P. G., & Iannarelli, J. E. (2022). A Vison for Management Education: The AACSB Perspective. In M. R. Fellenz, S. Hoidn, & M. Brady (Eds.), *The Future of Management Education* (pp. 225–243). Routledge. doi:10.4324/9781003095903-17

Charroin, L., Fortin, B., & Villeval, M. C. (2022). Peer effects, self-selection and dishonesty. *Journal of Economic Behavior & Organization*, *200*, 618–637. doi:10.1016/j.jebo.2022.06.024

Chedrawi, C., Howayeck, P., & Tarhini, A. (2019). CSR and legitimacy in higher education accreditation programs, an isomorphic approach of Lebanese business schools. *Quality Assurance in Education*, *27*(1), 70–81. doi:10.1108/QAE-04-2018-0053

Claro, P. B., & Esteves, N. R. (2021). Teaching sustainability-oriented capabilities using active learning approach. *International Journal of Sustainability in Higher Education*, *22*(6), 1246–1265. doi:10.1108/IJSHE-07-2020-0263

Deer, S., & Zarestky, J. (2017). Balancing profit and people: Corporate social responsibility in business education. *Journal of Management Education*, *41*(5), 727–749. doi:10.1177/1052562917719918

DeSombre, E. R. (2018). *Why good people do bad environmental things*. Oxford University Press. doi:10.1093/oso/9780190636272.001.0001

Dickson, M. A., Eckman, M., Loker, S., & Jirousek, C. (2013). A model for sustainability education in support of the PRME. *Journal of Management Development*, *32*(3), 309–318. doi:10.1108/02621711311318337

Driscoll, C., Price, S., McKee, M., & Nicholls, J. (2017). An assessment of sustainability integration and communication in Canadian MBA programs. *Journal of Academic Ethics*, *15*(2), 93–114. doi:10.1007/s10805-017-9275-0

Etse, D., & Ingley, C. (2016). Higher education curriculum for sustainability: Course contents analyses of purchasing and supply management programme of polytechnics in Ghana. *International Journal of Sustainability in Higher Education*, *17*(2), 269–280. doi:10.1108/IJSHE-07-2015-0121

Figueiró, P. S., & Raufflet, E. (2015). Sustainability in higher education: A systematic review with focus on management education. *Journal of Cleaner Production*, *106*, 22–33. doi:10.1016/j.jclepro.2015.04.118

Gal, A. (2023). From recycling to sustainability principles: The perceptions of undergraduate students studying early childhood education of an education for sustainability course. *International Journal of Sustainability in Higher Education*, *24*(5), 1082–1104. doi:10.1108/IJSHE-05-2022-0165

Galleli, B., Teles, N. E. B., Freitas-Martins, M. S., Semprebon, E., & Hourneaux, F. Junior. (2022). Sustainability in management undergraduate courses: Mapping the Brazilian higher education institutions. *International Journal of Sustainability in Higher Education*, *23*(7), 1628–1647. doi:10.1108/IJSHE-03-2021-0109

Gil-Doménech, D., Magomedova, N., Sánchez-Alcázar, E. J., & Lafuente-Lechuga, M. (2021). Integrating sustainability in the business administration and management curriculum: A sustainability competencies map. *Sustainability (Basel)*, *13*(16), 9458. doi:10.3390/su13169458

Gottardello, D., & Pàmies, M. M. (2019). Business School Professors' Perception of Ethics in Education in Europe. *Sustainability (Basel)*, *11*(3), 608. doi:10.3390/su11030608

Hay, R., & Eagle, L. (2020). Impact of integrated sustainability content into undergraduate business education. *International Journal of Sustainability in Higher Education, 21*(1), 131–143. doi:10.1108/IJSHE-05-2019-0174

Hill, L. M., & Wang, D. (2018). Integrating sustainability learning outcomes into a university curriculum: A case study of institutional dynamics. *International Journal of Sustainability in Higher Education, 16*(4), 699–720. doi:10.1108/IJSHE-06-2017-0087

Jose, P. D. (2016). Sustainability education in Indian business schools: A status review. *AD-Minister, 28*(28), 255–272. doi:10.17230/ad-minister.28.13

Jun, H., & Moon, S. (2021). An analysis of sustainability integration in business school curricula: Evidence from Korea. *Sustainability (Basel), 13*(5), 2779. doi:10.3390/su13052779

KanwalK.MalikM.AkhtarA.RehmanF.KhanS. H.TabbasumA.HussainM. (2023). Investigating the effectiveness of sustainability integration strategies in business schools: A comparative study. *Journal of Xi'an Shiyou, 66*(4), 147-176. http://dx.doi.org/ doi:10.17605/OSF.IO/RSBXH

Kasch, J., Schutjens, V. A. J. M., Bootsma, M. C., Van Dam, F. W., Kirkels, A. F., van der Molen, M. K., Rimac, A., & Rebel, K. T. (2023). Distance and presence in interdisciplinary online learning. A challenge-based learning course on sustainable cities of the future. *Journal of Integrative Environmental Sciences, 20*(1), 2185261. Advance online publication. doi:10.1080/1943815X.2023.2185261

Kemper, J. A., Ballantine, P. W., & Hall, C. M. (2019). Combining the 'why' and 'how' of teaching sustainability: The case of the business school academics. *Environmental Education Research, 25*(12), 1751–1774. doi:10.1080/13504622.2019.1667959

Kim, E. (2023). Sustainable New Product Development for Ten Thousand Villages, a Fair-Trade Social Enterprise: Empowering Women and Economic Development through Problem-Based Service Learning. *Sustainability (Basel), 15*(8), 6452. Advance online publication. doi:10.3390/su15086452

Leal Filho, W., Amaro, N., Avila, L. V., Brandli, L., Damke, L. I., Vasconcelos, C. R., ... Salvia, A. (2021). Mapping sustainability initiatives in higher education institutions in Latin America. *Journal of Cleaner Production, 315*, 128093. doi:10.3390/su141912289

Martins, F. P., Cezarino, L. O., Liboni, L. B., Botelho Junior, A. B., & Hunter, T. (2022). Interdisciplinarity-Based Sustainability Framework for Management Education. *Sustainability (Basel), 14*(19), 12289. doi:10.3390/su141912289

Mburayi, L., & Wall, T. (2018). Sustainability in the professional accounting and finance curriculum: An exploration. *Higher Education. Skills and Work-Based Learning, 8*(3), 291–311. doi:10.1108/HESWBL-03-2018-0036

Michael, F. L., Sumilan, H., Bandar, N. F. A., Hamadi, H. A. N. A., Jonathan, V., & Nor, N. N. M. (2020). Sustainable development concept awareness among students in higher education: A preliminary study. *Journal of Sustainability Science and Management, 15*(7), 113–122. doi:10.46754/jssm.2020.10.011

Mori, R. Junior, Fien, J., & Horne, R. (2019). Implementing the UN SDGs in universities: Challenges, opportunities, and lessons learned. Sustainability. *Sustainability (New Rochelle, N.Y.), 12*(2), 129–133. doi:10.1089/sus.2019.0004

Morshidi, S., & Wan, C. D. (2022). Higher Education in Malaysia. In L. P. Symaco & M. Hayden (Eds.), *International Handbook on Education in South East Asia* (pp. 1–23). Springer International Handbooks of Education. doi:10.1007/978-981-16-8136-3_14-1

Obrecht, M., Feodorova, Z., & Rosi, M. (2022). Assessment of environmental sustainability integration into higher education for future experts and leaders. *Journal of Environmental Management, 316*, 115223. doi:10.1016/j.jenvman.2022.115223 PMID:35569358

Painter-Morland, M., Sabet, E., Molthan-Hill, P., Goworek, H., & de Leeuw, S. (2016). Beyond the curriculum: Integrating sustainability into business schools. *Journal of Business Ethics, 139*(4), 737–754. doi:10.1007/s10551-015-2896-6

Palinkas, L. A., Horwitz, S. M., Green, C. A., Wisdom, J. P., Duan, N., & Hoagwood, K. (2015). Purposeful Sampling for Qualitative Data Collection and Analysis in Mixed Method Implementation Research. *Administration and Policy in Mental Health, 42*(5), 533–544. doi:10.1007/s10488-013-0528-y PMID:24193818

PRME. (2021). *Principles for responsible management education.* https://www.unprme.org/about

Saldana, J. (2013). *The coding manual for qualitative researchers* (2nd ed.). SAGE Publications Ltd.

Saleem, A., Aslam, S., Sang, G., Dare, P. S., & Zhang, T. (2023). Education for sustainable development and sustainability consciousness: Evidence from Malaysian universities. *International Journal of Sustainability in Higher Education, 24*(1), 193–211. doi:10.1108/IJSHE-05-2021-0198

Shanika, V. G., Jayawickrama, T. S., & Perera, B. A. K. S. (2023). The Adaptability of Sustainability Education in Quantity Surveying Degree Programs in Sri Lanka: The Industry Perception. *International Journal of Construction Education and Research, 19*(3), 299–320. doi:10.1080/15578771.2022.2100017

Sharma, U., & Stewart, B. (2022). Enhancing sustainability education in the accounting curriculum: An effective learning strategy. *Pacific Accounting Review, 34*(4), 614–633. doi:10.1108/PAR-02-2021-0029

Sidiropoulos, E. (2014). Education for sustainability in business education programs: A question of value. *Journal of Cleaner Production, 85*, 472–487. doi:10.1016/j.jclepro.2013.10.040

Stake, R. E. (2005). Qualitative Case Studies. In N. K. Denzin & Y. S. Lincoln (Eds.), *The SAGE Handbook of Qualitative Research* (3rd ed., pp. 443–466). Sage Publications.

Symaco, L. P., & Wan, C. D. (2017). Development of Higher Education in Malaysia: Issues and Challenges. In M. Samuel, M. Y. Tee, & L. P. Symaco (Eds.), *Education in Malaysia, Education in the Asia-Pacific Region: Issues, Concerns and Prospects* (pp. 53–66). Springer Singapore Pte. Limited. doi:10.1007/978-981-10-4427-4_4

Tijsma, G., Horn, A., Urias, E., & Zweekhorst, M. B. M. (2023). Training students in inter- and transdisciplinary sustainability education: Nurturing cross-faculty staff commitment and continuous community collaboration. *International Journal of Sustainability in Higher Education*, *24*(4), 765–787. doi:10.1108/IJSHE-02-2022-0049

Triana, N. (2022). The need for sustainability and CSR in undergraduate business education. *Journal of Undergraduate Research (Gainesville, Fla.)*, *24*. Advance online publication. doi:10.32473/ufjur.24.130792

Wan, C. D., & Abdullah, D. (2021). Internationalisation of Malaysian Higher Education: Policies, Practices and the SDGs. *International Journal of Comparative Education and Development*, *23*(3), 212–226. doi:10.1108/IJCED-08-2020-0052

Wersun, A., Dean, B. A., Mills, R., Perkiss, S., Acosta, P., Anastasiadis, S., Gibbons, B., Gonzalez-Perez, M. A., Heithaus, T., Jun, H., Mesicek, R. H., & Bayerlein, L. (2019). An exploration of student learning for sustainability through the WikiRate student engagement project. *International Journal of Management Education*, *17*(3), 100313. doi:10.1016/j.ijme.2019.100313

Weybrecht, G. (2017). Greening the business of business education. *Corporate Knights, 16*(4), 32-35. https://www.jstor.org/stable/26789227

Winter, J., Zhai, J., & Cotton, D. R. E. (2022). Teaching environmental sustainability in China: Opportunities and challenges for business and economics faculty in higher education. *Environmental Education Research*, *28*(2), 318–332. doi:10.1080/13504622.2021.2012560

Zguir, M. F., Dubis, S., & Koç, M. (2021). Embedding education for Sustainable Development (ESD) and SDGs values in curriculum: A comparative review on Qatar, Singapore and New Zealand. *Journal of Cleaner Production*, *319*, 128534. doi:10.1016/j.jclepro.2021.128534

Chapter 15
Integration of Conservation Studies in Library and Information Science Curriculum in the Eastern and Southern African Region

Thatayaone Segaetsho
https://orcid.org/0000-0001-7247-4377
University of Botswana, Botswana

ABSTRACT

Despite the eloquent strives to improve education in library and information sciences, the insidious challenges in preservation and conservation education and training continue to traumatize the world, especially in Africa. This chapter weaves together literature on preservation and conservation curriculum in library and information sciences education in the Eastern and Southern Africa Regional Branch of the International Council of Archives (ESARBICA) member states. The chapter then explores the challenges observed and proposes a sustainable framework for integrating conservation in the preservation curriculum in library and information studies education. The chapter concludes by offering recommendations for improvement.

INTRODUCTION

The preservation and conservation of classical information resources still continue to exist despite the emergent of disruptive technologies in the digital age. As Iyishu, Nkanu and Ogar (2013:36) pointed out "every generation desire to leave a mark of documented issues, concerns, ideas, discourse and events, and the only way to do this is by leaving a legacy in the form of documented matter". This implies that regardless of the new digital media, information holders still have to continue preserving the classical cultural heritage media such as graphic documented matter. Consequently, continuous education for

DOI: 10.4018/978-1-6684-9859-0.ch015

both classical and new digital information media is critical. Education platforms are, therefore, the key backbones of sustainable protection of our cultural and documentary resources. Preservation and conservation education is a well-established cadre in most developed countries such as the United States of America, Canada, United Kingdom, France, Germany, Spain, China and Australia. Robust research on preservation and conservation is observed with good research publications in the field. For example, in 2012 Krtalic and Hasenay (2012) explored a framework for comprehensive and successful preservation management in libraries. The study proposed a foundation that preservation should be based on five key components of 1) strategic and theoretical 2) economic and legal, 3) Educational, 4) technical and operational, and 5) cultural and social. The framework recognizes education as one of the key components of preservation management.

The history of preservation and conservation emerged from the craftsmanship of making and repairing objects. Traditionally that process was associated with engineering maintenance, and that involved the art of regular lubrication, replacement of worn parts, removal of damaged varnish or paint, repainting, and so on (Pye, 2016). Such prepositions were then mirrored by various challenges that prompted detailed understanding of those materials/objects that needed to be conserved. Consequently, this brought the need for shifting the practices and philosophy of conservation towards more scientific approaches. The current 21st century approaches on conservation are thus geared towards application of innovative Information and Communication Technologies (ICTs) in order to fully understand the chemical, mechanical, physical and biological agents of deterioration and address their effects.

Despite the long history and advancement in conservation, the insidious challenges on the lack of expertise still exist in the 21st Century. The challenges are now exacerbated by the newer defiance of technology and the new emerging industrial specialties. Consequently, educational platforms are continuously stumbling with how to address the increasing demands of industrial specialties. The demand on conservations kills and knowledge, especially in the field of Library and Information Science Education (LISE), is not an exception. Zervos (2012), observed that often the mistakes in the understanding and use of terminology, such as “non-rusting paper” (in analogy to non-rusting steel) instead of acid-free paper, is one of the key demonstrations of lack of teaching skills in conservation and preservation in LISE. In Africa, the lack of knowledge and skills in conservation is highly expressed in the context of the lack of preservation expertise in libraries and archives. Msibi (2015) highlighted the challenges on preservation and conservation education, among others, including the lack of skilled manpower resulting in backlogs of collections that need treatment, prioritization and financial constraints, poor and irrelevant curricula, and lack of proper structures to widen knowledge and skills on preservation. Of recent, the new trends of lack of digital preservation skills, lack of understanding and integration of digital information management strategies have been raised as a concern in education curricula (Segaetsho & Moloi, 2019). The emergent of digital era, therefore, even complicates the already existing classical challenges of education in preservation and conservation.

It is in common ground throughout the world that intellect is a legitimate force in the construction of erudition, and by extension, all vital realms of human endeavor are guided by intellect. The United Nations (UN) agenda for 2030, ‘Transforming our world: the 2030 Agenda for Sustainable Development’ (A/RES/70/1), places a great priority on education through Strategic Goal 4. The Strategic Goal 4 points out that it is critical to ‘ensure inclusive and equitable quality education and promote lifelong learning opportunities for all’. As part of the milestones in achieving the agenda for Strategic Goal 4, the United Nations held a Transforming Education Summit in New York, from the 16, 17 and 19 September 2022, in which the summit clearly expressed that *“education is a human right and a foundation for peace, tolerance, other human rights and sustainable development*" (UN, 2022:01). These prepositions and perceptions of the world on education are also critical in preservation and conservation field. The continued demands on education are explicitly raised in UNESCO Institute for Lifelong Learning (UIL) Policy Brief 14, 'Transforming higher educat1ion institutions into lifelong learning institutions' (UNESCO, 2022:01) expressing that

In the face of the challenges of the twenty-first century, including technological advancements, climate change, demographic shifts, globalization and the transforming world of work, higher education institutions(HEIs) must be willing to expand their LLL opportunities to reach a greater range of individuals and respond better to societal needs.

The trace of preservation training efforts in the ESARBICA are observed in the 1980s ESARBICA 1980/5 resolutions in which it was resolved for members to consider technical training in conservation and reprographics (Mnjama, 2007). In 2005 members of ESARBICA realized that there was still scarcity of preservation professionals in the region hence an urge for national archival institutions, stakeholders and educators to partner in setting regional conservation training facilities for the benefit of ESARBICA members. Members were further urged to integrate conservation in national archival policies (Mnjama, 2007). Despite these previous efforts made by ESABIRCA, the dire need for preservation and conservation trainingand capacity building inLISE in the Eastern and Sothern Africa Regional Branch of the International Council of Archives (ESARBICA), still remain unsolved as observed by Shameenda and Kanyengo, (2012) in three universities in Zambia. Post the efforts by ESARBICA, a significant number of solutions have been proposed by various scholars and professionals in the region, with an ultimate goal of addressing the various challenges in preservation. Among others, various studies proposed that to solve preservation issues, especially on education, academic institutions should increase the number of undergraduate and graduate studies in LISE, re-evaluate skills essential to success, include modern materials in curriculum to cover new aspects of digital equipment, audio visuals, and conduct continuous updating of curriculum (Murray, 2002: 2006; Ngulube, 2007; 2007b; Katuu, 2015; Shameenda & Kanyengo, 2012). Segaetsho (2018) recommended that tertiary institutions, universities and consultants should establish courses in preservation, and if need be, at least short courses could be designed to assist on revamping education in preservation.

Despite a tremendous increase in quality and Outcome Based Education (OBE) throughout the world, the LISE still trails behind when it comes to preservation and conservation studies in the ESARBICA. The key question is that; is it not time to start integrating conservation studies in LISE? If yes, then what are the frameworks that could be adopted for a development of a curriculum in LISE? In order to address this key question, this chapter explores content/principles relevant to conservation in LISE in the ESARBICA. Using the findings, the chapter then develops a framework for the integration of conservation studies in LISE in ESARBICA. The chapter ends by providing key conclusions and recommendations for improving conservation studies in curriculum development for LISE.

PRESERVATION AND CONSERVATION EDUCATION IN LIBRARIES AND ARCHIVES

The practice of conservation in libraries and archives worldwide can be linked with the origins of library, archives and records management profession. Application of conservation in education curricula varies from one country to another. The next subsections discusses the international outlook on the establishment of conservation education, followed by an African perspective focusing on ESARBICA.

Definitions on Preservation and Conservation

The history of preservation goes a long way back with a common definition that acknowledges the terms 'management' and 'technical operations' as key in preservation. A significant number of researchers in ESARBICA (Abdulrahim, 2016; Ngulube, 2007; Kootshabe & Mnjama, 2014) continuously define preservation by using the International Federation of Library (IFLA) (1998; 2019) definition that has been adopted from Harvey (1994). The definition proposes that preservation "includes all the managerial and financial considerations, including storage and accommodation provisions, staffing levels, policies, techniques and methods involved in preserving library and archive materials and information contained in them" (IFLA, 1998:05). It is a function of ensuring long the term access to collections through implementation of managerial and technical considerations of storage and accommodation provisions, staffing levels, policies, techniques, and interventive conservation treatments. However, the above definition of preservation from IFLA and Harvey seems not to be comprehensive enough given the innovative and technological advancements that have happened. The definition does not address all aspects of conservation and restoration work that takes places in libraries and archival institutions and often the term is confused or used interchangeably with the term conservation. To address this limitation, Segaetsho (2020:250) proposed that preservation could be defined broadly as "the management process that integrates both natural sciences, humanities, education and social sciences principles and concepts for the purpose of the protection of valuable library and archival resources/materials for the short and long-term access and use." The definitions of preservation and conservation still need further academic and professional discourse for clearer theoretical understanding and distinction between the two terms.

On the other hand, the concept of conservation emerged from the ability to fully understand the structure, content, materials, and the complexity of the science of mechanical, physical, chemical and biological composition. Pye (2016) noted that a number of literatures (Keene, 2008; Pye, 2008; Saunders et al, 2008) discuss the foundation of conservation in two key goals. The first goal is to limit the rate of natural decay, which may happen in heritage collections leading to loss of authenticity and information on objects. The second goal is to halt deterioration and make objects physically and intellectually accessible through the use of interventive measures. Internationally, conservation is understood as "those specific policies and practices involved in protecting library and archive materials from deterioration, damage and decay, including the methods and techniques devised by technical staff" (Harvey, 1994:6). Conservation in libraries and archives can thus be defined as a profession that is devoted to the preservation of cultural property through application of various interventive techniques/methods such as examination, documentation, treatment, and preventive care supported by research and education. Another term that is used loosely and interchangeably with conservation is restoration. Restoration means the action of repair that is conducted in materials to stabilize them for continued access. As indicated by Roberts (2021:01) restoration "attempts to return an artifact back to its original condition while in contrast, conservation attempts to preserve an artifact in its current condition." Australian Institute for the Conservation of Cultural Material (2022:01) expresses that restoration is all about "treatments that enhance the interpretation of cultural heritage, e.g. in painting losses in an oil painting so that the original appearance of the image is maintained". They further explain that involve the "reassembly of displaced components, removal of extraneous matter or integrating new materials or components in order to stabilise and strengthen the original artefact".

Using the above given broad definitions of preservation and conservation, the next section reviews an international outlook of preservation and conservation curriculum in library and information science field.

An International Outlook of Preservation and Conservation Curriculum

The practice of conservation education throughout the world is somewhat inconsistent. Countries develop conservation programmes as per their need and the programmes are spread across various departments or faculties. It is common practice that conservation curriculums are mostly developed under faculties of humanities, especially the arts and culture departments. For instance, in the UK, at the Camberwell College of Arts, West Dean College, the conservation curricula was developed within the Arts Departments, The Northumbria University, places conservation programmes within the department of arts as a standalone programme and the university does not have library and information studies offered (North Umbria University, 2023; West Dean College, 2023). The West Dean College of Arts and Conservation specializes in conservation programmes mainly at postgraduate level and short courses. The Cardiff University Postgraduate Prospectus 2024 showed that the university offers both undergraduate and graduate programmes in Conservation of Objects in Museums and Archaeology and they are offered under the College of Arts, Humanities and Social Sciences within the History, Archaeology and Religion (Cardiff University, 2023). From the observed information on UK universities obtained from their websites, it is evident that most UK conservation programmes do not necessarily fall within the library and information studies programmes. However, a list of library and information studies courses in UK is provided by the Historical Association (2022), among others including University College London, Leeds Beckett University, Loughborough University, Department for Information Science, Manchester Metropolitan University, Northumbria University, University of Sheffield, University of the West of England, City University London etc. Some of these universities have preservation courses named differently such as collections care, collection management, or preservation as part of the modules that deals with preservation.

Another typical example on conservation education is a study by Zervos (2012) on 'Teaching Conservation and Preservation conducted at the Department of Library Science and Information Systems of the TEI of Athens' in Greece. The findings of that study revealed that the conservation programme is taught in the 6th semester of the 8 semesters that comprise the length of studies, with 3-hour theory and 2-hourworkshop components (65 hours is total). The study further revealed that the curriculum covered topics on an introductory to preservation, conservation and restoration, objects and materials, aging and degradation, conservation and preservation. The conservation aspect is a well detailed programme that covers:

- Definition of terms- preservation, restoration, conservation, aim sand principles of conservation, methods and stages of paper conservation treatments

- Examination- methodology and criteria for the evaluation of the suitability of conservation treatments, insect extermination, environmental monitoring of carbon dioxide and nitrogen treatments, obsolete methods, sterilization and fumigation.
- Treatment- Cleaning dry and wet, methods of chemical stabilization, chemistry of deacidification; various experimental methods, treatment of iron gall ink corrosion, consolidation and strengthening, mechanical stabilization, drying of wet books and archival material.
- Conservation of bookbinding- restoration of functionality, conservation of leather and parchment, conservation of modern media objectives, cleaning and consolidation, specific issues (baking of tape, mechanical restoration of film and tape, etc.) and limitations and drawbacks of conservation.
- Paper and book conservation workshop infrastructure- preservation quality materials, and health considerations
- Documentation of conservation- rationale, methods, and principles of conservation management.

Other European universities that cover LISE include the University of Malta in Malta which offers a preservation and conservation courses in its information and knowledge management degrees as an elective named LIS2611 Preservation Studies. In Austria the Danube University Krems offers a specific programme on 'Interdisciplinary Methods in Graphic Art, Book and Document Conservation' (University for Continuing Education, Krems, 2016). According to the university's website, the programme is specifically targeted to restaurators, archivists, librarians, collection managers, public funding agencies, art historians, researchers in the fields of codicology, historians and relevant natural sciences professionals.

Preservation history in Australia goes as far as the early 1970s. For example in AICCM, Pearson, Lyall, Sloggett, and Cook, (2011) expressed that presentations in preservation and conservation goes back to the first AICCM conference which was held in Canberra in 1976 at the Australian National University. When it comes to libraries and archives, Paul Wilson (1992) stated that the first course specific to libraries preservation in Australia was offered through The Bachelor of Information Studies (Conservation) with a devotion to the preservation administration and conservation of library and archive materials. The International Education Specialist (2022) provides a list of international courses in Information Science and Librarianship courses across Australia and elsewhere. Specific courses observed in the internet that deals with conservation mainly fall within the cultural heritage management sector such as the Master of Museum and Heritage Practice at Victoria University of Wellington and the graduate Certificate in Heritage Materials Conservation at University of Canberra. The Curtin University also offers both undergraduate and graduate specific course on preservation through the CUR-INFM335-2022 Conservation and Preservation module.

Similar trends were also observed in the United States of America with universities placing conservation as an entity within the humanities or applied science divisions. In addition, a significant number of specifically established institutions for preservation and conservation profession and training are observed across the various states. For example, the Northeast Document Conservation Center (NEDCC), which was initially an independent conservation laboratory, specializes exclusively in the conservation and preservation of paper- and film-based collections. In addition to professional conservation treatment of various graphic materials, the NEDCC offers training programmes, named 'Preservation 101', which cover sessions on; introduction to preservation; building and environment; caring for collections, care and handling of photographs, media collections, and reformatting and digitization. The course also covers the emergency preparedness, building a preservation program, and conducting preservation assessment reviews (NEDCC, 2021). The introductory course thus bridges knowledge gaps in preservation and conservation through provision of basics on theory and practice of preserving library and archival collections. Although some of the conservation programmes in USA are developed under departments such as archaeology and history, students from library and information sciences often take the courses as electives. For instance, in 2008, a conservator, Ellen, recruited students from library and archive materials into her preservation and management classes after joining the UCLA Depart of Information Studies (Center for Curating the Archives, 2021).

A significant number of institutions in Canada offer training in LISE. For example the Algonquin College in Ottawa, the University of British Columbia iSchool, University of Manitoba, University of Toronto iSchool, McGill University and many others offers graduate archival education programmes (The Association of Canadian Archivist, 2022). The programmes offered by the universities does emphasizes on preservation. For example the University of Toronto emphasizes that the prgramme learning outcomes in the areas of study are for long term preservation. Other universities in Canada offer preservation as part of collections management course learning outcomes. For instance the joint collaboration between Algonquin College and Carleton University offer training on Information Resource Management. The programme structure covers collection management but silent on whether that covers preservation. Similarly the Southern Alberta Institute of Technology (SAIT) offer programmes on 'Information and Records Management' and the programme structures have modules on 'Management of Vital Records'. This does not clearly pronounce if this is preservation and conservation aspect.

Preservation and Conservation Curriculum in ESARBICA

Preservation education in Africa can be traced concomitant with the emergent of Archives and Records Management (ARM) education following the need to preserve records that were initially managed by colonial administrators after a significant number of African countries who began to gain independence and realized the need to preserve their cultural heritage (Katuu, 2015). Since the development of ARM education in the early 1960s (Katuu, 2015), preservation and conservation education is a concern in most African academic institutions.

Prior to the establishment of preservation programmes in African universities most archivists from the ESARBICA region obtained their training in through International Centre for the Study of the Preservation and the Restoration of Cultural Property (ICCROM) through the first Prevention in Museums of Africa (PREMA) course of 1986-87 programmes (Jokilehto, 1988). Of recent, the outlook of preservation was summarized by Segaetsho and Mnjama (2019), specifically looking at ICTs and their applications to preservation, revealing a web of innovation and technology. The key principles and concepts in preservation are embracing, covering environmental considerations, care and handling, education, conservation, disasters management, pest management, binding, security, preservation risk assessments and documentations. A conceptual outlook of preservation functions developed by Segaetsho (2020) showed a pretty much complex outlook of preservation through a conceptual visualization of relationships and disciplines that contribute to preservation (See Figure 1, pp254). The figure is a web of functionalities that streams from both humanities, social sciences, education, and applied sciences.

To ascertain the outlook of preservation and the integration of conservation curriculum in ESARBICA, this chapter presents a survey of LISE institutions in Southern Africa as presented in summary in Table 1. The information has been obtained the universities' websites.

Although a significant number of the programme tittles have the word conservation, very little of the institutions offer practical conservation practice. Only the University of Cape Town, Center for

Table 1. Courses on preservation and conservation in LISE

Program Title	Level	Institution	Course Description/Coverage
Preservation of information resources, Advanced preservation,	Diploma, Bachelor's Degree and Masters	University of Botswana, Department of library and information studies, Botswana	Theoretical teaching on preservation topics where possible with industrial visits. • Introduction to preservation covering the role of preservation and conservation in library and archive management • Nature and deterioration of documentary materials including environmental considerations • Collections care • Disaster preparedness • Preservation policy
Conservation and Preservation, Preservation and Disaster Management,	Diploma and Bachelor's degree	Institute of Development Management (IDM), Botswana	• Conservation and Preservation • Preservation and Disaster Management,
LS 311 Preservation and Conservation of Library Resources	Bachelor of Library and Library Studies (BLIS)	University of Dar ES Salaam, College of Social Science, Tanzania	Optional/Elective course on Preservation and Conservation of Library Resources • No details
IRA4201 Conservation and Preservation Management IRA 6305 Advanced Conservation and Preservation Management	Bachelor of Science Honours Degree in Records and Archives Management and Master of Science Degree in Records and Archives Management	National University of Science and Technology, Zimbabwe	• Undergraduate- The module is an in-depth study of the main threats to the survival of different types of information media. It includes general protection and preventative measures against hazards like fire, flooding and protection units. Strategies to resume operations following disasters, including temporary or alternative facilities are also examined. • Graduate- The module presents and examines modern approaches to the conservation and preservation of various archival material and restoration methods for deteriorated materials. Students will be expected to practically use microphotography, dehumidifying and binding equipment for various preservation and restoration tasks under the guidance of an expert conservator.
Collection Management in Library and Information Services	Degree, Postgraduate diploma, masters	University of Western Cape, South Africa.	Collection preservation • No details
Preserving Records and Archives	Higher Certificatein Archives and Records Management	University of South Africa, South Africa,	A compulsory course that equips students with a sound understanding of the: • Theories and practices relating to the preservation of records and archives. • apply the basic principles and techniques of preservation to safeguard records and archival material
Preventive conservation, materiality, History and current practice of preventive conservation	Honors in Curatorship	University of Cape town, Center for Curating Archives (CCA)	Offers short courses on various topics, internships, compulsory and elective courses. Example of workshop for 2016 • Preventive conservation- the courses have practical demonstration in conservation studios. • Workshop on materiality- Speakers from various academic fields ranging from art history, science, classical studies, digital media, design history, feminism, film and television studies, history, philosophy, psychoanalysis, religious studies and visual culture will share their insights during the morning sessions. • History and current practice of preventive conservation
Preserving archives, preservation and conservation	Diploma and Bachelors' Degree	University of Namibia, Namibia	Preserving archives, preservation and conservation • No details
Paper Making And Reprography	Degree in Records and Archives Management	Makerere University, The East African School of Library and Information Science, Kenya	Paper Making And Reprography • No details

Source: Conceptualized by Author.

Curating Archives (CCA) offer practical teaching through the use of professional workshops. Specialist programmes that are not necessarily within LISE are the ones offering practical hands-on in preservation and conservation such as the South African Institute of Object Conservation. Despite the little information on preservation in LISE a significant number of universities do offer programmes in LISE almost in every country in ESARBICA region such as Botswana, Zambia, South Africa, Namibia, Swaziland, Lesotho, Zimbabwe and others.

CONTEMPORARY CHALLENGES IN PRESERVATION AND CONSERVATION CURRICULUM

A cloud of challenges on preservation and conservation curriculum is observed throughout the world. The challenges in education, termed by Adetimirin (2019) as unethical behaviour (academic dishonesty, plagiarism, cheating, copying and information infringement) still stands as challenges in academic institutions. The need to teach students ethical behaviour that is acceptable for ideal operation following their graduations becomes critical as key issues to address in every disciple including conservation and preservation. It is, therefore, critical to take note of key ethics to inculcate in students in preservation and conservation curriculum. The questions that still need theoretical exploitation are; which ethics are important for preservation and conservation in LISE? How do these ethics link/interrelate with other ethical expectation of other specializations in LISE?

The present shift from print to electronic records management equally importantly brings a mirrored challenge on digital preservation. Indeed, at some point these new technologies and innovations culminating in various technical objects such as computers, mobile phones, tablets audiovisuals etc., will need to be preserved in perpetuity as cultural heritages. The diversity of preservation and conservation grows on daily bases hence the need for continuous revamping of education development in LISE. In discussing preservation challenges in Nigeria, a study by Uluocha, (2019), highlighted issues on changes in software and hardware, lack of policies, and inadequate technology infrastructures. Although the Lagos State Library recently did not observe lack of skilled manpower, inadequate funding, technological obsolesces, administrative bottlenecks and lack of trained curators as challenges, these challenges are significant elsewhere such as on studies conducted by Ngulube (2007a) and Nsibirwa (2012; 2015) in South Africa, Kootshabe and Mnjama (2015) in Botswana, and Msibi (2015) in Swaziland and Kanyengo, (2009) in Zambia. A study by Murray (2006) analyzed the situation of preservation education in South African library and archives degree programmes and observed preservation awareness in South Africa came late in the 1980s. The study by Murray (2006) observed that the basic underlying topics of preservation were critical in library and archives, and on average preservation courses were not yet well established.

Despite the undeniable benefits brought about by ICTs, Segaetsho and Mnjama (2019) observed that these technocratic measures are often expensive to purchase. Concomitantly, given obsolesces in technology, academic institutions therefore have to revamp their teaching resources frequently. These bring in the challenge of the sustainability of the education programmes. The major achievements realized in the implementation of Agenda 2063 over the past ten years (2021-2023) are still vague in academic universities and often discussed as slow on implementation of SDGs, especially in developing countries (El-Jardali et al., 2018; Atlhopheng, Mbongwe & Segaetsho, 2020). It is therefore critical that academic institutions should take into cognizant the key pillars in SDGs for integrated, indivisible and balance three dimensions of sustainable development: the economic, social and environmental. The SDGs, Strategic goal 4, emphasizes that education development should by 2030, "substantially increase the number of youth and adults who have relevant skills, including technical and vocational skills, for employment, decent jobs and entrepreneurship" (United Nations, 2015:21).

A FRAMEWORK FOR INTEGRATING CONSERVATION CURRICULUM IN LISE

The confusion on conservation education platforms is a living reality amongst developing countries, especially in Africa. As would Hedstrom and King (nd: 02) attest, "confusion arises at the intersection of the known and the unknown and is an essential feature of the expanding frontier of knowledge." The confusion brought about by the known and unknown in terms of preservation and conservation platforms of education to be put in place in LISE, is increasing with changes brought about by ICT. For this confusion to be resolved, education systems need to create knowledge by managing the ambiguity inherent in the unknown of conservation studies in LISE. Consequently, the more the integration of conservation studies in curriculum development for LISE, the more the rapid and effective knowledge is acquired, and progress would be sustained in preservation and conservation of heritage collections. This section, therefore, proposes a framework for integrating conservation curriculum in LISE.

The Role of Higher Education Institutions in Implementing the SDGs is clearly discussed in various literatures, among others, including the rate of inclusion of SDGs in research, outreach, stewardship, participation, collaboration and commitment by higher education associations such as the International Association of Universities (IAU) (Atlhopheng, Mbongwe & Segaetsho, 2020). In line with these commitments in education, preservation programmes have to make contribution to SDGs in various ways, such as revamping LISE teaching curricula. The development of conservation concept needs to be based on the realities of industrial demand and realize the UN Agenda 2063 on SDGs. The content curriculum development in conservation should be based on the principles of preservation coverage (preservation sustainability, care and handling, environmental considerations, disaster management, security, pest management, preservation risk assessment/surveys etc.), conservation documentation, technical analysis, treatment, environmental recommendation, research and restoration in the form of binding.

Figure 1. A framework for integration of conservation concepts in LISE

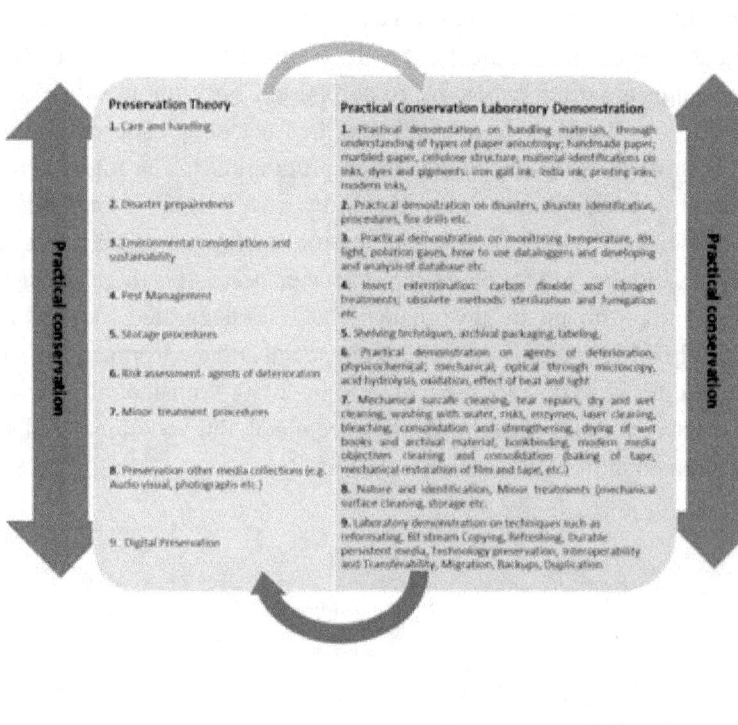

The preservation and conservation curriculum in LISE need to consider integrating the following restoration, preservation and conservation teaching topics as shown in Figure 1. The model is based on the premise that theoretical preservation management activities can be strengthened by technical practical demonstrations which require demonstrations by conservators in conservation laboratory setup. For example, theoretical teaching on 'Care and Handling' could be augmented by introducing practical laboratory activities such as demonstration on handling of materials through the understanding of types of paper (e.g., anisotropy, handmade paper, marbled paper) using microscopy to view cellulose structure, material identifications on inks, dyes and pigments, iron gall ink, printing inks and modern inks.

The list given in Figure 1 is not exhaustive, the various topics covered in various levels in LISE could be added as needed. Such developments should be guided by possibilities of sustainability and various strategies for implementing the programme. For sustainability purposes, the design of preservation and conservation content could serve as a logical connection between preservation and information science fields offered in LISE. In cases where there are no conservation studios, practical workshops could be arranged with local conservation studios in national archives and museums.

SUMMARY AND CONCLUSIONS

This chapter focused on the integration of conservation principles in preservation curriculum in Library and Information Science Education. This chapter reveals that knowledge development in conservation could only be achieved by the application of a systematic protocol abstracted from the effort to understand the curriculum applicable to LISE. Common principles both for preservation and conservation are identified. The principles were then used to map possible conservation teaching concepts that can be integrated in preservation curriculum development for Library and Information Science Education. The chapter proposes that conservation principles/topics on understanding natural decay, nature of information resources, environmental considerations, basic conservation treatments, restoration in the form of book binding, and documentation principles are critical for practical conservation demonstration in LISE. Going forward, it is critical to conduct comprehensive survey studies on education and training, education sustainability and impact on preservation and conservation. Such research should assist LISE to transform themselves. The LISE should offer courses that are focused on preservation of documentary collections, and other media such as audio visual materials, photographic and digital collections.

RECOMMENDATIONS

In addition to the proposed framework on integration of conservation concepts in LISE, the following recommendations are furthered proposed for improvement.

- As much as the conservation functions and the general preservation activities could be integrated in LISE curriculum, it is still recommended that the education curriculum should develop conservation streams as part of specialties for obtaining degrees in conservation.
- Cooperation, partnerships and participation between universities, public and private entities is critical. The diverse partnership in programmes can be used for resource sharing given that con-

servation equipment is expensive to purchase especially during the initial development of conservation studios.

- To execute a robust skills and knowledge sharing culture to develop conservation curriculum in LISE, it is critical to adopt a force field analysis that encourages group discussions, agreements and reflections on the underlying education problems.
- Specific basic conservation treatments on other media beyond graphic materials such as digital preservation and audio visuals also need to be further investigated, course outlines' being developed and integrated for demonstrations in LISE curriculum.
- Given the expensiveness of running conservation programmes throughout the world, African Universities should develop specific funding schemes to address educational issues such that courses offered on preservation and conservation are sustainable and comply with SDGs indicators.

REFERENCES

Abdulrahim, A. (2016). *Preservation and conservation of records in the national archives and national museum*. Academic Press.

Adetimirin, A. (2019). Educating library and information science students for an ethical information age. In W. M. Olatokun, A. O. Aremu, & A. Adetimirin (Eds.), Research on contemporary issues in media resources and information and communication technology use (pp. 99-112). Department of Library, Archival and Information Studies, University of Ibadan.

Atlhopheng, J. R., Mbongwe, B., & Segaetsho, T. (2020). Implementing Sustainable Development Goals at Institutional Level: The Case of University of Botswana. In S. O. Keitumetse, L. Hens, & D. Norris (Eds.), *Sustainability in Developing Countries*. Springer. doi:10.1007/978-3-030-48351-7_13

Australian Institute for the Conservation of Cultural Material. (2022). *Conservation*. Available at: https://aiccm.org.au/conservation/

Cardiff University. (2023). *Cardiff University Postgraduate Prospectus 2024*. Cardiff University. Available at https://www.cardiff.ac.uk/__data/assets/pdf_file/0009/2713941/Cardiff-University-Postgraduate-Prospectus-2024.pdf

Center for Curating the Archives. (2021). *2016 Workshops, The history and current practices of preventive conservation.* University of Cape Town. Available at: http://www.cca.uct.ac.za/cca/honours-in-curatorship/2016-2/workshops

El-Jardali, F., Ataya, N., & Fadlallah, R. (2018). Changing roles of universities in the era of SDGs: Rising up to the global challenge through institutionalising partnerships with governments and communities. *Health Research Policy and Systems*, *16*(1), 38. doi:10.1186/s12961-018-0318-9 PMID:29743112

Harvey, R. (1994). *Preservation in libraries: principles, strategies and practices*. Bowker.

Hedstrom, M., & King, J.L. (n.d.). On the LAM: Library, Archive, and Museum Collections in the Creation and Maintenance of Knowledge Communities. *School of Information University of Michigan*, (2), 139–148.

Historical Association. (2022). *The voice of history, Library and Information studies.* Available at: https://www.history.org.uk/student/resource/2790/library-and-information-studies

International Education Specialist. (2022). *Information Science and Librarianship courses.* Available at: https://www.idp.com/australia/search/information-science-and-librarianship/

International Federation of Library Association. (2019). *Preservation and conservation.* Available at https://www.ifla.org/preservation-and-conservation

International Federation of Library Association. (1998). *Principles for the Care and Handling of Library Material, International Preservation Issues, Number One.* Available at: https://repository.ifla.org/bitstream/123456789/1183/1/ipi1-en.pdf

Iyishu, V. A., Nkanu, W. O., & Ogar, F. O. (2013). Preservation and conservation of library materials in digital age. *Journal Information and Knowledge Management, 04*(02).

Kanyengo, C. W. (2009). Preservation and Conservation of Information Resources in the University of Zambia Library. *Journal of Archival Organization, 7*(3), 116–128. doi:10.1080/15332740903126736

Katuu, S. (2015). The development of archives and records management education and training in Africa–challenges and opportunities. *Archives and Manuscripts, 43*(2), 96–119. doi:10.1080/01576895.2015.1050677

Keene, S. (2008). *Collections for People: the stored collections of UK museums as a public resource.* UCL Institute of Archaeology. Available at https://www.collectionslink.org.uk/assets/userfiles/use_my_collection/collections_for_people.pdf

Kootshabe, T. J., & Mnjama, N. (2014). Preservation of government records in Botswana. *Journal of the South African Society of Archivists, 47*, 26–49.

Krtalic, M., & Hasenay, D. (2011). Exploring a framework for comprehensive and successful preservation management in libraries. *The Journal of Documentation, 68*(3), 353–377. doi:10.1108/00220411211225584

Msibi, N. M. (2015). *Preservation of public records and archives in Swaziland government ministries and the Department of Swaziland National Archives, Master of Information Studies* [Dissertation]. University of KwaZulu-Natal, Pietermaritzburg, South Africa.

Murray, K. (2002). *Preservation Education and Training for South African Library and Archive Professionals and Students, MBibl.* University of Cape Town.

Murray, K. (2006). Preservation education in South African library and archive degree programmes. *South African Journal of Library and Information Science, 72*(02), 98–107. doi:10.7553/72-2-1130

National University of Science and Technology (NUST). (2019). *Faculty of-Communication Information and Science Prospectus, Department of Records and Archives.* Retrieved from: https://www.nust.ac.zw/cis/images/documents/faculty-of-comm-info-science-prospectus-2019.pdf

Ngulube, P. (2007a). Preserving South Africa's paper trail and making public records available for present and future generations. *ESARBICA Journal, 26*(1), 35–50. doi:10.4314/esarjo.v26i1.31014

Ngulube, P. (2007b). The Achilles Heel of the Preservation of Documentary Materials in Sub Saharan Africa: Knowledge and Skills or Funding? *Restaurator (Copenhagen)*, *28*(3), 158–168. doi:10.1515/REST.2007.159

North Umbria University. (2023). *Preventive Conservation MA, Newcastle.* Available at https://www.northumbria.ac.uk/search/?q=conservation&trigger=

Northeast Document Conservation Center (NEDCC). (2021). *About Preservation Training at NEDCC.* Available at https://www.nedcc.org/about/overview

Nsibirwa, B. Z. (2012). *Preservation of, and access to, legal deposit materials in South Africa* [PhD Dissertation]. University of Kwazulu-Natal, South Africa.

Nsibirwa, Z. (2015). Staff education and the preservation of legal deposit materials in South Africa. Innovation. *Journal of Appropriate Librarianship and Information Work in Southern Africa*, *2015*(50), 1263–1145.

Pearson, C., Lyall, J., Sloggett, R., & Cook, I. (2011). Preserving the past: how to ensure the development of conservation profession in Australia is preserved for the future. *Australian Institute for the Conservation of Cultural Material, A.I.C.C.M., National Conference 2011.*

Pye, E. (2008). The benefits of access through handling outweigh the risks. *Conservation and Access Preprints of the 22nd IIC Congress*, 162–165. 10.1179/sic.2008.53.Supplement-1.162

Pye, E. (2016). Challenges of conservation: Working objects. *Science Museum Group Journal.* Advance online publication. doi:10.15180/160608

RobertsC. (2021). *Conservation VS Restoration: What is the difference.* National Museum of the Pacific War. Retrieved from; https://www.pacificwarmuseum.org/about/news/conservation-vs-restoration-what-is-the-difference#

Segaetsho, T. (2018). *Environmental Considerations- Preservation of library and archival papers.* LAP Lambert Publishers.

Segaetsho, T. (2020). *Transcending Theories of Preservation in Library and Archives: The Perspective in the ESARBICA, Cases on Electronic Record Management in the ESARBICA Region.* IGI Global. doi:10.4018/978-1-7998-2527-2.ch013

Segaetsho, T., & Mnjama, N. M. (2019). Information and communication technology use in the preservation of paper-based materials in the Eastern and African Region. Academic Press.

Segaetsho, T., & Moloi, J. (2019). Integration of Digital Preservation Knowledge, Skills and Competencies in the Teaching Curricula of Library and Information Studies at the University of Botswana. *Mousaion*, *37*(1). Advance online publication. doi:10.25159/2663-659X/6436

Shameenda, K. L., & Kanyengo, W. K. (2012). An Audit of Skills and Qualifications in Preservation and Conservation Techniques: The Case of the University of Zambia Libraries. *Journal of Archival Organization*, *10*(02), 137–149. doi:10.1080/15332748.2012.736769

Study Portal. (2022). *Master's degrees in Library Science in Europe.* Available at: https://www.mastersportal.com/search/master/library-science/europe

The Association of Canadian Archivist. (2022). *Graduate archival education in Canada.* Available at: https://archivists.ca/Archival-Education-Programs

The United Nations (UN). (2016). *Transforming our world: the 2030 Agenda for Sustainable Development* (A/RES/70/1). UN.

Uluocha, A. (2019). Assessment of preservation practices at Lagos State House of Assembly Library. In W. M. Olatokun, A. O. Aremu, & A. Adetimirin (Eds.), Research on contemporary issues in media resources and information and communication technology use (pp. 465-479). Department of Library, Archival and information studies, University of Ibadan.

UNESCO. (2022). *Transforming higher education institutions into lifelong learning institutions.* UIL POLICY BRIEF 14, UNESCO Institute for Lifelong Learning.

United Nations. (2022). *Transforming Education Summit 2022 - Concept and Programme Outline, convened by the United Nations Secretary-General.* UN.

United Nations. (2015). *Transforming our world: the 2030 Agenda for Sustainable Development.* Available at https://sustainabledevelopment.un.org/post2015/transformingourworld

University for Continuing Eduacation. Krems. (2016). Interdisciplinary Methods in Graphic Art, Book and Document Conservation. *Auszug aus Mitteilungsblatt, 20*(98). Available at https://www.iiconservation.org/sites/default/files/news/attachments/7420-krems_university_course_details.pdf

West Dean College. (2023). *Master of Arts Conservation Studies Programme Specification and Unit Descriptors.* West Dean College, University of Sussex. Available at https://assets.westdean.ac.uk/pdfs/original/60954-ma-conservation-studies-programme-specification-june23.pdf

Wilson, P. (1992). *Library conservation course in Australia, Conservation DistList.* https://cool.culturalheritage.org/byform/mailing-lists/cdl/1992/0236.html

Zervos, S. (2012). *Teaching Conservation and Preservation at the Department of Library Science and Information Systems of the TEI of Athens.* Restorator.

Chapter 16
Using Mother Tongue in Teaching of STEM for Sustainable Development in Zimbabwe

William Chakabwata
https://orcid.org/0000-0002-4224-5239
University of South Africa, South Africa

ABSTRACT

This chapter submits that the teaching of science, technology, engineering, and mathematics (STEM) is perceived as one of the key to the economic development of Africa. Africa is grappling with the challenge of economic development, and strengthening the teaching of STEM across the curriculum can have a positive effect on the continent. The chapter submits that teaching STEM in a foreign language hinders the sustainable development and the opportunity for learners to acquire knowledge in mother tongue and also to innovate in critical areas that are desperately needed in Zimbabwe as they spent a lot of time learning English in STEM at the expense of acquiring scientific knowledge. The chapter advocates the use of tranlanguaging and multilingualism in order to improve student performance in the teaching of STEM.

INTRODUCTION

Language remains central to human society for interactions at any level either academic or social (Ozoemena, Ngweke, & Nwoko, 2021). Language has been defined as engagement among people which involve symbols and where auditory and oral features are used (de Castillo, 2015). Language provides human beings with a facility for thought and creativity.Language is situated in culture and is passed on from one generation to the next. Language is intimately linked to issues of identity of a people,their history, and aspirations and potential for creativity. Because language plays a central role in education, it becomes very political in terms which language can be used in education and which language is mar-

DOI: 10.4018/978-1-6684-9859-0.ch016

ginalised.Human beings are acknowledged to be social beings and this connotes that they share beliefs, knowledge and customs. It because of the primacy of mother tongue in education that United Nations, Scientific and Cultural Organisations has been a chief advocate on the use of mother tongue as instrument for learning in schools for early childhood education since 1953. It is apparent that if sustainable development is to occur among the different nations, it is imperative to ensure that instruction occurs in mother tongue throughout the education system with other foreign languages being taught as subject even up to University.

Mensa (2019) avers that the term sustainable development is being used in as a mantra in day to day engagements. The term has continued to evade many people and the history of sustainable development has remained obscure for many people. The chapter will present a brief background of sustainable development in terms of the origins of the

The purpose of this chapter is to contribute to sustainable development discourse especially in terms educational practices in the teaching of Sciences, Technology, Engineering and Mathematics (STEM). Sustainable Development and Teaching of STEM. The chapter will also argue that using mother tongue in teaching of STEM, is the key to ascertaining sustainable development for developing nations.

BACKGROUND TO ZIMBABWEAN EDUCATION

Peresuh & Masuku (2002) asserted that was occupied by Britan and it attained political autonomy in 1980. From the inception of the colonial state missionaries of European origin, played a dominant role in the financing and provision of education to the indigenous population. The major socio-lingustic composition in Zimbabwe are the Shona constituting making up 70% of the population followed by the Ndebele who make up more than 15% of the population. Other minority lingustic categories are Venda, Tonga, Kalanga, Ndau and Sotho which make up approximately 12% of the population. Unfortunately English a language that is utilised by less than 1% of the population is the official lingua franca of the country (Peresuh & Masuku, 2002).

Chivhunga & Chimhenga (2013) noted that the current teaching practices in Zimbabwe promotes the use of English language as the offcial lingua in schools. English has tended to exercise hegemony in areas such as in spheres such as business,education, administration and politics. This dominance of English has affected the development of African languages in terms of being utilised as a language of instruction in STEM in schools.This has a negative impact on the need to promote sustainable teaching of STEM in Zimbabwean schools. Sustainable education means that the mother tongue has to gain priority in terms of being the language of instruction in schools in line with learning and development theories such as socialcultural theory by Vygotsky which articulates the way cognitive development occurs in socialcultural context.

More than fourty years after political independence Zimbabwe has not yet developed a language policy. Indigenous languages have not been endorsed as a language of instruction although empirical evidence support the use of mother tongue for meaningful learning (Chivhunga & Chimhenga, 2013). In the next section, the coloniality of language is explored in more detail.

ORIGINS OF THE CONCEPT SUSTAINABLE DEVELOPMENT

The term sustainable development has its genesis in the discursive discourse on sustainable development. Sustainability as a paradigm originated from the three pillars namely environment, society and economy which are intertwined. These concepts tends to display tension, complementarity and also tends to shape the way people acts at national and international level (Mensa, 2019). The terms have also had an effect on practices in many domains such as development and education.

Du Pisani (2007) asserts that the genesis of sustainable development may be traced to the twentieth century particularly the Brandt Report that has been hailed as a chief milestone in sustainable development. The Brandt Report was christened after the Chairperson of the World Commission on Environment and Development (WECD) (World Commission on Economic Development (WCED,1987; Cayley, 2023). The concept was foregrounded in the awareness of an impending ecological peril that was an outcome of modenization and rapacious capitalism. Sustainable development has roots in western thoughts of linear progress that was built on industrialization and modernity which would culminate in the final utopia of a scoiety that was able to address all human needs through science and technological innovation. Human progress was conceptualised in terms of material and economic advancement (Du Pisani, 2007).

There was an overwhelming blind belief that science would grant mankind mastery over nature ushering humanity into some kind of golden age remiscient of christian utopia as represented by the eschatalogical belief. However as history attests the economic expansion that transpired in the western world ended up benefitting certain nations at the expense of other nations creating a world that was characterised by plenty in the nations of the North and destitution in the nations of the South. It was becoming apparent that the world was facing a challenge of resource depletion due to wanton use of resources and an imminent ecological disaster. Although sustainable development may be linked to developments prior to industrialization, it is apparent that the massive demand of natural resources tended to exacebate the threat on the exploitation of natural resources and accompaying environmental pollution.

Du Pisani (2007:91) notes that the term sustainable development was already in common currency in the 1970s and was coined by Babra Ward and it was consolidated and employed to connote a society that evinces stability, while providing material needs of all people in the future. There was a growing realization in the 1970s,that development had to be sustainable by ensuring that the resources are used prudently in order to benefit future generations. In the 1980s United Nations, tasked 23 people from the developed and developing countries to explore the needs to ensure global equity for posterity by ensuring that there is social equality and environmental protection and economic growth were addressed in a compartible way. The report that they compiled was termed the Brandt Report and it was named the Our Common Future to the United Nations in 1987.

EDUCATION AS A SUSTAINABLE ACTIVITY

Education itself can be said to be a sustainable activity when it is designed and conducted in a manner that aligns with principles of sustainability. This means that the educational process, institutions, and systems should consider the social, environmental, and economic dimensions of sustainability (United Nations Educational Scientific and Cultural Organizations UNESCO, 2012). Education does not only provide individuals with the knowledge and skills needed to address sustainability challenges but it also serves as a model for sustainable practices in society.

Education must respond to the environmental dimension by establishing green schools by adopting sustainable practices such as energy-efficient buildings, waste reduction, sustainable transportation options, and green spaces. Likewise, integrating environmental and sustainability topics across the curriculum helps raise awareness and build knowledge about environmental issues.

The relationship between the education and economic systems can described as mutual. The economic system provides the materials used for education purpose. Education responds to the economic dimension through efficient resource allocation. Schools use materials such as papers, chalks, board markers etc (UNESCO, 2012). Schools should manage these resources efficiently to reduce waste and maximise the impact of their investments. Moreover, education provides human capital needed by the economic system. Thus, education provides skills for sustainable careers. Preparing students for careers in sustainable industries can contribute to economic sustainability by promoting the growth of green jobs and sustainable businesses.

In response to the social dimension, education should promote inclusivity. Education should be accessible to all, irrespective of background or ability. Inclusivity ensures that education benefits society as a whole. Likewise, education should promote equity. Efforts should be made to reduce educational disparities and ensure that all individuals have equal opportunities to learn and succeed (United Nations Educational Scientific and Cultural Organization UNICEF, 2014). Didham & Ofel-Manu (2015) avers that quality education is linked to the provision of sustainable education. Education is perceived as a key strategy for realising SDGs.

LANGUAGE AND COLONIALITY

In this chapter the nomenclature that was used by Ogbodo et al.(2015:p6) " where the first language is the language a person acquires with his contact with his environment. It is the only language a monolingual person acquires in his native environment…". This first language which is also termed as mother tongue is assumes priority when it comes to communication in a person's speech repository. It is generallny presuposed that a child leans this language from a parent of a gurdian. The second language is acquired after the first language and is termed the target language.In the case of Zimbabwe a fourteen languages are identified as official languages and each of them has the capacity to be identified as a mother tongue. By virtue of the colonial history for the majojority of the people English is a second language.

Hsu (2017) explained that the use of English as an official language of instruction was a part and parcel of the western ideology based on racial superiority driven by a motive to civlise babarians they had subjugated in their empire. This connotes that English as a language of instruction carries with it huge colonial overtones. English as an official language contributed to the racial and lingusitic ordering of people that still has a negative effect today in Zimbabwe and other parts of the world that were a part of the British empire.

The colonial construction of the self and the other remains dominant in Zimbabwe discourse to use indigenous languages as official languages of instructions. The ruling elite and industry, are known to defy efforts to upset this lingusitic and racial ordering. It is for that reason that learners continue to struggle in STEM subjects. This runs contrary to sustainable education which advocates literacy and the development of thinking skill. Katukula, Set, & Nyambe, (2023) conducted a study in Namibia on the language policy used in schools. The study noted that teaching in junior primary school, grade 0-3, is required to be done in mother tongue. The same stigma that as noted in Zimbabwe was also observed in

Namibia,where parents and gurdians are reluctant to enrol their children in schools that teach in mother tongue. Coloniality is still so dominant in most states that were colonised in Africa, that instructional in colonial lingua is associated with quality, while instruction in mother tongue is perceived as inferior (Mufwene, 2006; Mendoza, 2016; Lugones, 2008).

Katukula et al (2023) explicate the attitude of the parents in Namibia as supporting the use of English only as a medium of instruction to the exclusion of all other languages such as the mother tongue, despite the reality that Namibia used Afrikaans before independence and the majority of teachers still struggle with proficiency in the English. The use of English as an official language in schools also pervades the thinking of school principals and educational administrators.

There is a growing critique of the use of English or other foreign languages as a medium of instruction at the expense of local languages. Mufwene (2006) lamented that the 19th century ideology based on social Darwinism has remained recalcitrant in many states where European language and culture are still highly esteemed. The perception that is dominant was that languages that were that (Indo) European languages which display fusional morphology are more developed than the other languages that those with isolating morphosyntax.

THE LANGUAGE POLICY IN ZIMBABWE

Zimbabwe does not have a written single language policy of education document, the language policy is rather implied or inferred from various documents (Nhongo, 2020; Sibanda, 2019). Seven years after Zimbabwe attained its independence, the government launched a language policy as part the education act of 1987 in pursuit for linguistic decolonialisation. The act pronounced English, Shona, and Ndebele as the three main languages to be used in schools in Zimbabwe. Mother tongue, Shona or Ndebele may be used as a medium of instruction from grade 1-3. English shall be the medium of instruction from grade 4 with Shona or Ndebele taught as subjects on equal time allocation as the English language. The Minister may authorise the teaching of such languages in primary schools in areas where minority languages exist (Government of Zimbabwe, 1987). Revision of the Education Act in 2006 added Tonga and Kalanga languages to be used as medium of instructions. The revision also allowed the teaching of Shona, Ndebele, and English up to form 2 with the same instructional time in all districts in Zimbabwe.

Presidential commission of inquiry into education and training in Zimbabwe known as the Nziramasanga commission was an educational initiative to assess and make recommendations on the country's education system. The Nziramasanga commission report on education in Zimbabwe addressed the issue of language in education. Language has been a significant and complex issue in Zimbabwe due to its multilingual and multicultural society. The report recognised the importance of language in education and made several recommendations regarding language policy and its implementation. The Nziramasanga report reiterated some of the language policy fundamentals of the Education Act of 1987. The report stressed multilingual education. The report acknowledged the linguistic diversity of Zimbabwe, with English, Shona, Sindebele (Ndebele), and other languages being spoken. It recognised that multilingualism is an essential aspect of Zimbabwean society and should be reflected in the education system. The report promoted mother tongue-based bilingual education at the primary level. This means that students should be educated in their native language, while also learning English as a second language in the early years of primary education. A gradual transition from the use of the mother tongue to English as the medium of instruction as students progressed through the education system was recommended. The report emphasised the need to promote and preserve indigenous languages and cultures as part of the education system. Teacher training programs should prepare educators to teach in multiple languages. The report stressed development of curriculum materials in multiple languages to support mother tongue-based education.

Zimbabwe's new constitution of 2013 recognises English and 15 local language as official languages. In addition to English, the following are official recognised languages of Zimbabwe: ChiNambya, ChiKalanga, ChiVhenda (XiVhenda), ChiShangani (Tsonga), Sign Language, ChiTonga, ChiChewa, ChiBarwe, English, Sotho, Koisan/ Tshawo, Tswana, Xhosa, ChiNdau, Shona and Ndebele (Mazuruse,2018). The government and all its institutions and agencies are required to ensure that all officially recognised languages are treated equitably. English has continued to overshadow indigenous languages in most official communication (Mazuruse, 2018) including communication in schools in Zimbabwe. Sibanda (2019) found that the "footprints" of colonial language policy fundamentals still prevail in the current language policy. English remains a dominant language of instruction used in schools. Indigenous languages are considered inferior. The policy failed to create indigenous ways of thinking, knowing, seeing, doing, and responding to the world.

Nhongo (2013) avers that the in Zimbabwe an Education Act was introduced in 1987,which Shona, English and Ndebele were supposed to be taught in schools in primary schools in areas where majority of residents were Shona or Ndebele.From fourth grade learners would start learning instruction in English, and Shona or Ndebele could be taught as subject with equal time allocation as English. The education act was reviewed in 2009, with no distingushable changes.

INTERNATIONAL DECLARATIONS ON THE USE OF LANGUAGE IN AFRICA

Significant declarations that were made at a continental level regarding regarding the language used in schools were *The Language Plan of Action for Africa by* African Union, The Harare Commonwealth Declaration 1991 and Asmara Declaration (African Union Academy of Languages, 2005; Blommaert, 2001;Nhongo, 2013).

THE LANGUAGE PLAN OF ACTION FOR AFRICA

The language plan of Action for Africa was mooted by African intellectuals who were discussing the use African languages as medium of instruction at universities and in tertiary education by promoting master degrees and doctoral programmes in African languages. These concepts culminated in the Language Plan of Action for Africa, which was endorsed in Cameroon in Yaonde in 2004. African Union Academy of Languages (ACALAN) is an arm of the African Union that has a mandate to help member states in developing language policy in Africa and other associated programmes.

Pan African Master's and PhD project in African languages and Applied Linguistics (PAMAPAL) for the purpose of developing African linguists with the appropriate linguistic competencies. The same platform acknowledges that African culture and languages are at risk from the English Language on the continent.

THE HARARE COMMONWEALTH DECLARATION 1991

The Harare Commonwealth declaration 1991 asserted that sound and sustainable development can help address the issues of poverty on the continent. There are numerous challenges that are confronting Africa such as human trafficking, poverty, environmental degradation among many other challenges. All these challenges require the use of technology to address them as well as sustainable educational practices.

ASMARA DECLARATION

(Blommaert, 2001) noted that a declaration was made by African scholars in Asmara Eritrea pronouncing the rights of African languages. It is noted that colonialism created a number of impediments for Africa languages. The same conference conceded that African languages must be viewed as an instrument to promote African unity and also that an African child must have the right to enroll for school and to learn in mother tongue. Asmara declaration also declared that every African child has a right to learn in mother tongue at all levels of learning. The same declaration also added that the development of science and technology depends on the development of local languages.

THEORETICAL FRAMEWORK

Vygotsky explained ways in which human beings develop under the influence of socialcultural conditions (Toom,2015). The key concepts are that human beings use of cultural tools, higher in contrast to lower pscychological tools, effective use of cultural tools,gaining competency over one's psychological processes, and behaviour. The social cultural theory postulated that all human processes consists of processes. The lower psychological processes consists of processes such as memory and basic attention are involuntary and inate. In contrast higher psychological processes develop due to mediation of cultural tools (Abtahi, 2018). Abtahi (2018) asserted that the relationship between learning and development is central to Vygotsky theorising.

Vygotsky (1978) postulate how cognitive development is central to cognitive development (Vygotsky, 1978; Alharbi, 2023). The social cultural theory posited by Vygotsky was a reaction to the theories that were dominant in the 1970s on language development such as behaviorist and social interactionist theories. Deng & Zou (2016) asserts that behaviourists submit that language acquisition begin around the age of three years as children acquire the mother tongue or primary language. Any other language that a child acquires after that is termed a second language acquisition. In the case of children in Zimbabwe second language acquisition would tend to be English which most children encounter at school while the primary language can be an indigenous language. Behaviorist theory attribute language acquisition to internal and external processes.

The behaviorist theories on language acquisition can be placed into two categories, namely work of early behaviorists and behaviorism as epitomised by the work of Skinner. Watson work is associated with the enunciation of stimulus-response theory that was founded on the premises of classical conditioning theory. Skinner categorised behaviour into two categories namely the respondent behaviour and operational behaviour. The conditioned response is further categorised into respondent behaviour and operational behaviour. Skinner theory has been perceived as shedding some light on how adults acquire a second language (Deng & Zou, 2016). Skinner asserted that language acquisition occurs through the process of reinforcement which works through a system of rewards and punishment.

Vygotsky (1978) argued that there are two symbolic functions to the use of language which are He maintained that language acquisition contrary to the view by Piaget doesn't depend on maturation, but social and cultural learning takes priority over maturation. Arshad & Chen (2009,p319) noted that " Vygotsky concept of appropriation provides a springboard for a discussion on how children from imitating adults". Chidlren acquire the psychological tool of language by modelling the behaviour of adults. Vygotsky (1978) utilised terms such as semiotic mediation, Zone of Proximal Development and internalization. In socio-cultural theory the term semiotic mediation was utilised to account for how children learn by modelling adults behaviour. Internalization has connotation in terms of how children learn reading and writing and consequently develop literacy competence. Vygotsky postulated that any function on the child's psychological development occurs at two planes firstly at the social and then at the psychological dimensions.

Mediation of language occurs as mutual as opposed to being unidirectional when children engage with adults and other people who are more competent than them. The child social and cultural context of the child is interrelated to language learning outcomes (Toom, 2015). When viewed from the perspective of modelling, it is submitted that children acquire competencies to read and write by interacting with others. Vygotsky submitted that internalisation refers to the way that the language that is used is transformed into inner speech and then turned into higher levels of thinking. Other key concepts in the social-cultural theory are the Zone of Proximal Development and scaffolding. Zone of proximal development refers to the level that a child can engage in problem solving on their own without the help of an adult and the level where they can develop with the help of a competent adult or peer.

Scaffolding is very crucial for child development as it promotes high levels of the child's development. Children engage in private speech at different levels such as social, inner and private speech. The first level of speech is termed social and it serves to articulate short sentences and to demonstrate emotions. The second level is termed private speech and it displays the child egocentric nature at the age of around 3 years (Yuniar, 2017). The private speech displays a child's efforts to come to terms with a complex world and the desire for self direction. Private speech helps the child to gain competence in communication, and those that originate from the middle income families tends to become more proficient in the use of the language due to the stimulating family environment. Internalisation of speech occurs during the period when the child engages in inner thought, which means that the child is able to structure their own thoughts and to think in more abstract terms.

THE IMPORTANCE OF STEM

There is no debate on the importance of teaching of STEM in Africa as one strategy to realise development (Office of Science and Technology Policy, 2021 (Kocabas, Ozfidan, & Burlbaw, 2020; Nhongo & Tshotsho,2020). Countries such as the United States and other developed nations have taken a strategic interests in the teaching of STEM. It is acknowledged that STEM as an importance field that contributed to the industrial revolution especially with the invention of the bulb among many other inventions. The concept of STEM which coalesces four disciplines namely science, technology, engineering and mathematics has been linked to the development of critical thinking competencies, employment and productivity. People are who have been trained in STEM develop the competencies to enhance the world around them. STEM is has potential to realise the following goals in many states:

- Develop teachers who are capable of facilitating the development of foundational skills in learnes in a holistic way
- Develop scientists, engneers and digital specialists who are capable of conducting research at national level and global level
- Develop workers who are able to create, design and apply complex forms of technology and innovations
- Generate citizens who have empirical skills to reflect on issues and address issues in their environment such as lack of clean water, environmental degradation and climate change (Siekmann, 2016, p. 3).

Fomunyam (2020,p1516) averred that there are " 14 million students in higher education in Africa, representing 6,4% of the global tetiary education enrolments, but less than 25% of these students are in STEM fields with the majority of these students studying humanities and social sciences. Africa generates a paltry 1,1% of scientific knowledge. The contemporary demand for skilled people in STEM in industry in Africa does not match he level that tertiary institutions are producing skilled people. African Union also underscores the importance of STEM in its agenda 2063 which asserted that it intends to " develop Africa into a powerful house of the future" (African Union, 2023, p. 1). The AU agenda 2063 is linked to the SDGs and also explicate as goal 2 as the " a well educated citizens and skills revolution, underpinned by science, technology and innovation". This means that there is a convergence on the importance and necessity of STEM for sustainable development in Africa in general and Zimbabwe in particular. The use of mother tongue can therefore play a critical role in meeting the need for people with skills in STEM in Zimbabwe.

TRANSLANGUAGING AND CODE SWITCHING

Flores (2019) averred that interest in translanguaging has expanded in contemporary times. It was initially mooted to indicate a pedagogical approach, but Ofelia Garcia widened the concept to embrace numerous discursive practices employed by bilingual communities. The expanded conceptualisation led to the argument that translanguagung is the right approach to challenging the dominance of certain languages such as English in academic space. Garcia & Lin, (2016) explicated translanguging is a term that is used to refer to how bilingual students navigate in a sophisticated and dynamic way to express themselves by using alternative languages and also to the pedagogical practices.

The work of Garcia has led to a call to incorporate translanguaging in the teaching in the classes (Garcia & Lin, Translanguaging in Bilingual Education, 2016). Translanguaging enables students to transverse the culturally determined linguistic constrictions as they converse. This approach has challenged the tendency which has characterised teaching of STEM in Zimbabwe across the education spectrum where the use of English is perceived as the norm. Bilingualism has also been described in terms such as additive bilingualism and substractive bilingualism. Subtrative bilingualism refers to a process where children when they acquire a second language, they loss competence in the first language. In additive bilingualism a student is able to acquire a second language and then sustain the use of mother tongue at home (Garcia & Lin, Translanguaging in Bilingual Education, 2016). Research shows that additive bilingualism when practiced in education when compared to substrative bilingualism, it can lead to better learning outcomes among students. Additive bilingualism helps students to develop a sense of self esteem, as they feel connected to their culture and are able to express themselves in the language of the school and also language they use at home (Garcia,2008).

Nhongo and Tshotsho (2020) advocates the use of translanguaging as one option to resolve the students learning challenges they face in STEM. The proposition to use translanguaging in the teaching of STEM in Zimbabwe and other developing nations has been based on points which include the multiplicity of languages that are in Zimbabwe and other African countries. Another counter point to translanguing in STEM in Zimbabwe is that African languages are not highly developed to meet the demand of the terminology needed teaching STEM (Ozoemena, Ngweke, & Nwoko 2021; Ezeh, Umeh, & Anynwu, 2022). These counterpoints are based on the presupposition that STEM has to be taught in a multilingual classes where none of them is neglected in the class (Nhongo & Tshotsho,2020). Practical considerations means that the use of African language doesn't have to be undertaken in an exclusionary way where no other foreign language may be used, but the use of mother tongue in learning environment that is inclusive.

Lemmi & Perez (2023) explicated translanguaging as a linguistic practice which allow students to extract from their language repository which may be in mother tongue although they are taught in another language. Practices of this nature which takes into account the prior language competence a student brings to the learning engagement is just beginning to be appreciated in the teaching of sciences (Pierson, Clark, & Brady, 2021;Charamba, 2020) . Daniel & Pacheno (2015) expounds that language is a resource for construction of meaning. In translanguaging a student can manoeuvre from one across languages and registers of speech and are able to generate meaning in many languages. Dougherty (2021) asserted that students and instructors of learning can strengthen their metalinguistic awareness by summarising and expanding the vocabulary that they use through the medium of translanguaging.

Translanguaging practices encompass the utilisation of "code switching, translating, language brokering or interpreting between culturally and linguistically diverse individuals" (Dougherty, 2021, p. 12). In order to implement translanguaging successfully in the teaching of STEM, it is imperative that the instructors in STEM be conscious of the ideologies that are dominant in the learning context regarding the use of language, especially those languages that are considered powerful and those that are marginalised. These perceptions have an impact on perceptions that people have regarding the languages that are thought to promote the capacity of learners to secure jobs and those that are seen as a liability. Ideologies are not static but are fluid across populations and may demonstrate a conflict even within an individual. It is critical that instructors are aware of reasons that make students use language and then appropriate advantage of that in the teaching and learning context (Daniel & Pacheno, 2015).

In Zimbabwe the use of English is perceived to give students an advantage in terms of job opportunity and also chances of pursuing higher education. It is for that reason that teachers, parents and administrators tend to habour negative sentiments regarding the use of mother tongue for instruction beyond grade 4 (Daniel & Pacheno, 2015; Nziramasanga, 1999). This chapter contends that if instructors take attention of the students capacity for translanguaging after grade 4, in the learning of STEM, it can help enhance their chances of progressing in such subjects. Of course, in this chapter it is also acknowledged that translanguaging comes with a number of challenges for teachers as well such as the challenge on their part when it comes to using multiple languages and also the challenges regarding the learning resources to make such efforts successful.

Academic research demonstrates that the ability of learners to master a second language relies on their capacity to function in mother tongue (Daniel & Pacheno, 2015). Ezeh, Umeh, & Anynwu (2022) averred that Code switching and Code mixing are some of the tools at the disposal of the instructor in STEM that can help enhance learning. However the use of Code switching and Code mixing has been perceived as a sign of inability to communicate in a foreign language like English (Ezeh, Umeh, & Anynwu, 2022). Code switching and Code mixing helps an instructor to facilitate learning in a class where learners have knowledge of other languages which is different from the official language of instruction.

Benson (2001) and Panhwar & Buriro (2020) expounded that the history of codeswitching can be traced to the 1970s and late 1960s by Albarrilo and early 1970s. Benson (2001) explicate code switching was originally used in the 1950s, although its etymology is largely not known due to interdisciplinary nature of the term. By 1972 the term codeswitching was well established in linguistics. Panhwar & Buriro (2020) asserted that Codeswitching is commonly in use in bilingual contexts where students can engage in mother tongue and then a second language. In Codeswitching the elements of language are altered in order to communicate in different languages and also different forms of engagements.

USING MULTILINGUALISM FOR SUSTAINABLE DEVELOPMENT

Given the multiplicity of languages in Zimbabwe and the central role of language in communicating ideas, it is vital that the country implements a multilingual approach in order to promote sustainable development. English by virtue of its colonial heritage has tended to occupy a central role in didactic processes at all levels of learning in Zimbabwe, in spite of the fact that the majority of the learners across the educational spectrum are from a diverse linguistic background. This means that children from other linguistic backgrounds have to communicate in mother tongue at home while using English at school, a development that interferes with their capacity to engage effectively in STEM.

Zimbabwe introduced a new constitution in 2013, which acknowledged promulgated in 2013 averred that the official languages that can be used in Zimbabwe in line with the ethnic composition are Chewa, Chibarwe, English, Kalanga, Koisan, Nambya, Ndau, Ndebele, Ndebele, Shangani, Shona, sign language, Sotho, Tonga, Tswana, Venda, and Xhosa (Republic of Zimbabwe, 2018). The Constitution added that all state institutions including schools and tertiary institutions, to identify that the use of languages is done equitably. Moreover, the 6 (4) is explicit in terms of schools recognizing the equal value of languages that are used in Zimbabwe that:

The state must promote and advance the use of all languages used

in Zimbabwe, including sign language and must create conditions for the

development of those languages.

This clause in the constitution of Zimbabwe is explicit in recognizing the importance of language for sustainable development especially the use of mother tongue in institutions of learning. However, these noble intentions normally demonstrate noble intentions that may not reach fruition due to the recalcitrant nature of schools cultures in Zimbabwe and the continued veneration of English above other languages.

The proposition to use mother tongue as a medium of instruction has not been without its own critiques. Ozoemena, Ngwoke, & Nwoke (2021) avers that in a country that has multiplicity of indgenous languages, the implementation of instruction in mother tongue can prove to be very challenging.

MULTILINGUAL PRACTICE IN CHINA

China is home to a wide array of ethnicities, each with its own languages and traditions. Mandarin, the standard Chinese variant known as Putonghua, is the official language of education in China and is used as the medium of instruction in most schools across the country. The languages in China adds up to 80 and they are all derived from five categories which are Sino-Tibetan, Altaic, Austronesian, South Asian and Indo-European. However, China recognises the linguistic and cultural diversity of its population, and there are efforts to preserve and promote indigenous languages in certain regions and among specific ethnic groups, thus bilingual education programmes were implemented in some regions. For example, in Tibetan schools, Tibetan language is used as a medium of instruction alongside Mandarin. Shan (2018) avers that China constitute a multiplicity of languages numbering 55. Minority groups in China are dispersed over the entire country and they display a cultural and linguistic diversity. The majority of the population is made of hans who constitute 91,5% of the population and the rest of the population making up 8,5% of the population constituting the minority group (Shan, 2018).

The government of China has followed a multilingual educational framework that seeks to promote linguistic and cultural diversity. Minority groups use mother tongue to communicate and a minute number can speak Chinese. A significant part of the minority of the population are multilingual and have the competence to converse in Chinese, mother tongue and possibly two other languages. China has adopted a multilingual education system that allows for the learning of three languages in a formal setting. The ethnic minority language is however, utilized as the medium of instruction. China uses four distinct models in its educational system which are the use of mother tongue, use of Chinese, bilingual education which involves the use of Chinese and English and use of Chinese and minority languages.

DISCUSSION

It is apparent that many previously colonized states such as Zimbabwe, South Africa and Namibia utilize colonial languages as medium of instruction in primary schools. The sentiments of the generality of the population is for the monolingual practice which is based on the use of one colonial language. The presupposition of the use of a colonial language was that it would promote unity in African states that have numerous language and a wide range of cultural diversity (Katukula, Set, & Nyambe, 2023). It was also acknowledged in the foregoing discussion that local languages suffered from neglect during the colonial era and post independent period that they do not seem to be adequate for (Chivhunga & Chimhenga, 2013; Didham & Ofel-Manu, 2015). However the effect of using a foreign language when teaching STEM, has been noted as devaluing the self -esteem of the learner and also a negative effect when it comes to engaging in learning. The social cultural theory acknowledges the importance of language for transmitting culture and for cognitive development of the learning.

It is noteworthy that the success of teaching STEM in mother tongue depends on a transformation of attitudes (Beka, 2016). UNESCO is a firm advocate for the use of mother tongue instruction. In spite of UNESCO disposition towards the use of mother tongue from early childhood and primary schools since 1953, the practices in many states upholds monolingualism.

It is also important to take into account the principle of translanguaging in the teaching of STEM. This means that in place of adopting an inflexible language policy that expunges the teaching in English in STEM, mother tongue has to be included in the class in any inclusive way. Not only does the use of mother tongue alongside English in Zimbabwe across the educational spectrum promote the self -esteem of the students, it also builds democratic classes (Daniel & Pacheno, 2015). The training of instructors in STEM in the use of code switching can also help promote sustainable learning.

RECOMMENDATIONS

In this chapter it was noted that there is a perception among stakeholders such as parents, guardians and administrators habour the sentiments that the use of a colonial language such as English is associated with the provision of quality education. Presupposition that the use of a foreign language such as English has capacity to unite people in culture that are very diverse as in Nigeria were also explored. Local language have been underdeveloped over time that they are not perceived as being well developed for effective teaching of STEM. The chapter has also noted that such sentiments have little empirical value and based more on coloniality. Based on the foregoing observation it is submitted that African governments and other stakeholders in education in Zimbabwe must launch programmes that target parents, guardians and other stakeholders in education on the importance of using mother tongue for sustainable development. It also of essence to ensure that the government of Zimbabwe begin to invest resources in the development of local languages that are acknowledged in the constitution for the purpose of utilizing them for teaching of STEM, in line with earlier recommendations by Nziramasanga Presidential Commission of Inquiry into Education (Nziramasanga, 1999). The use of mother tongue needs also to sustained up to university level with the use of other languages that have international value such as English in a multilingual model. The use of mother tongue as a tool for instruction has the potential of ensuring that "educational, technological, cultural levels and spiritual levels of Zimbabweans" could be lifted up (Nziramasanga, 1999,p156).

The use of mother tongue can help to accelerate national development and the development of technology in Zimbabwe. The government needs to review the language policy in Zimbabwean schools in order to align it to the use of mother tongue and a multilingual model. It is apparent that students come to class with linguistic resources that can help them to learn through the process of translanguaging. Given the ideological nature regarding the perceptions that instructors and parents have towards languages it is pre-eminent that people change their perceptions and allow for translanguaging to occur in STEM classes. Tranlanguaging helps facilitate learning for students in STE by allowing students to appropriate linguistic advantages they bring to school when they learn STEM. In this chapter it is recommended that instructors of STEM be trained in the use of translanguaging for the sake of sustainable education in Zimbabwe.

CONCLUSION

The paper submits that although there is a widespread practice in using English as an official language of instruction in many states, it is a practice that can impeded the learning of students in STEM. It is a practice that is not supported by research, but is based more on coloniality. Such practices therefore do not support sustainable development of learners in Zimbabwe and other previously colonized states. The implication of this observation is that stakeholders must endeavor to launch educational outreach to parents and administrators on the importance of leaning in mother tongue in order to shift educational policy in the direction of multilingual educational practices which gives mother tongue primacy in schools. It is multilingual educational practices that can lead to sustainable education in Zimbabwe.

REFERENCES

Abtahi, Y. (2018). Vygotsky: A learning theory or a theory of development. *FLM*, *38*(1), 21–25.

African Union. (2023). *African Union*. Retrieved from Agenda 2063: The Africa that we want: https://au.int/en/agenda2063/overview

African Union Academy of Languages. (2005). PanAfrican Masters and PhD Project in African Languages and Applied Lingustics (PAMAPAL). Addis Ababa: African Union Academy of Languages.

Alharbi. J, M. (2023). Insight into the role of interaction in language acquisition: Vygotsky's Interactionist Theory. Arab World English Journal, 4(2), 281-294.

Arshad, M., & Chen, W. (2009). Vygotsky Social Cultural Theory of Literacy: Scafolding Children to Read and Write at an Early Age. *Wacana*, *11*(2), 319–334. doi:10.17510/wjhi.v11i2.164

Beka, M. (2016). Mother tongue as medium of instruction: Benefits and challenges? *International Journal of Innovative Language, Literature & Art Studies*, *4*(1), 16–26.

Benson, E. (2001). The neglected early History of Codeswitching. *Language & Communication*, *21*(1), 23–36. doi:10.1016/S0271-5309(00)00012-4

Blommaert, J. (2001). The Asmara Declaration as a sociolingustic problem:Reflections on scholarstic and lingustic rights. *Journal of Sociolinguistics*, *5*(1), 131–155. doi:10.1111/1467-9481.00142

Cayley, M. (2023). *The Brundtland Report: A short Critique*. Retrieved from https://www.academia.edu/7757839/The_Brundtland_Report_A_Short_Critique

Charamba, E. (2020). Tranalanguaging in a multilngual class: A study of the relation between students' languages and epistemological access in science. *International Journal of Science Education*, *42*(11), 1779–1798. doi:10.1080/09500693.2020.1783019

Chivhunga, E., & Chimhenga, S. (2013). Language planning in Zimbabwe: The use indegenous Languages Shona as a medium of instruction in primary schools. *IOSR Journal of Humanities and Social Science, 12*(5), 58-65.

Daniel, S., & Pacheno, M. (2015). Translanguaging practices and perspectives of four Multilingual teens. *Journal of Adolescent & Adult Literacy*, 1–11.

de Castillo, J. M. (2015). Meaning and Language. *International Journal of Language and Linguistics*, 1-6.

Deng, F., & Zou, Q. (2016). A study on whether the adults the adults second language acquisition is easy or not from the perceptive of children native language acquisition. *Theory and Practice in Language Studies*, *6*(4), 776–780. doi:10.17507/tpls.0604.15

Didham, R., & Ofel-Manu, P. (2015). *The role of Education in Sustainable Development Agenda: Empowering a learning Society for sustainability through quality education.* http://data.myworld2015.org/

Dougherty, J. (2021). Translanguaging in action: Pedagogy that elevates. *ORTESOL Journal*, *38*, 19–31.

Du Pisani, J. (2007). Sustainable Development: Historical roots of the concept. *Environmental Sciences (Lisse)*, *3*(2), 83–96. doi:10.1080/15693430600688831

Ezeh, N., Umeh, I. A., & Anynwu, E. C. (2022). Code Switching and Code Mixing in Teaching and Learning of English as a second language: Building on knowledge. *English Language Teaching*, *15*(9), 106–113. doi:10.5539/elt.v15n9p106

Flores, N. (2019). Translanguaging into raciolingistic Ideologies: A personal reflection on legacy of Ofelia Garcia. *Journal of Multilingual Education Research*, *9*, 1–18.

Fomunyam, K. (2020). Towards Enhancing Science, Technology, Engneering, Mathematics (STEM) Education: A case for higher Education in Africa. *International Journal of Engineering Research & Technology (Ahmedabad)*, *13*(7), 1516–1524. doi:10.37624/IJERT/13.7.2020.1516-1524

Garcia, O. (2008). *Bilingual Education in the 21Century: A Global Perspective*. Wiley-Blackwell.

Garcia, O., & Lin, A. (2016). Translanguaging in Bilingual Education. In O. Garcia, A. Lin, & S. May (Eds.), *Bilingual and Multilingual Education, Encyclopedia of Language* (pp. 1–14). Springer International Publishing. doi:10.1007/978-3-319-02324-3_9-1

Katukula, K., Set, B., & Nyambe, J. (2023). Language ideologies and the use of mother tongue as the medium of instruction and learning in junior primary schools: A case of parents and teachers in a Namibian Schools. *Ivannore Journal of Education*, *11*(4), 70–79. doi:10.22159/ijoe.2023v11i4.48164

Kocabas, S., Ozfidan, B., & Burlbaw, L. (2020). American STEM Education in its global, National, and Linguistic Contexts. *EURASIA STEM Journal of Mathematics, Sciences and Technology and Technology Education*, *16*(1), 1–23.

Lemmi, C., & Perez, G. (2023). Translanguaging in Elementary Science. *International Journal of Science Education*, 1–27. doi:10.1080/09500693.2023.2185115

Lugones, M. (2008). The coloniality of gender. *Worlds and Knowledge Otherwise*, (Spring), 1–17.

Mendoza, B. (2016). Coloniality of Gender and Power: From Postcoloniality to Deconiality. In L. Disch & M. Hawkesworth (Eds.), *The Oxford Handbook of Feminist Theory* (pp. 1–24). Oxford University Press.

Mensa, J. (2019). Sustainable Development: Meaning, history, principles, and implications for action: Literature review. *Cogent Social Sciences*, *36*(1), 1–21. doi:10.1080/23311886.2019.1653531

Mufwene, S. (2006). Race, racialism, and the study of language evolution in America. In M. Picone & K. Davis (Eds.), *Language Variety in the South: Historical and Contemporary Perspectives* (pp. 1–49). University of Alabama Press.

Nhongo, R. (2013). A National Language Policy for Zimbabwe in the Twenty first Century:Myth or reality? *Journal of Language Teaching and Research*, *4*(6), 1208–1215. doi:10.4304/jltr.4.6.1208-1215

Nhongo, R., & Tshotsho, B. (2020). Is terminology the real problem in failure to implement STEM in African Languages? Translanguaging as an intevention strategy. *Multicultural Education*, *6*(2), 78–89.

Nziramasanga, D. (1999). *Zimbabwe Report of the Presidential Commission of Inquiry into Education and Training*. Harare: Republic of Zimbabwe.

Office of Science and Technology Policy. (2021). *Progress Report on the Implementation of the Federal STEM education Strategic Plan*. Office of Science and technology.

Ogbodo, J., Otaguruagu, E., Ngonebu, C., & Ogenyi, L. (2015). *New Horizons in Communication Skills: A Use of English Course Text*. Nsukka: Afro-Orbis Publishing Ltd.

Ozoemena, J., Ngwoke, F., & Nwoke, B. (2021). Prospects of mother tongue as a medium of instruction in Nigerian primary level education. *English Language Teaching, 14*(4), 1-8.

Ozoemena, T., Ngweke, F., & Nwoko, B. (2021). Prospects of mother tongue as a medium of instruction in Nigerian primary level education. *English Language Teaching*, *14*(4), 1–8. doi:10.5539/elt.v14n4p1

Panhwar, F., & Buriro, G. (2020). An overview of theories and approaches Code-Swtiching. *International Research Journal of Arts and Humanities*, *48*(48), 265–281.

Peresuh, M., & Masuku, J. (2002). The role of primary language in the bilingual and bicultural education in Zimbabwe. *Zambezia*, *29*(1), 27–37.

Pierson, A., Clark, D., & Brady, C. (2021). Scientific Modeling and Translanguaging: A multilingual and multimodal approach to support science learning and engagement. *Science Education*, *150*(4), 776–813. doi:10.1002/sce.21622

Shan, F. (2018). Multilingual Education in China: Taking the situation of Guizhou minority area as examples. *Theory and Practice in Language Studies*, *8*(2), 197–202. doi:10.17507/tpls.0802.03

Siekmann, G. (2016). *What is STEM? The need for unpacking its definitions and applications*. NCVER.

Toom, A. (2015). Teaching Leo Vygotsky's Theory of Socialcultural Theory Development with the use of literature and Information Technology. In A. Toom (Ed.), *Academic research of SSaH 2015* (pp. 60–79). Czech Technical Republic.

Toom, A. (2015). Teaching Leo Vygotsky's theory of sociocultural development with the use of literature and information technology. In *Academic research of SSaH*. Czech Technical University.

United Nations Educational Scientific and Cultural Organization (UNICEF). (2014). *Sustainable Development post 2015 begins with Education*. UNICEF.

United Nations Educational Scientific and Cultural Organizations (UNESCO). (2012). *Educational for Scientific and Cultural Development Source Book*. UNESCO.

Vygotsky, L. (1978). Mindset in society: The Development of higher psychological processes. Havard University Press.

World Commission on Economic Development (WCED). (1987). *From One Earth to One World: An Overview of Our Common Future: Report of the World Commission on Environment and Development*. Oxford University Press.

Yuniar, G. K. (2017). Story-Reading as a Teaching method for young Children: Teacher's Challenges. In *3rd International Conference on Early Childhood (ICECE-16)* (pp. 179-186). Atlantis Press.

KEY TERMS AND DEFINITIONS

Bilingual: A knowledge to speak or write in two languages.

Coloniality: A mindset that is programmed to function using a Eurocentric mindset in the previously colonized states.

Conditioning: A process used to prepare a person to accept certain conditions.

Eschatological: Dealing with events at the end of the world.

Monolingualism: A practice of speaking or writing in a single language.

Mother Tongue: The first dominant language a child is exposed to at birth before learning any other language.

Second Language: The second language a child is exposed to after learning the first language.

Translanguaging: The competency to move easily between two languages and it can also be used as a teaching approach that uses two languages.

Zone of Proximal Development: An area between what a student already knows and what the student is able to learn with support from another person.

Compilation of References

AAACSB. (2023). *Let's Lead Boldly: Elevating the Value of Business Schools.* Available online: https://www.aacsb.edu/insights/briefings/lets-lead-boldly-elevating-the-value-of-business-schools

Abdulrahim, A. (2016). *Preservation and conservation of records in the national archives and national museum.* Academic Press.

Abtahi, Y. (2018). Vygotsky: A learning theory or a theory of development. *FLM, 38*(1), 21–25.

Acisli, S., Yalcin, A., & Turgut, U. (2009). The influence of teaching based on 5e learning model on undergraduates' skills of scientific process and their attitudes towards physics laboratory in the electric topics. *Bayburt University Education Faculty Journal, 4*(1-2), 80–92.

Adetimirin, A. (2019). Educating library and information science students for an ethical information age. In W. M. Olatokun, A. O. Aremu, & A. Adetimirin (Eds.), Research on contemporary issues in media resources and information and communication technology use (pp. 99-112). Department of Library, Archival and Information Studies, University of Ibadan.

Aeschlimann, B., Herzog, W., & Makarova, E. (2016). How to foster students' motivation in mathematics and science classes and promote students' STEM career choice. A study in Swiss high schools. *International Journal of Educational Research, 79*, 31–41. doi:10.1016/j.ijer.2016.06.004

Affolderbach, J. (2022). Translating green economy concepts into practice: Ideas pitches as learning tools for sustainability education. *Journal of Geography in Higher Education, 46*(1), 43–60. doi:10.1080/03098265.2020.1849063

African Union Academy of Languages. (2005). PanAfrican Masters and PhD Project in African Languages and Applied Lingustics (PAMAPAL). Addis Ababa: African Union Academy of Languages.

African Union. (2023). *African Union.* Retrieved from Agenda 2063: The Africa that we want: https://au.int/en/agenda2063/overview

Agostini, L., Bigliardi, B., Filippelli, S., & Galati, F. (2021). Seller reputation, distribution and intention to purchase refurbished products. *Journal of Cleaner Production, 316*, 128296. doi:10.1016/j.jclepro.2021.128296

Aguirre, M. S., Aldamiz-Echevarría González de Durana, C., Charterina, J., & Vicente-Molina, M. (2003). El consumidor ecológico. *Distribución y consumo, 13*(67), 41-54.

Ajzen, I., & Fishbein, M. (1991). The Influence of Attitudes on Behavior. The Handbook of Attitudes, 32.

Ajzen, I., & Fishbein, M. (1980). *Understanding attitudes and predicting social behavior.* Prentice-Hall.

Albareda-Tiana, S., García-González, E., Jiménez-Fontana, R., & Solís-Espallargas, C. (2019). Implementing pedagogical approaches for ESD in initial teacher training at Spanish Universities. *Sustainability (Basel), 11*(18), 4927. doi:10.3390/su11184927

Albareda-Tiana, S., Vidal-Raméntol, S., Pujol-Valls, M., & Fernández-Morilla, M. (2018). Holistic approaches to develop sustainability and research competencies in pre-service teacher training. *Sustainability (Basel), 10*(10), 3698. doi:10.3390/su10103698

Albus, P., Vogt, A., & Seufert, T. (2021). Signaling in virtual reality influences learning outcome and cognitive load. *Computers & Education, 166*, 104154. doi:10.1016/j.compedu.2021.104154

Alharbi. J, M. (2023). Insight into the role of interaction in language acquisition: Vygotsky's Interactionist Theory. Arab World English Journal, 4(2), 281-294.

Al-Kuwari, M. M., Du, X., & Koç, M. (2021). Performance assessment in education for sustainable development: A case study of the Qatar education system. *Prospects, 52*(3–4), 513–527. doi:10.1007/s11125-021-09570-w

Allcoat, D., & von Mühlenen, A. (2018). Learning in virtual reality: Effects on performance, emotion and engagement. *Research in Learning Technology, 26*(0), 1–13. doi:10.25304/rlt.v26.2140

Allport, G. (1935). Attitudes. In Handbook of social psychology. Clark University Press.

Amabile, T. M. (1983). *The Social Psychology of Creativity*. doi:10.1007/978-1-4612-5533-8

Anastasiadis, S., Perkiss, S., Dean, B. A., Bayerlein, L., Gonzalez-Perez, M. A., Wersun, A., Acosta, P., Jun, H., & Gibbons, B. (2021). Teaching sustainability: Complexity and compromises. *Journal of Applied Research in Higher Education, 13*(1), 272–286. doi:10.1108/JARHE-02-2020-0029

Anđić, D., & Šuperina, L. (2021). How important is future teachers' "Connectedness to Nature"? Adaptation and validation of the connectedness to nature scale. *Education Sciences, 11*(5), 250. doi:10.3390/educsci11050250

Andrási, G., Körtvési, D., & Szegedi, K. (2022). Developing Ethical Future Managers Embracing Sustainability: Case Study of Budapest Business School. *International Journal of Web-Based Learning and Teaching Technologies, 17*(8), 1–19. doi:10.4018/IJWLTT.300786

Annan-Diab, F., & Molinari, C. (2017). Interdisciplinarity: Practical approach to advancing education for sustainability and the Sustainable Development Goals. *International Journal of Management Education, 15*(2), 73–83. doi:10.1016/j.ijme.2017.03.006

Antil, J. (1984). Socially Responsible Consumers: Profile and Implications for Public Policy. *Journal of Macromarketing, 4*(2), 18–39. doi:10.1177/027614678400400203

Arias, B. (2016). El consumo responsable: educar para la sostenibilidad ambiental. *Aibi Revista de Investigación, Administración e Ingeniería, 4*, 29-34. doi:10.15649/2346030X.385

Arshad, M., & Chen, W. (2009). Vygotsky Social Cultural Theory of Literacy: Scafolding Children to Read and Write at an Early Age. *Wacana, 11*(2), 319–334. doi:10.17510/wjhi.v11i2.164

Arslantas, E. E., & Yesilirmak, E. (2020). Changes in the climatic growing season in western Anatolia, Turkey. *Meteorological Applications.*

Association of Commonwealth Universities. (1993, August). Retrieved from https://www.iau-hesd.net/sites/default/files/documents/the_swansea_declaration.pdf

Atlhopheng, J. R., Mbongwe, B., & Segaetsho, T. (2020). Implementing Sustainable Development Goals at Institutional Level: The Case of University of Botswana. In S. O. Keitumetse, L. Hens, & D. Norris (Eds.), *Sustainability in Developing Countries*. Springer. doi:10.1007/978-3-030-48351-7_13

Augoustinos, M., Walker, I., & Donaghue, N. (2014). Social cognition: An integrated introduction. *Sage (Atlanta, Ga.)*.

Australian Institute for the Conservation of Cultural Material. (2022). *Conservation*. Available at: https://aiccm.org.au/conservation/

Awasthy, R., Flint, S., Sankarnarayana, R., & Jones, R. (2020). A framework to improve university–industry collaboration. *Journal of Industry-university Collaboration*, *2*(1), 49–62. doi:10.1108/JIUC-09-2019-0016

Babbie, E. (2010). *The practice of social research* (12th ed.). Wadsworth.

Badea, L., Şerban-Oprescu, G., Dedu, S., & Piroşcă, G. (2020). The impact of education for sustainable development on romanian economics and business students' behavior. *Sustainability (Basel)*, *12*(19), 8169. doi:10.3390/su12198169

Baer, H. A. (2022). How Environmentally Sustainable Is the Internationalisation of Higher Education? A View from Australia. In K. Bjørkdahl & A. S. Franco Duharte (Eds.), *Academic Flying and the Means of Communication* (pp. 103–132). Palgrave Macmillan. doi:10.1007/978-981-16-4911-0_5

Bagozzi, R. P., & Warshaw, P. R. (1990). Trying to consume. *The Journal of Consumer Research*, *17*(2), 127–140. doi:10.1086/208543

Bai, B., Shen, B., & Wang, J. (2021). Impacts of social and emotional learning (SEL) on English learning achievements in Hong Kong secondary schools. *Language Teaching Research*, •••, 13621688211021736. doi:10.1177/13621688211021736

Bailenson, J. (2018). *Experience on demand: What virtual reality is, how it works, and what it can do*. W. W. Norton & Company.

Bakal, I., & Akcali, E. (2009). Effects of Random Yield in Remanufacturing with Price-Sensitive Supply and Demand. *Production and Operations Management*, *15*(3), 407–420. doi:10.1111/j.1937-5956.2006.tb00254.x

Barbarossa, C., Beckmann, S. C., De Pelsmacker, P., Moons, I., & Gwozdz, W. (2015). The impact of different social media channels on consumers' buying decision. *Journal of Marketing Management*, *31*(1-2), 136–154. doi:10.1080/0267257X.2014.965148

Barber, N. A., Wilson, F., Venkatachalam, V., Cleaves, S. M., & Garnham, J. (2014). Integrating sustainability into business curricula: University of New Hampshire case study. *International Journal of Sustainability in Higher Education*, *15*(4), 473–493. doi:10.1108/IJSHE-06-2013-0068

Barnett, W. S. (2003). *Better teachers, better preschools: Student achievement linked to teacher qualifications* (Preschool Policy Matters No. 2). NIEER. Retrieved from https://nieer.org/resources/policybriefs/2.pdf

Başkan, Z., Alev, N., & Kara, I. S. (2010). Physics and mathematics teachers' ideas about topics that could be related or integrated. *Procedia: Social and Behavioral Sciences*, *2*(2), 1558–1562. doi:10.1016/j.sbspro.2010.03.235

Bass, F. M., & Wayne Talarzyk, W. (1972). An Attitude Model for the Study of Brand Preference. *JMR, Journal of Marketing Research*, *9*(February), 93–96. doi:10.1177/002224377200900121

Bass, F., & Wilkie, W. (1973). A Comparative Analysis of Attitudinal Predictions of Brand Preference. *JMR, Journal of Marketing Research*, *10*(3), 262–269. doi:10.1177/002224377301000306

Bates, T. (2019). *Teaching in a Digital Age: Guidelines for designing and learning* (2nd ed.). Tony Bates Associates LTD.

Baumber, A. (2021). Transforming sustainability education through transdisciplinary practice. *Environment, Development and Sustainability, 24*(6), 7622–7639. doi:10.1007/s10668-021-01731-3 PMID:34393621

Beckem, J. M., & Watkins, M. (2012). Bringing life to learning: Immersive experiential learning simulations for online and blended courses. *Journal of Asynchronous Learning Networks, 16*(5), 61–70.

Beka, M. (2016). Mother tongue as medium of instruction: Benefits and challenges? *International Journal of Innovative Language, Literature & Art Studies, 4*(1), 16–26.

Ben-Eliyahu, A. (2019). Academic Emotional Learning: A critical component of self-regulated learning in the emotional learning cycle. *Educational Psychologist, 54*(2), 84–105. doi:10.1080/00461520.2019.1582345

Benson, E. (2001). The neglected early History of Codeswitching. *Language & Communication, 21*(1), 23–36. doi:10.1016/S0271-5309(00)00012-4

Berglund, T., Gericke, N., Pauw, J. B., Olsson, D., & Chang, T. (2019). A cross-cultural comparative study of sustainability consciousness between students in Taiwan and Sweden. *Environment, Development and Sustainability, 22*(7), 6287–6313. doi:10.1007/s10668-019-00478-2

Bergmann, M., Schäpke, N., Marg, O., Stelzer, F., Lang, D. J., Bossert, M., & Sußmann, N. (2021). Transdisciplinary sustainability research in real-world labs: Success factors and methods for change. *Sustainability Science, 16*(2), 541–564. doi:10.1007/s11625-020-00886-8

Berkowitz, L., & Daniels, L. R. (1964). Affecting the salience of the social responsibility norm: Effects of past help on the response to dependency relationships. *Journal of Abnormal and Social Psychology, 68*(3), 275–281. doi:10.1037/h0040164 PMID:14126841

Berkowitz, L., & Lutterman, K. G. (1968). The Traditional Socially Responsible Personality. *Public Opinion Quarterly, 32*(2), 169–185. doi:10.1086/267597

Berliner, D. C. (1987). Simple views of effective teaching and a simple theory of classroom instruction. In D. C. Berliner & B. Rosenshine (Eds.), *Talks to teachers* (pp. 93–110). Random House.

Bermejo, B., Juiz, C., Cortes, D., Oskam, J., Moilanen, T., Loijas, J., Govender, P., Hussey, J., Schmidt, A. L., Burbach, R., King, D., O'Connor, C., & Dunlea, D. (2023). AR/VR Teaching-Learning Experiences in Higher Education Institutions (HEI): A Systematic Literature Review. *Informatics (MDPI), 10*(2), 1–16. doi:10.3390/informatics10020045

Bernardes, J., Ferreira, F., Marques, A. D., & Nogueira, M. (2018). "Do as I say, not as I do" - a systematic literature review on the attitude-behaviour gap towards sustainable consumption of Generation Y. *IOP Conference Series. Materials Science and Engineering, 459*, 012089. doi:10.1088/1757-899X/459/1/012089

Bertão, R. A., Jung, C. H., Chung, J., & Joo, J. (2023). Design thinking: A customized blueprint to train R & D personnel in creative problem-solving. *Thinking Skills and Creativity, 48*, 101253. doi:10.1016/j.tsc.2023.101253

Bettman, J. R., Capon, N., & Lutz, R. J. (1975). Multiattribute measurement models and multiattribute attitude theory: A test of construct validity. *The Journal of Consumer Research, 1*(4), 1–15. doi:10.1086/208602

Bianchi, C., & Bianchi, C. (2023). YPF Sustainability: Renovation of Service Stations and YPF Full Stores in Responsible Consumption and Sustainability. In *Case Studies from Corporate Social Responsibility, Social Marketing, and Behavioral Economics*. Springer Business Cases.

Bielawska, K., & Grębosz-Krawczyk, M. (2021). Consumers' Choice Behaviour Toward Green Clothing. *European Research Studies, 24*(2), 238–256. doi:10.35808/ersj/2124

Biggs, J. B., & Collins, K. F. (1982). *Evaluating the Quality of Learning: the SOLO Taxonomy*. Academic Press.

Bisquerra, R. (2003). Educación emocional y competencias básicas para la vida. *Revista de Investigación Educacional, 21*(1), 7–43.

Blackwell, R. D., Miniard, P. W., & Engel, J. F. (1986). *Consumer Behavior. Better World Books*. Estados Unidos de America.

Blair, T. R., Rupley, W. H., & Nichols, W. D. (2007). The effective teacher of reading: Considering the "what" and "how" of instruction. *The Reading Teacher, 60*(5), 432–438. doi:10.1598/RT.60.5.3

Blommaert, J. (2001). The Asmara Declaration as a sociolingustic problem:Reflections on scholarstic and lingustic rights. *Journal of Sociolinguistics, 5*(1), 131–155. doi:10.1111/1467-9481.00142

Borich, G. D. (2014). *Effective teaching methods* (8th ed.). Pearson Prentice Hall.

Borowski, T. (2019). CASEL's framework for systemic social and emotional learning. *Measuring SEL: Using Data to Inspire Practice, 8*, 1–7.

Boss, S., & Larmer, J. (2018). *Project-based teaching: How to create rigorous and engaging learning experiences*. Buck Institute for Education.

Boudon, R., & Bourricaud, F. (1989). *Dictionnaire critique de la sociologie*. Presses universitaires de France.

Bovill, C. (2014) An investigation of co-created curricula within HE in the UK, Ireland and the USA. *Innovations in Education and Teaching International, 51*(1), 15-25.

Bowen, G. A. (2009). Document Analysis as a Qualitative Research Method. *Qualitative Research Journal, 9*(2), 27–40. doi:10.3316/QRJ0902027

Bower, M., Lee, M. J. W., & Dalgarno, B. (2018). Collaborative learning across physical and virtual worlds: Factors supporting and constraining learners in a blended reality environment. *British Journal of Educational Technology, 48*(2), 407–430. doi:10.1111/bjet.12435

Brackenreg, J. (2004). Issues in reflection and debriefing: How nurse educators' structure experiential activities. *Nurse Education in Practice, 4*(4), 264–270. doi:10.1016/j.nepr.2004.01.005 PMID:19038168

Braga, M., Paccagnella, M., & Pellizzari, M. (2014). Evaluating students' evaluations of professors. *Economics of Education Review, 41*, 71–88. doi:10.1016/j.econedurev.2014.04.002

Brand, S., Felner, R., Shim, M., Seitsinger, A., & Dumas, T. (2003). Middle school improvement and reform: Development and validation of a school-level assessment of climate, cultural pluralism, and school safety. *Journal of Educational Psychology, 95*(3), 570–588. doi:10.1037/0022-0663.95.3.570

Bridgeland, J., Bruce, M., & Hariharan, A. (2013). *The missing piece: A national teacher survey on how social and emotional learning can empower children and transform schools*. A Report for CASEL. Civic Enterprises.

Brodin, E., & Avery, H. (2020). Cross-disciplinary collaboration and scholarly independence in multidisciplinary learning environments at the doctoral level and beyond. *Minerva, 58*(3), 409–433. doi:10.1007/s11024-020-09397-3

Brophy, J., & Alleman, J. (1991). Activities as instructional tools: A framework for analysis and evaluation. *Educational Researcher, 20*(4), 9–23. doi:10.2307/1176471

Brown, T. (2008). Design thinking. *Harvard Business Review, 86*(6), 84. PMID:18605031

Brown, T. (2009). *Change by Design: How Design Thinking Transforms Organizations and Inspires Innovation*. HarperCollins Publishers.

Brown, T., & Katz, B. (2011). Change by design. *Journal of Product Innovation Management*, *28*(3), 381–383. doi:10.1111/j.1540-5885.2011.00806.x

Brown, T., & Wyatt, J. (2010). Design thinking for social innovation. *Development Outreach*, *12*(1), 29–43. doi:10.1596/1020-797X_12_1_29

Bruce, B. C., & Bloch, N. (2013). Pragmatism and community inquiry: A case study of community-based learning. *Education and Culture*, *29*(1), 27–45. doi:10.1353/eac.2013.0004

Brundtland Commission. (1987). *Our Common Future, Chapter 2: Towards Sustainable Development*. Geneva, United Nation, World Commission on Environment and Development (WCED). http://www.un-documents.net/ocf-02.htm#I

Brunnquell, C., Brunstein, J., & Jaime, P. (2015). Education for sustainability, critical reflection and transformative learning: Professors' experiences in Brazilian administration courses. *International Journal of Innovation and Sustainable Development*, *9*(3-4), 321–342. doi:10.1504/IJISD.2015.071858

Bryant, S. M., Cullen, P. G., & Iannarelli, J. E. (2022). A Vison for Management Education: The AACSB Perspective. In M. R. Fellenz, S. Hoidn, & M. Brady (Eds.), *The Future of Management Education* (pp. 225–243). Routledge. doi:10.4324/9781003095903-17

Bryhn, A. C., & Belgrano, A. (2023). Teaching, learning and assessment methods for sustainability education on the land–sea interface. *Discover Sustainability*, *4*(1), 3. Advance online publication. doi:10.1007/s43621-023-00120-2 PMID:36694724

Bucea-Manea-Ţoniş, R., Bucea-Manea-Ţoniş, R., Simion, V. E., Ilic, D., Braicu, C., & Manea, N. (2020). Sustainability in Higher Education: The Relationship between Work-Life Balance and XR E-Learning Facilities. *Sustainability (Basel)*, *12*(14), 1–19. doi:10.3390/su12145872

Buchanan, R. (1992). Wicked Problems in Design Thinking. *Design Issues*, *8*(2), 5–21. doi:10.2307/1511637

Burbules, N., Fan, G., & Repp, P. (2020). Five trends of education and technology in a sustainable future. *Geography and Sustainability*, *1*(2), 93–97. doi:10.1016/j.geosus.2020.05.001

Burgin, X., Coli, S., & Daniel, M. C. (2021). Exploratory Study of Ecuadorian Teachers' Understanding of Social Emotional Learning: An Examination of Primary School Teachers. *GIST Education and Learning Research Journal*, *22*(22), 125–146. doi:10.26817/16925777.964

Bürgi, P. T., Jacobs, C. D., & Roos, J. (2005). From metaphor to practice in the crafting of strategy. *Journal of Management Inquiry*, *14*(1), 78–94. doi:10.1177/1056492604270802

Burr, V. (2015). *Social constructionism*. Routledge. doi:10.4324/9781315715421

Butler, M. G., Church, K. S., & Spencer, A. W. (2019). Do, reflect, think, apply: Experiential education in accounting. *Journal of Accounting Education*, *48*, 12–21. doi:10.1016/j.jaccedu.2019.05.001

Büyüköztürk, Ş., Kılıç Çakmak, E., Akgün, Ö. E., Karadeniz, Ş., & Demirel, F. (2013). *Bilimsel araştırma yöntemleri* [Scientific research methods]. Pegem Press.

Bybee, R. (2010). Advancing STEM education: A 2020 vision. *Technology and Engineering Teacher*, *70*(1), 30–35.

Bybee, R. W., Taylor, J. A., Gardner, A., Van Scotter, P., Powell, J. C., Westbrook, A., & Landes, N. (2006). The BSCS 5E instructional model: Origins and efec- tiveness. *Colorado Springs Co: BSCS*, *5*, 88–98.

Çakmak, M. (2009). Prospective teachers' thoughts on characteristics of an "effective teacher". *Education in Science, 34*, 74–82.

Calvert, F., Crowe, T. & Grenyer, B. (2016). Dialogical reflexivity in supervision: An experiential learning process for enhancing and relational competencies. *Journal of Supervision in Psychotherapy and Mental Health, 35*(1), 1-21. . do i:10.1080/07325223.2015.1135840

Calvo, P., & Lévy, J. P. (2020). *Foundations of augmented cognition: Directing the future of adaptive systems*. Springer.

Camacho-Otero, J., Boks, C., & Pettersen, I. (2018). Consumption in the Circular Economy: A Literature Review. *Sustainability (Basel), 10*(8), 2758. doi:10.3390/su10082758

Campbell, D. T., & Stanley, J. (1963). *Experimental and quasi-experimental designs for research*. Rand McNally.

Canavari, M., Dimech, M., & Caputo, V. (2011). Attitudes of Maltese Consumers Towards Quality in Fruit and Vegetables in Relation to Their Food-Related Lifestyles. *The International Food and Agribusiness Management Review*, 14.

Cantor, J. A. (1995). ASHEERIC Higher Education Report: Vol. 7. Experiential Learning in Higher Education. Academic Press.

Capgemini Research Institute. (2020). *Consumer products and retail. How sustainability is fundamentally changing consumer preferences*. Author.

Carayannis, E. G., & Morawska-Jancelewicz, J. (2022). The futures of Europe: Society 5.0 and Industry 5.0 as driving forces of future universities. *Journal of the Knowledge Economy, 13*(4), 3445–3471. doi:10.1007/s13132-021-00854-2

Cardiff University. (2023). *Cardiff University Postgraduate Prospectus 2024*. Cardiff University. Available at https://www.cardiff.ac.uk/__data/assets/pdf_file/0009/2713941/Cardiff-University-Postgraduate-Prospectus-2024.pdf

Carnegie Melon University. (n.d.). Retrieved from https://www.cmu.edu/about/mission.html

Carroll, M., Goldman, S., Britos, L., Koh, J., Royalty, A., & Hornstein, M. (2010). Destination, Imagination and the Fires within: Design Thinking in a Middle School Classroom. *International Journal of Art & Design Education, 29*(1), 37–53. doi:10.1111/j.1476-8070.2010.01632.x

Castaño, L., Perdomo-Ortiz, J., Dueñas-Ocampo, S., & Durán, W. (2016). Socially responsible consumption: An application in Colombia. *Business Ethics (Oxford, England), 25*(4), 460–481. doi:10.1111/beer.12128

Castañón, J. (2018). *El consumo sostenible en España. Influencia de la información sobre la acción ecológica entre los consumidores millennial*. Universidad Pontificia de Comillas.

Castillo, R., Fernández-Berrocal, P., & Brackett, M. A. (2013). Enhancing teacher effectiveness in Spain: A pilot study of the RULER approach to social and emotional learning. *Journal of Education and Training Studies, 1*(2), 263–272. doi:10.11114/jets.v1i2.203

Cavaliero, T. (2015). 'Creative blocs': Action research study on the implementation of Lego as a tool for reflective practice with social care practitioners. *Journal of Further and HE*, 1-10.

Cayley, M. (2023). *The Brundtland Report: A short Critique*. Retrieved from https://www.academia.edu/7757839/The_Brundtland_Report_A_Short_Critique

CCR. (2015). *Skills for the 21st Century: What should students learn?* Center for Curriculum Redesign. Available at https://4dedu.org/wp-content/uploads/2022/10/CCR-Skills-for-the-21st-Century-5-2015.pdf

CCR. (2019). *Competencies/Subcompetencies Framwork, Rev 1.0*. Available at https://curriculumredesign.org/framework/

Çelebi, N. (2015). Bir meslek olarak öğretmenlik [Teaching as a profession]. In A. Bakioğlu (Ed.), *Eğitim bilimine giriş* [Introduction to Education] (pp. 115–145). Nobel Akademik Yayıncılık.

Center for Curating the Archives. (2021). *2016 Workshops, The history and current practices of preventive conservation.* University of Cape Town. Available at: http://www.cca.uct.ac.za/cca/honours-in-curatorship/2016-2/workshops

Cerreta, M., & La Rocca, L. (2021). Urban Regeneration Processes and Social Impact: A literature review to explore the role of evaluation. In Lecture Notes in Computer Science (pp. 167–182). doi:10.1007/978-3-030-86979-3_13

Chamorro, A., & Palacios-González, M. (2020). Analysis of Socially Responsible Consumption: A Segmentation of Spanish Consumers. *Sustainability (Basel)*, *12*(20), 8418. Advance online publication. doi:10.3390/su12208418

Chancel, L., Piketty, T., Saez, G., & Zucman, G. (2022). *World Inequality Report 2022*. Academic Press.

Chang, J., & Le, T. N. (2010). Multiculturalism as a dimension of school climate: The impact on the academic achievement of Asian American and Hispanic youth. *Cultural Diversity & Ethnic Minority Psychology*, *16*(4), 485–492. doi:10.1037/a0020654 PMID:21058811

Charamba, E. (2020). Tranalanguaging in a multilngual class: A study of the relation between students' languages and epistemological access in science. *International Journal of Science Education*, *42*(11), 1779–1798. doi:10.1080/09500693.2020.1783019

Charroin, L., Fortin, B., & Villeval, M. C. (2022). Peer effects, self-selection and dishonesty. *Journal of Economic Behavior & Organization*, *200*, 618–637. doi:10.1016/j.jebo.2022.06.024

Chedrawi, C., Howayeck, P., & Tarhini, A. (2019). CSR and legitimacy in higher education accreditation programs, an isomorphic approach of Lebanese business schools. *Quality Assurance in Education*, *27*(1), 70–81. doi:10.1108/QAE-04-2018-0053

Chen, C. J., & Teh, C. S. (2013). Enhancing an instructional design model for virtual reality-based learning. *Australasian Journal of Educational Technology*, *29*(5), 699–716. doi:10.14742/ajet.247

Cheng, J. C. H., & Monroe, M. C. (2012). Connection to nature: Children's affective attitude toward nature. *Environment and Behavior*, *44*(1), 31–49. doi:10.1177/0013916510385082

Chen, J. (2007). Flow in games (and everything else). *Communications of the ACM*, *50*(4), 31–34. doi:10.1145/1232743.1232769

ChenY.ChenY.LinW.ZhengY.XueT.ChenC.ChenG. (2022). Application of Active Learning Strategies in Metaverse to Improve Student Engagement: An Immersive Blended Pedagogy Bridging Patient Care and Scientific Inquiry in Pandemic. https://ssrn.com/abstract=4098179 doi:10.2139/ssrn.4098179

Chen, Y., Ren, X., Xu, X., Wang, R., & Xu, Y. (2023). Evaluation of Living Environment Quality in Urban Residential Areas under the Concept of Urban Renewal—A Case Study of Binjiang District, Hangzhou, China. *Sustainability (Basel)*, *15*(19), 14104. doi:10.3390/su151914104

Chiang, C., Wells, P. K., & Xu, G. (2021). How does experiential learning encourage active learning in auditing education? *Journal of Accounting Education*, *54*, 100713. doi:10.1016/j.jaccedu.2020.100713

Chin, C., & Chia, L. G. (2004). Problem-based learning: Using students' questions to drive knowledge construction. *Science Education*, *88*(5), 707–727. doi:10.1002/sce.10144

Chivhunga, E., & Chimhenga, S. (2013). Language planning in Zimbabwe: The use indegenous Languages Shona as a medium of instruction in primary schools. *IOSR Journal of Humanities and Social Science, 12*(5), 58-65.

Christ University. (n.d.). *Centre for Soccial Action.* Retrieved from Christ University: https://christuniversity.in/center/C/Centre-for-Social-Action

Churcher, K. M. A., Downs, E., & Tewksbury, D. (2014). "Friending" Vygotsky: A social constructivist pedagogy of knowledge building through classroom social media use. *The Journal of Effective Teaching, 14*(1), 33–50. Retrieved January 2nd, 2023, from https://uncw.edu/cte/et/articles/Vol14_1/Churcher.pdf

Citysem. (2023). *Marketing sostenible, estrategias y ejemplos para un impacto positivo.* https://citysem.es/marketing-sostenible-estrategias-y-ejemplos/

Clarke, A. E., & Friese, C. (2007). Grounded theorizing using situational analysis. The Sage handbook of grounded theory, 363-397.

Clark, W., & Dickson, N. (2003). Sustainability science: The emerging research program. *Proceedings of the National Academy of Sciences of the United States of America, 100*(14), 8059–8061. doi:10.1073/pnas.1231333100 PMID:12794187

Claro, P. B., & Esteves, N. R. (2021). Teaching sustainability-oriented capabilities using active learning approach. *International Journal of Sustainability in Higher Education, 22*(6), 1246–1265. doi:10.1108/IJSHE-07-2020-0263

Claxton, G. (2015). *Intelligence in the Flesh: Why Your Mind Needs Your Body Much More Than It Thinks.* Yale University Press.

Clewell, A. F., & Aronson, J. (2013). Ecological references. In Island Press/Center for Resource Economics eBooks (pp. 137–153). doi:10.5822/978-1-59726-323-8_7

CMI. (2021). *Work ready graduates: Building employability skills for a hybrid world.* The Chartered Management Institute. Available online: https://www.managers.org.uk/wp-content/uploads/2021/09/employability-skills-research_work-ready-graduates.pdf

Cochrane, T., & Munn, J. (2016). EDR and Design Thinking: Enabling Creative Pedagogies. In *Proceedings of EdMedia 2016—World Conference on Educational Media and Technology* (pp. 315–324). Vancouver, BC, Canada: Association for the Advancement of Computing in Education (AACE). Retrieved August 1, 2023 from https://www.learntechlib.org/p/172969/

Comisión Mundial sobre el Medio Ambiente y el Desarrollo. (1987). *Nuestro futuro común.* Fondo de Cultura Económica.

Conley, C. S., Durlak, J. A., & Kirsch, A. C. (2015). A meta-analysis of universal mental health prevention programs for HE students. *Prevention Science, 16*(4), 487–507. doi:10.1007/s11121-015-0543-1 PMID:25744536

Cooper, L., Orrell, J., & Bowden, M. (2010). *Work integrated learning: A guide to effective practice.* Routledge. doi:10.4324/9780203854501

Corballis, M. C. (2017). The evolution of lateralized brain circuits. *Frontiers in Psychology, 8*, 1021. doi:10.3389/fpsyg.2017.01021 PMID:28670296

Corlu, S., Capraro, R. M., & Caprar, M. M. (2014). Introducing STEM education: Implications for educating our teachers for the age of innovation. *Education in Science, 39*(171), 74–85.

Corporation, B. (2023b). *Find a B Corp: Search here to buy from, work with, or learn more about Certified B Corporations around the world.* https://www.bcorporation.net/en-us/find-a-b-corp/?refinement%5BbftwCategories%5D%5B0%5D=Community

CorporationB. (2023a). *About B Lab.* https://www.bcorporation.net/en-us/movement/about-b-lab/

Corral-Verdugo, V., & Pinheiro, J. (2004). Aproximaciones al estudio de la conducta sustentable. *Medio Ambiente y Comportamiento Humano*, *5*, 1–26.

Costes-Onishi, P., & Kwek, D. (2023). Technical skills vs meaning-making: Teacher competencies and strength of inquiry-based learning in aesthetic inquiry. *Teaching and Teacher Education*, *130*, 104152. doi:10.1016/j.tate.2023.104152

Crabbé, A. (2019). Interdisciplinarity and sustainable development. In Springer eBooks (pp. 998–1006). doi:10.1007/978-3-030-11352-0_215

Crawford-Garrett, K. (2017). 'The problem is bigger than us': Grappling with educational inequity in TeachFirst New Zealand. *Teaching and Teacher Education*, *68*, 91–98. doi:10.1016/j.tate.2017.08.010

Creswell, J. W. (2014). *Research design. Qualitative, quantitative and mixed method approaches* (4th ed.). Sage.

Cruickshank, D. R., & Haefele, D. (2001). Good teachers, plural. *Educational Leadership*, *58*(5), 26–30.

Cuadra, E., & Moreno, J. M. (2005). Expanding opportunities and building competencies for young people: Anew agenda for secondary education. https://www.researchgate.net/publication/260517903_Introducing_STEM_Education_Implications_for_Educating_Our_Teachers_For_the_Age_of_Innovation

d.school. (2023). *Design thinking bootleg*. Retrieved August 1, 2023, from https://dschool.stanford.edu/resources/design-thinking-bootleg

Dagiliute, R., & Liobikiene, G. (2015). University contributions to environmental sustainability: Challenges and opportunities from the Lithuanian case. *Journal of Cleaner Production*, *108*, 891–899. Advance online publication. doi:10.1016/j.jclepro.2015.07.015

Daher, E., Maktabifard, M., Kubicki, S., Decorme, R., Pak, B., & Desmaris, R. (2021). Tools for citizen engagement in urban planning. In Springer eBooks (pp. 115–145). doi:10.1007/978-3-030-85566-6_7

Dalgarno, B., & Lee, M. J. (2010). What are the learning affordances of 3-D virtual environments? *British Journal of Educational Technology*, *41*(1), 10–32. doi:10.1111/j.1467-8535.2009.01038.x

Damgaard, M., & Nielsen, H. (2018). Nudging in education. *Economics of Education Review*, *64*, 313–342. Advance online publication. doi:10.1016/j.econedurev.2018.03.008

Daniel, S., & Pacheno, M. (2015). Translanguaging practices and perspectives of four Multilingual teens. *Journal of Adolescent & Adult Literacy*, 1–11.

Dann, S. (2018). Facilitating co-creation experience in the classroom with Lego Serious Play. *Australasian Marketing Journal*, *26*(2), 121–131. doi:10.1016/j.ausmj.2018.05.013

Danone. (2023a). *Danone Goals*. https://www.danone.es

Danone. (2023b). *Raison D'Être*. https://www.danone.com/about-danone/sustainable-valuecreation/danone-entreprise-a-mission.html

Day, C., Sammons, P., Stobart, G., Kington, A., & Gu, Q. (2007). *Teachers matter*. Open University Press.

De Bono, E. (1970). *Lateral Thinking: Creativity step by step*. Perennial Library.

de Castillo, J. M. (2015). Meaning and Language. *International Journal of Language and Linguistics*, 1-6.

de Freitas Netto, S. V., Sobral, M. F. F., Ribeiro, A. R. B., & Soares, G. R. D. L. (2020). Concepts and forms of greenwashing: A systematic review. *Environmental Sciences Europe*, *32*(1), 1–12. doi:10.1186/s12302-020-0300-3

De Jesus, A., & Mendonna, S. (2018). Lost in Transition? Drivers and Barriers in the Eco-Innovation Road to the Circular Economy. *Ecological Economics*, *145*, 75–89. doi:10.1016/j.ecolecon.2017.08.001

Deer, S., & Zarestky, J. (2017). Balancing profit and people: Corporate social responsibility in business education. *Journal of Management Education*, *41*(5), 727–749. doi:10.1177/1052562917719918

DeGiancomo, J. A. (2002). Experiential learning in higher education. *Forestry Chronicle*, *78*(2), 245–247. doi:10.5558/tfc78245-2

Dengel, A., Auer, A., Urlbauer, P., & Läufer, T. (2022). Game-Based Teaching of Basic Hardware Components With an Educational Virtual Reality at Different Levels of Immersion. ITiCSE 2022: Innovation and Technology in Computer Science Education, Dublin, Ireland.

Deng, F., & Zou, Q. (2016). A study on whether the adults the adults second language acquisition is easy or not from the perceptive of children native language acquisition. *Theory and Practice in Language Studies*, *6*(4), 776–780. doi:10.17507/tpls.0604.15

Denham, D., Rozance, M. A., Malone, M., & Goodling, E. (2020). Sustaining future environmental educators: Building critical interdisciplinary teaching capacity among graduate students. *Journal of Environmental Studies and Sciences*, *11*(1), 101–114. doi:10.1007/s13412-020-00611-y

Derakhshan, A., Coombe, C., Arabmofrad, A., & Taghizadeh, M. (2020). Investigating the effects of English language teachers' professional identity and autonomy in their success. *Issue Lang. Teach*, *9*, 1–28. doi:10.22054/ilt.2020.52263.496

Design Council. (2023). *The double diamond.* Retrieved August 1, 2023 from https://www.designcouncil.org.uk/our-resources/the-double-diamond/

DeSombre, E. R. (2018). *Why good people do bad environmental things.* Oxford University Press. doi:10.1093/oso/9780190636272.001.0001

Dewaele, J.-M., & MacIntyre, P. D. (2014). The two faces of Janus? Anxiety and enjoyment in the foreign language classroom. *Studies in Second Language Learning and Teaching*, *4*(2), 237–274. doi:10.14746/ssllt.2014.4.2.5

Dewey, J. (1933). *How We Think: A Restatement of the Relation of Reflective Thinking to the Educative Process.* DC Health.

Díaz-Chao, Á., & Torrent-Sellens, J. (2010). ¿ Pueden el uso de las TIC y los activos intangibles mejorar la competitividad? Un análisis empírico para la empresa catalana. *Estudios de Economía Aplicada*, *28*(2), 1–22.

Dickson, M. A., Eckman, M., Loker, S., & Jirousek, C. (2013). A model for sustainability education in support of the PRME. *Journal of Management Development*, *32*(3), 309–318. doi:10.1108/02621711311318337

Didham, R., & Ofel-Manu, P. (2015). *The role of Education in Sustainable Development Agenda: Empowering a learning Society for sustainability through quality education.* http://data.myworld2015.org/

Ding, X., & Li, Z. (2022). A review of the application of virtual reality technology in higher education based on Web of Science literature data as an example. *Frontiers in Education*, *7*, 1048816. doi:10.3389/feduc.2022.1048816

Dirkx, J. M. (2001). The power of feelings: Emotion, imagination, and the construction of meaning in adult learning. *New Directions for Adult and Continuing Education*, *2001*(89), 63–72. doi:10.1002/ace.9

Dix, K. L., Slee, P. T., Lawson, M. J., & Keeves, J. P. (2012). Implementation quality of whole school mental health promotion and students' academic performance. *Child and Adolescent Mental Health*, *17*(1), 45–51. doi:10.1111/j.1475-3588.2011.00608.x PMID:22518095

Dougherty, J. (2021). Translanguaging in action: Pedagogy that elevates. *ORTESOL Journal*, *38*, 19–31.

Dowdy, E. T., Dever, B. V., DiStefano, C., & Chin, J. K. (2011). Screening for Emotional and Behavioral Risk among Students with Limited English Proficiency. *School Psychology Quarterly*, *26*(1), 14–26. doi:10.1037/a0022072

Dresner, M., & Gill, M. (1994). Environmental education at summer nature camp. *The Journal of Environmental Education*, *25*(3), 35–41. doi:10.1080/00958964.1994.9941956

Driscoll, C., Price, S., McKee, M., & Nicholls, J. (2017). An assessment of sustainability integration and communication in Canadian MBA programs. *Journal of Academic Ethics*, *15*(2), 93–114. doi:10.1007/s10805-017-9275-0

Du Pisani, J. (2007). Sustainable Development: Historical roots of the concept. *Environmental Sciences (Lisse)*, *3*(2), 83–96. doi:10.1080/15693430600688831

Dubovi, I. (2022). Cognitive and emotional engagement while learning with VR: The perspective of multimodal methodology. *Computers & Education*, *183*, 104495. doi:10.1016/j.compedu.2022.104495

Dunne, D., & Martin, R. (2006). Design thinking and how it will change management education: An interview and discussion. *Academy of Management Learning & Education*, *5*(4), 512–523. doi:10.5465/amle.2006.23473212

Durlak, J. A., Weissberg, R. P., Dymnicki, A. B., Taylor, R. D., & Schellinger, K. (2011). The impact of enhancing students' social and emotional learning: A meta-analysis of school-based universal interventions. *Child Development*, *82*(1), 405–432. doi:10.1111/j.1467-8624.2010.01564.x PMID:21291449

Earle, A. G., & Leyva-de la Hiz, D. I. (2021). The wicked problem of teaching about wicked problems: Design thinking and emerging technologies in sustainability education. *Management Learning*, *52*(5), 581–603. doi:10.1177/1350507620974857

Earnest, D., Rosenbusch, K., Wallace-Williams, D., & Keim, A. (2016). Study abroad in Psychology: Increasing cultural competencies through experiential learning. *Teaching of Psychology*, *43*(1), 75–79. doi:10.1177/0098628315620889

Echegoyen-Sanz, Y., & Martín-Ezpeleta, A. (2021). A holistic approach to education for sustainability: Ecofeminism as a tool to enhance sustainability attitudes in pre-service teachers. *Journal of Teacher Education for Sustainability*, *23*(1), 5–21. doi:10.2478/jtes-2021-0002

Efstratia, D. (2014). Experiential education through project-based learning. *Procedia: Social and Behavioral Sciences*, *152*, 1256–1260. doi:10.1016/j.sbspro.2014.09.362

El-Jardali, F., Ataya, N., & Fadlallah, R. (2018). Changing roles of universities in the era of SDGs: Rising up to the global challenge through institutionalising partnerships with governments and communities. *Health Research Policy and Systems*, *16*(1), 38. doi:10.1186/s12961-018-0318-9 PMID:29743112

Elkington, J., & Hailes, J. (1988). *The green consumer guide*. Grafton Books.

Elsbach, K. D., & Stigliani, I. (2018). Design thinking and organizational culture: A review and framework for future research. *Journal of Management*, *44*(6), 2274–2306. doi:10.1177/0149206317744252

Enell-Nilsson, M., Jaskari, M.-M. & Borisov, P. (2023). Students' Perception of the Development of Generic Competencies in a User Innovation Course Context – A Longitudinal Study Comparing Different Teaching Modes in Higher Education. Nordic Journal of Business, 72(4), 218-236.

Enell-Nilsson, M., Jaskari, M. M., & Kantola, J. (2019). How to use interdisciplinary team innovation to foster crucial engineering competencies? In *Global Advances in Engineering Education* (pp. 101–115). CRC Press. doi:10.1201/b22184-5

Engel, J. F., Kollat, D. T., & Blackwell, R. D. (1968). *Consumer Behavior*. Holt, Rinehart & Winston.

Erdogan, M. (2009). *Fifth grade students' environmental literacy and the factors affecting students' environmentally responsible behaviors* [Unpublished doctoral dissertation]. Middle East Technical University, Turkey.

Erdogan, M., Erentay, N., Balaban, D., Çakır, U., Şen, C., Aydogan, B., & Nechita, A. (2010). Expanding the horizons through field trips: Developing global action plan for saving endangered species and threatened environments. Academic Press.

Erginsoy, G. (2022). *Antroposen çağ.* https://ansiklopedi.tubitak.gov.tr/ansiklopedi/antro posen_cag

Eriksson, P., & Kovalainen, A. (2008). *Qualitative Methods in Business Research.* Sage. doi:10.4135/9780857028044

Erkuş, A. (2009). *Davranış bilimleri için bilimsel araştırma süreci* [Scientific research process for behavioral sciences]. Seçkin Press.

Esigul, T. (2021). Innovative approaches to the ancient timbre of Anatolian kabak kemane. *Online Journal of Music Sciences*, *6*(2), 195–221. doi:10.31811/ojomus.1026560

Etse, D., & Ingley, C. (2016). Higher education curriculum for sustainability: Course contents analyses of purchasing and supply management programme of polytechnics in Ghana. *International Journal of Sustainability in Higher Education*, *17*(2), 269–280. doi:10.1108/IJSHE-07-2015-0121

Europea, C. (2015). *Comunicación de la comisión al parlamento europeo, al consejo, al comité económico y social europeo y al comité de las regiones. Cerrar el círculo: un plan de acción de la UE para la economía circular.* Bruselas.

European Commission. (2018). *Behavioural study on consumers' engagement in the circular economy: executive summary.* https://data.europa.eu/doi/10.2818/921596

Ezeh, N., Umeh, I. A., & Anynwu, E. C. (2022). Code Switching and Code Mixing in Teaching and Learning of English as a second language: Building on knowledge. *English Language Teaching*, *15*(9), 106–113. doi:10.5539/elt.v15n9p106

Fader, P. (2020). *Customer centricity: Focus on the right customers for strategic advantage.* University of Pennsylvania Press.

Fazelian, P., Ebrahim, A. N., & Soraghi, S. (2010). The effect of 5E instructional design model on learning and retention of sciences for middle class students. *World Conference on Psychology, Counselling and Guidance (WCPCG 2010), 5*, 140143. 10.1016/j.sbspro.2010.07.062

Fazio, R. H., Sanbonmatsu, D. M., Powell, M. C., & Kardes, F. R. (1986). On the automatic activation of attitudes. *Journal of Personality and Social Psychology*, *50*(2), 229–238. doi:10.1037/0022-3514.50.2.229 PMID:3701576

Ferguson, R., Coughlan, T., Egelandsdal, K., Gaved, M., Herodotou, C., Hillaire, G., ... Misiejuk, K. (2019). *Innovating Pedagogy 2019.* Open University Innovation Report 7.

Ferguson, T., Roofe, C., & Cook, L. D. (2021). Teachers' perspectives on sustainable development: The implications for education for sustainable development. *Environmental Education Research*, *27*(9), 1343–1359. doi:10.1080/13504622.2021.1921113

Ferreira, A. J. D., Lopes, M. A. R., & Morais, J. P. F. (2006). Environmental management and audit schemes implementation as an educational tool for sustainability. *Journal of Cleaner Production*, *14*(9-11), 973–982. doi:10.1016/j.jclepro.2006.01.003

Fiel'ardh, K., Fardhani, I., & Fujii, H. (2023). Integrating perspectives from Education for Sustainable Development to foster plant awareness among trainee science teachers: A mixed methods study. *Sustainability (Basel)*, *15*(9), 7395. doi:10.3390/su15097395

Fien, J., & Maclean, R. (2000). Teacher Education for Sustainability. II. Two Teacher Education Projects from Asia and the Pacific. *Journal of Science Education and Technology*, *9*(1), 37–48. doi:10.1023/A:1009420923692

Figueiró, P. S., & Raufflet, E. (2015). Sustainability in higher education: A systematic review with focus on management education. *Journal of Cleaner Production*, *106*, 22–33. doi:10.1016/j.jclepro.2015.04.118

Findler, F., Schönherr, N., Lozano, R., & Stacherl, B. (2018). Assessing the Impacts of Higher Education Institutions on Sustainable Development—An Analysis of Tools and Indicators. *Sustainability (Basel)*, *11*(1), 59. doi:10.3390/su11010059

Finger, D. C., Drăghici, C., Perniu, D., Smederevac-Lalić, M., Hălbac-Cotoară-Zamfir, R., Sehic, A., & Solomun, M. K. (2021). The importance of international collaboration to enhance education for environmental citizenship. *Sustainability (Basel)*, *13*(18), 10326. doi:10.3390/su131810326

Finlay, L., & Gough, B. (Eds.). (2008). *Reflexivity: A practical guide for researchers in health and social sciences*. John Wiley & Sons.

Flores, N. (2019). Translanguaging into raciolingistic Ideologies: A personal reflection on legacy of Ofelia Garcia. *Journal of Multilingual Education Research*, *9*, 1–18.

Florida State University. (n.d.). https://sustainablecampus.fsu.edu/about/mission

Fogarassy, C., & Finger, D. (2020). Theoretical and Practical Approaches of Circular Economy for Business Models and Technological Solutions. *Resources*, *9*(6), 76. doi:10.3390/resources9060076

Fomunyam, K. (2020). Towards Enhancing Science, Technology, Engineering, Mathematics (STEM) Education: A case for higher Education in Africa. *International Journal of Engineering Research & Technology (Ahmedabad)*, *13*(7), 1516–1524. doi:10.37624/IJERT/13.7.2020.1516-1524

Forgas, J. P., & Williams, K. D. (2014). *The social self: Cognitive, interpersonal and intergroup perspectives*. Psychology Press. doi:10.4324/9781315800516

França, C. L., Broman, G., Robert, K. H., Basile, G., & Trygg, L. (2017). An approach to business model innovation and design for strategic sustainable development. *Journal of Cleaner Production*, *140*, 155–166. doi:10.1016/j.jclepro.2016.06.124

Fransen, J., Peralta, D. O., Vanelli, F., Edelenbos, J., & Olvera, B. C. (2021). The Emergence of Urban Community Resilience Initiatives During the COVID-19 Pandemic: An International Exploratory Study. *European Journal of Development Research*, *34*(1), 432–454. doi:10.1057/s41287-020-00348-y PMID:33456209

Freedman, N. (1983). On psychoanalytic listening: The construction, paralysis, and reconstruction of meaning. *Psychoanalysis and Contemporary Thought*.

Freemana, S., Eddy, S. L., McDonough, M., Smith, M. K., Okoroafor, N., Hannah Jordt, H., & Wenderoth, M. P. (2014). Active learning increases student performance in science, engineering. *Proceedings of the National Academy of Sciences of the United States of America*, *111*(23), 8410–8415. PMID:24821756

Freina, L., & Ott, M. (2015). A literature review on immersive virtual reality in education: State of the art and perspectives. *Proceedings of the 11th International Scientific Conference eLearning and Software for Education*.

Frick, E., Tardini, S., & Cantoni, L. (2013). *LEGO® SERIOUS PLAY®*. Academic Press.

Friedman, F., & Goldbaum, C. (2016). Experiential learning: Developing insights about working with older adults. *Clinical Social Work Journal*, *44*(2), 186–197. doi:10.1007/s10615-016-0583-4

Fuchs, M., Fizek, S., Ruffino, P., & Schrape, N. (2014). *Rethinking gamification*. Meson Press.

Fuchs, T. (2019). The Experience of Time and its Disorders. In *The Oxford Handbook of Phenomenological Psychopathology* (p. 431). Oxford University Press.

Furner, J. M., & Kumar, D. D. (2007). The mathematics and science integration argument: A stand for teacher education. *Eurasia Journal of Mathematics, Science and Technology Education, 3*(3), 185–189. doi:10.12973/ejmste/75397

Gal, A. (2023). From recycling to sustainability principles: The perceptions of undergraduate students studying early childhood education of an education for sustainability course. *International Journal of Sustainability in Higher Education, 24*(5), 1082–1104. doi:10.1108/IJSHE-05-2022-0165

Gallardo-Vázquez, D., Barroso-Méndez, M., Pajuelo-Moreno, M., & Sánchez-Meca, J. (2019). Corporate Social Responsibility Disclosure and Performance: A Meta-Analytic Approach. *Sustainability (Basel), 11*(4), 1115. doi:10.3390/su11041115

Galleli, B., Teles, N. E. B., Freitas-Martins, M. S., Semprebon, E., & Hourneaux, F. Junior. (2022). Sustainability in management undergraduate courses: Mapping the Brazilian higher education institutions. *International Journal of Sustainability in Higher Education, 23*(7), 1628–1647. doi:10.1108/IJSHE-03-2021-0109

Garcia, O. (2008). *Bilingual Education in the 21Century: A Global Perspective.* Wiley-Blackwell.

Garcia, O., & Lin, A. (2016). Translanguaging in Bilingual Education. In O. Garcia, A. Lin, & S. May (Eds.), *Bilingual and Multilingual Education, Encyclopedia of Language* (pp. 1–14). Springer International Publishing. doi:10.1007/978-3-319-02324-3_9-1

Gardner, H. (1983). *Frames of mind.* Basic Books.

Gatti, L., Seele, P., & Rademacher, L. (2019). Grey zone in–greenwash out. A review of greenwashing research and implications for the voluntary-mandatory transition of CSR. *International Journal of Corporate Social Responsibility, 4*(1), 1–15. doi:10.1186/s40991-019-0044-9

Gauntlett, D. (2014). *The LEGO System as a tool for thinking, creativity, and changing the world.* Academic Press.

Geissdoerfer, M., Vladimirova, D., & Evans, S. (2018). Sustainable business model innovation: A review. *Journal of Cleaner Production, 198*, 401–416. Advance online publication. doi:10.1016/j.jclepro.2018.06.240

Genovese, A., Acquaye, A., Figueroa, A., & Koh, S. C. L. (2015). Sustainable Supply Chain Management and the transition towards a Circular Economy: Evidence and some Applications. *Omega, 66*, 344–357. Advance online publication. doi:10.1016/j.omega.2015.05.015

Georgiou, I., Zahn, C., & Meira, B. J. (2008). A systemic framework for case-based classroom experiential learning. *Systems Research and Behavioral Science, 25*(6), 807–819. doi:10.1002/sres.858

Gerlach, J. M. (1994). Is this collaboration? In K. Bosworth & S. J. Hamilton (Eds.), Collaborative Learning: Underlying Processes and Effective Techniques, New Directions for Teaching and Learning No. 59. Academic Press.

Ghisellini, P., Cialani, C., & Ulgiati, S. (2016). A review on circular economy: The expected transition to a balanced interplay of environmental and economic systems. *Journal of Cleaner Production, 114*, 11–32. doi:10.1016/j.jclepro.2015.09.007

Giannakos, M. N., Mikalef, P., & Pappas, I. O. (2021). Systematic Literature Review of E-Learning Capabilities to Enhance Organizational Learning. *Information Systems Frontiers, 24*(2), 619–635. doi:10.1007/s10796-020-10097-2 PMID:33551669

Giefer, J., Peterson, M., & Chen, Y. (2019). Understanding perceived corruption in business: Views from the United States and China. *Journal of International Business Ethics, 2*(1), 24–34. doi:10.18352/joibe.37

Gilbert, D. T., Pinel, E. C., Wilson, T. D., Blumberg, S. J., & Wheatley, T. P. (1998). Immune neglect: A source of durability bias in affective forecasting. *Journal of Personality and Social Psychology, 75*(3), 617–638. doi:10.1037/0022-3514.75.3.617 PMID:9781405

Gil-Doménech, D., Magomedova, N., Sánchez-Alcázar, E. J., & Lafuente-Lechuga, M. (2021). Integrating sustainability in the business administration and management curriculum: A sustainability competencies map. *Sustainability (Basel), 13*(16), 9458. doi:10.3390/su13169458

Girvan, C., Conneely, C., & Tangney, B. (2016). Extending experiential learning in teacher professional development. *Teaching and Teacher Education, 58*, 129–139. doi:10.1016/j.tate.2016.04.009

Glasson, J., & Wood, G. (2009). Urban regeneration and impact assessment for social sustainability. *Impact Assessment and Project Appraisal, 27*(4), 283–290. doi:10.3152/146155109X480358

Glen, R., Suciu, C., & Baughn, C. (2014). The need for design thinking in business schools. *Academy of Management Learning & Education, 13*(4), 653–667. doi:10.5465/amle.2012.0308

Global Resources Outlook. (2019). *Natural Resources for the Future We Want*. United Nations Environment Programme.

Goh, D., Muskat, B., & Tan, A. H. T. (2017). The nexus between sustainable practices in hotels and future Gen Y hospitality students' career path decisions. *Journal of Teaching in Travel & Tourism, 17*(4), 237–253. Advance online publication. doi:10.1080/15313220.2017.1362971

Goi, C., & Ng, P. Y. (2008). E-learning in Malaysia: Success factors in implementing e-learning program. *International Journal on Teaching and Learning in Higher Education, 20*(2).

Göksoy, S., & Arıcan, K. (2018). Öğretmenlik mesleği [Teaching profession]. In K. Kıroğlu & C. Elma (Eds.), *Eğitime giriş* [Introduction to Education] (pp. 260–277). Pegem Akademi.

Goldmark, S., Raucher, L., & Cardenas, A. (2023). Building a Circular Campus: Consumption, Net Zero Emissions, and Environmental Justice at Barnard College. In M. R. Maulicci, S. Pfirman, & H. Callahan (Eds.), *Transforming education for sustainability* (pp. 347–376). Springer. doi:10.1007/978-3-031-13536-1_20

González, J. S., & Sellens, J. T. (2009). Knowledge, networks and economic activity. Revisiting the network effects in the knowledge economy. *UOC Papers, e-Journal on the Knowledge Society,* (8).

Goodman, A., Joshi, H., Nasim, B., & Tyler, C. (2015). *Social and emotional skills in childhood and their long-term effects on adult life*. Institute of Education.

Gottardello, D., & Pàmies, M. M. (2019). Business School Professors' Perception of Ethics in Education in Europe. *Sustainability (Basel), 11*(3), 608. doi:10.3390/su11030608

Grabinger, S., & Dunlap, J. C. (2002, October). Problem-based learning as an example of active learning and student engagement: Invited talk. In *International Conference on Advances in Information Systems* (pp. 375–384). Springer Berlin Heidelberg. https://link.springer.com/chapter/10.1007/3-540-36077-8_39

Graça, J., Roque, L., Guedes, D., Campos, L., Trüninger, M., Godinho, C., & Vinnari, M. (2022). Enabling sustainable food transitions in schools: A systemic approach. *British Food Journal, 124*(13), 322–339. doi:10.1108/BFJ-11-2021-1188

Greenacre, L., Freeman, L., Jaskari, M. M., & Cadwallader, S. (2017). Editors' Corner: The "work-ready" marketing graduate. *Journal of Marketing Education, 39*(2), 67–68. doi:10.1177/0273475317712411

Greene, H. (2011). Freshmen Marketing: A first-year Experience with Experiential Learning. *Marketing Education Review, 21*(1), 79–88. doi:10.2753/MER1052-8008210111

Greening, H., Heck, K. L., McKinney, L., Diefenderfer, H. L., Boynton, W. R., Kleiss, B. A., Mishra, D. R., George, A. II, Kraft, B. C., Kling, C. A., & Windecker, L. A. (2022). Assessing the Effectiveness of Large-Scale Environmental Restoration: Challenges and Opportunities. *Estuaries and Coasts*, *46*(2), 293–301. doi:10.1007/s12237-022-01149-8

Green, R. A., Conlon, E. G., & Morrissey, S. M. (2017). Task values and self-efficacy beliefs of undergraduate psychology students. *Australian Journal of Psychology*, *69*(2), 112–120. doi:10.1111/ajpy.12125

Groh, F. (2012). Gamification: State of the art definition and utilization. *Institute of Media Informatics Ulm University*, *39*, 31.

Groth, C. (2017). *Making sense through hands: Design and craft practice analysed as embodied cognition.* Aalto University publication series. Doctoral Dissertations 1/2017. Available online: https://aaltodoc.aalto.fi/handle/123456789/24839

Grots, A., & Creuznacher, I. (2016). Design Thinking: Process or Culture? In Design Thinking for Innovation (pp. 183–191). Springer.

Guaman-Quintanilla, S., Everaert, P., Chiluiza, K., & Valcke, M. (2023). Impact of design thinking in higher education: A multi-actor perspective on problem solving and creativity. *International Journal of Technology and Design Education*, *33*(1), 217–240. doi:10.1007/s10798-021-09724-z

Günkör, C. (2017). Exploration of the relationship between educationand development. *Journal of International Social Sciences Education*, *3*(1), 14–32.

Gunnarsson, B. L., Linell, P., & Nordberg, B. (Eds.). (2014). *The construction of professional discourse*. Routledge. doi:10.4324/9781315842820

Guo, F., Yao, M., Wang, C., Yang, W., & Zong, X. (2016). The effects of service-learning on student problem solving: The mediating role of classroom engagement. *Teaching of Psychology*, *43*(1), 16–21. doi:10.1177/0098628315620064

Guo, Q., Tao, J., & Gao, X. (2019). Language teacher education in system. *System*, *82*, 132–139. doi:10.1016/j.system.2019.04.001

Guo, R., Tao, L., & Gao, P. (2014). The research on greenwashing brands' rebuilding strategies and mechanism of brand trust after biochemical and other pollutions. *Biotechnology*, *10*(9), 3270–3279.

Guo, S. Y., Wu, W. C., & Cai, A. T. (2020, November). A study on promoting teachers' professional quality to achieve sustainable development goals. *IOP Conference Series. Earth and Environmental Science*, *576*(1), 012013. doi:10.1088/1755-1315/576/1/012013

Gursoy, I. T. (2021). Spring celebration, Hıdrellez: Myth and play. *Leisure Sciences*, *43*(6), 593–605. doi:10.1080/01490400.2020.1870593

Hajirasouli, A., Banihashemi, S., Sanders, P., & Rahimian, F. (2023). BIM-enabled virtual reality (VR)-based pedagogical framework in architectural design studios. Smart and Sustainable Built Environment, ahead-of-print.

Hamilton, D., McKechnie, J., Edgerton, E., & Wilson, C. (2021). Immersive virtual reality as a pedagogical tool in education: A systematic literature review of quantitative learning outcomes and experimental design. *Journal of Computers in Education*, *8*(1), 1–32. doi:10.1007/s40692-020-00169-2

Hart, A., & Akhurst, J. (2017). Community-based learning and critical community psychology practice: Conducive and corrosive aspects. *Journal of Community & Applied Social Psychology*, *27*(1), 3–15. doi:10.1002/casp.2287

Harvey, M., Coulson, D., & McMaugh, A. (2016). Towards a theory of the Ecology of Reflection: Reflective practice for experiential learning in higher education. *Journal of University Teaching & Learning Practice, 13*(2), 1–20. doi:10.53761/1.13.2.2

Harvey, R. (1994). *Preservation in libraries: principles, strategies and practices*. Bowker.

Hawes, D., & Arya, A. (2022). *VR-Based Context Priming to Increase Student Engagement and Academic Performance.* 8th International Conference of the Immersive Learning Research Network (iLRN), Vienna, Austria.

Hayes, C. (2018). Case Study Research. In Methodologies for Practice Research: Approaches for Professional Doctorates. SAGE Publications Limited.

Hayes, C. (2015). Meaning-making through transformative learning for HCA education. *British Journal of Healthcare Assistants, 9*(7), 354–361. doi:10.12968/bjha.2015.9.7.354

Hay, R., & Eagle, L. (2020). Impact of integrated sustainability content into undergraduate business education. *International Journal of Sustainability in Higher Education, 21*(1), 131–143. doi:10.1108/IJSHE-05-2019-0174

Hayter, C. S., & Cahoy, D. R. (2018). Toward a strategic view of HE social responsibilities: A dynamic capabilities approach. *Strategic Organization, 16*(1), 12–34. doi:10.1177/1476127016680564

Hedstrom, M., & King, J.L. (n.d.). On the LAM: Library, Archive, and Museum Collections in the Creation and Maintenance of Knowledge Communities. *School of Information University of Michigan*, (2), 139–148.

Helyer, R., & Lee, D. (2014). The role of work experience in the future employability of higher education graduates. *Higher Education Quarterly, 68*(3), 348–372. doi:10.1111/hequ.12055

Henje, C., & Jaskari, M.-M. (2013). Student and teacher experiences of multidisciplinary cross-border co-operation. *Reformation, Revolution, Evolution : Universitetslärandet Ur Ett Tidsperspektiv : Konferensrapport*, 153–154.

Herod, L. (2012). *Adult learning: From theory to practice*. Available at http://en.copian.ca/library/learning/adult_learning/adult_learning.pdf

Herranen, J., Tolppanen, S., Vesterinen, V., & Aksela, M. (2020). Challenges and tensions in collaborative planning of a student-led course on sustainability education. *Nordic Studies in Science Education, 16*(1), 18–36. doi:10.5617/nordina.6583

Herrera, F., Bailenson, J., Weisz, E., Ogle, E., & Zaki, J. (2018). Building long-term empathy: A large-scale comparison of traditional and virtual reality perspective-taking. *PLoS One, 13*(10), e0204494. doi:10.1371/journal.pone.0204494 PMID:30332407

Hidalgo-Carvajal, D., Carrasco-Gallego, R., & Morales, G. (2021). From Goods to Services and from Linear to Circular: The Role of Servitization's Challenges and Drivers in the Shifting Process. *Sustainability (Basel), 13*(8), 4539. doi:10.3390/su13084539

Hill, L. M., & Wang, D. (2018). Integrating sustainability learning outcomes into a university curriculum: A case study of institutional dynamics. *International Journal of Sustainability in Higher Education, 16*(4), 699–720. doi:10.1108/IJSHE-06-2017-0087

Historical Association. (2022). *The voice of history, Library and Information studies.* Available at: https://www.history.org.uk/student/resource/2790/library-and-information-studies

Hobson, K., Holmes, H., Welch, D., Wheeler, K., & Wieser, H. (2021). Consumption Work in the circular economy: A research agenda. *Journal of Cleaner Production, 321*, 128969. doi:10.1016/j.jclepro.2021.128969

Hodgson, P., Lee, V. W. Y., Chan, J. C. S., Fong, A., Tang, C. S. Y., Chan, L., & Wong, C. (2019). Immersive Virtual Reality (IVR) in Higher Education: Development and Implementation. In Augmented Reality and Virtual Reality, Progress in IS. Springer.

Hoeckel, K. (2008). *Costs and benefits in vocational education and training*. OECD Publishing. https://www.oecd.org/education/innovation-education/41538706.pdf

Hoffmann, J. D., Brackett, M. A., Bailey, C. S., & Willner, C. J. (2020). Teaching emotion regulation in schools: Translating research into practice with the RULER approach to social and emotional learning. *Emotion (Washington, D.C.)*, *20*(1), 105–109. doi:10.1037/emo0000649 PMID:31961187

Holland, J. H. (2006). Studying complex adaptive systems. *Journal of Systems Science and Complexity*, *19*(1), 1–8. doi:10.1007/s11424-006-0001-z

Holst, J. (2022). Towards coherence on sustainability in education: A systematic review of Whole Institution Approaches. *Sustainability Science*, *18*(2), 1015–1030. doi:10.1007/s11625-022-01226-8

Horn, A., Scheffelaar, A., Urias, E., & Zweekhorst, M. B. (2023). Training students for complex sustainability issues: A literature review on inter-and transdisciplinary higher education design. *International Journal of Sustainability in Higher Education*, *24*(1), 1–27. doi:10.1108/IJSHE-03-2021-0111

Howard, J., & Sheth, J. (1969). *The Theory of Buyer Behavior*. . doi:10.2307/2284311

Hoyer, W., Macinnis, D., & Pieters, R. (2016). *Consumer Behavior*. Academic Press.

Hsu, T. (2011). *Skepticism grows over products touted as eco-friendly*. https://www.latimes.com/business/la-xpm-2011-may-21-la-fi-greenwash-20110521-story.html

Huambachano, M., Arulingam, I., Bowness, E., Korzenszky, A., Mungai, C., Termine, P., & Wittman, H. (2022). Knowledge networks to support youth engagement in sustainable food systems. *Frontiers in Sustainable Food Systems*, *6*, 867344. Advance online publication. doi:10.3389/fsufs.2022.867344

Huang, H. M., Rauch, U., & Liaw, S. S. (2010). Investigating learners' attitudes toward virtual reality learning environments: Based on a constructivist approach. *Computers & Education*, *55*(3), 1171–1182. doi:10.1016/j.compedu.2010.05.014

Huang, K. T., Ball, C., Francis, J., Ratan, R., Boumis, J., & Fordham, J. (2019). Augmented Versus Virtual Reality in Education: An Exploratory Study Examining Science Knowledge Retention When Using Augmented Reality/Virtual Reality Mobile Applications. *Cyberpsychology, Behavior, and Social Networking*, *22*(2), 105–110. doi:10.1089/cyber.2018.0150 PMID:30657334

Huang, R., & Jiang, L. (2021). Authentic assessment in Chinese secondary English classrooms: Teachers' perception and practice. *Educational Studies*, *47*(6), 633–646. doi:10.1080/03055698.2020.1719387

IMF. (2023). *Danone, una empresa 100% concienciada con la sostenibilidad*. https://blogs.imf-formacion.com/blog/mba/danone-empresa-sostenible/

Inglehart, R. (2011). *Modernization, cultural change, and democracy: The human development sequence*. Cambridge University Press.

International Education Specialist. (2022). *Information Science and Librarianship courses*. Available at: https://www.idp.com/australia/search/information-science-and-librarianship/

International Federation of Library Association. (1998). *Principles for the Care and Handling of Library Material, International Preservation Issues, Number One.* Available at: https://repository.ifla.org/bitstream/123456789/1183/1/ipi1-en.pdf

International Federation of Library Association. (2019). *Preservation and conservation.* Available at https://www.ifla.org/preservation-and-conservation

International Labour Organisation (ILO). (2009). *Rural Skills Training: A Generic Manual on Training for Rural Economic Empowerment (TREE).* ILO.

Iyishu, V. A., Nkanu, W. O., & Ogar, F. O. (2013). Preservation and conservation of library materials in digital age. *Journal Information and Knowledge Management, 04*(02).

Jacobs, C. D. (2016). Making Is Thinking: The Design Practice of Crafting Strategy. In Design Thinking for Innovation (pp. 131–140). Springer.

Jagers, R. J., Rivas-Drake, D., & Borowski, T. (2018). *Equity & social and emotional learning: A cultural analysis.* CASEL Assessment Work Group Brief Series.

Jagers, R. J., Rivas-Drake, D., & Williams, B. (2019). Transformative social and emotional learning (SEL): Toward SEL in service of educational equity and excellence. *Educational Psychologist, 54*(3), 162–184. doi:10.1080/00461520.2019.1623032

Jagers, R. J., Skoog-Hoffman, A., Barthelus, B., & Schlund, J. (2021). Transformative Social Emotional Learning: In Pursuit of Educational Equity and Excellence. *American Educator, 45*(2), 12.

James, A. (2015). Learning in Three Dimensions: Using Lego Serious Play for Creative and Critical Reflection Across Time and Space. In Global Innovation of Teaching and Learning in HE (pp. 275-294). Springer International Publishing.

Jankowska, D. M., & Omelańczuk, I. (2018). Potential Mechanisms Underlying the Impact of Imaginative Play on Socio-Emotional Development in Childhood. *Creativity. Theories–Research-Applications, 5*(1), 84–103. doi:10.1515/ctra-2018-0006

Jaskari, M.-M. (2012). Using pictorial collages in user-centred construction of a creative and effective learning environment. In Veer, Ballantine, P. W. & Ozanne, L. K. (Eds). Asia-Pacific Advances in Consumer Research Vol. 10, (pp. 284–292). Association for Consumer Research.

Jaskari, M.-M. (2013b). *Teaching the future marketers through experiential client-based projects: marketing knowledge and skills in the context of early phases of service and product development.* Acta Wasaensia, 290. https://osuva.uwasa.fi/handle/10024/7377

Jaskari, M.-M. (2015). Teaching the Fuzzy Front End of Innovation: Real-Life Application With Cross-Functional And International Teams. In *Marketing Dynamism & Sustainability: Things Change, Things Stay the Same.... Developments in Marketing Science: Proceedings of the Academy of Marketing Science.* Springer. 10.1007/978-3-319-10912-1_106

Jaskari, M. M. (2013a). The Challenge of Assessing Creative Problem Solving in Client-based Marketing Development Projects: A SOLO Taxonomy Approach. *Journal of Marketing Education, 35*(3), 231–244. doi:10.1177/0273475313485586

Jaskari, M.-M., & Syrjälä, H. (2023). A Mixed-Methods Study of Marketing Students' Game-Playing Motivations and Gamification Elements. *Journal of Marketing Education, 45*(1), 38–54. doi:10.1177/02734753221083220

Jensen, L., & Konradsen, F. (2018). A review of the use of virtual reality head-mounted displays in education and training. *Education and Information Technologies, 23*(4), 1515–1529. doi:10.1007/s10639-017-9676-0

Jeronen, E. (2022). Sustainable Education. In Springer eBooks (pp. 1–10). doi:10.1007/978-3-030-02006-4_237-1

Jiménez-Marín, G., Bellido-Pérez, E., & López-Cortés, Á. (2019). Marketing sensorial: El concepto, sus técnicas y su aplicación en el punto de venta. *Vivat Academia*, (148), 121–147. doi:10.15178/va.2019.148.121-147

Jnr, B. A. (2021). Green campus paradigms for sustainability attainment in higher education institutions – A comparative study. *Journal of Science and Technology Policy Management.*

Johansen, K., & Werner, S. (2022). Something is sustainable in the state of Denmark: A review of the Danish district heating sector. *Renewable & Sustainable Energy Reviews*, *158*, 112117. doi:10.1016/j.rser.2022.112117

Jonassen, D. H., & Hung, W. (2012). Problem-Based learning. In Springer eBooks (pp. 2687–2690). doi:10.1007/978-1-4419-1428-6_210

Jones, S., Bailey, R., Brush, K., & Kahn, J. (2018). *Preparing for effective SEL implementation.* Harvard Graduate School of Education Easel Lab. Available from Wallace Foundation website: https://www. wallacefoundation. org/knowledge-center/Documents/Preparing-for-Effective-SEL-Implementation. pdf

Jose, P. D. (2016). Sustainability education in Indian business schools: A status review. *AD-Minister*, *28*(28), 255–272. doi:10.17230/ad-minister.28.13

Joseph, C., & Said, R. (2020). Community-Based education: a participatory approach to achieve the sustainable development goal. In Encyclopedia of the UN sustainable development goals (pp. 101–111). doi:10.1007/978-3-319-95870-5_67

Jun, H., & Moon, S. (2021). An analysis of sustainability integration in business school curricula: Evidence from Korea. *Sustainability (Basel)*, *13*(5), 2779. doi:10.3390/su13052779

Juvonen, J., Nishina, A., & Graham, S. (2006). Ethnic diversity and perceptions of safety in urban middle schools. *Psychological Science*, *17*(5), 393–400. doi:10.1111/j.1467-9280.2006.01718.x PMID:16683926

Kalamas Hedden, M., Worthy, R., Akins, E., Slinger-Friedman, V., & Paul, R. C. (2017). *Teaching sustainability using an active learning constructivist approach: Discipline-specific case studies in higher education.* doi:10.3390/su9081320

Kalfa, S., & Taksa, L. (2017). Employability, managerialism, and performativity in HE: A relational perspective. *HE*, *74*(4), 687-699.

Kalyuga, S., Renkl, A., & Pass, F. (2010). Facilitating flexible problem-solving: Cognitive load perspective. *Educational Psychology Review*, *22*(2), 175–186. doi:10.1007/s10648-010-9132-9

KanwalK.MalikM.AkhtarA.RehmanF.KhanS. H.TabbasumA.HussainM. (2023). Investigating the effectiveness of sustainability integration strategies in business schools: A comparative study. *Journal of Xi'an Shiyou*, *66*(4), 147-176. http://dx.doi.org/ doi:10.17605/OSF.IO/RSBXH

Kanyengo, C. W. (2009). Preservation and Conservation of Information Resources in the University of Zambia Library. *Journal of Archival Organization*, *7*(3), 116–128. doi:10.1080/15332740903126736

Kapagan, E. (2019). Gelenekten geleceğe "Nevruz". *Karabük Türkoloji Dergisi*, *1*, 46–60.

Karwowski, M., & Jankowska, D. M. (2016). Four faces of creativity at school. *Nurturing creativity in the classroom*, 337-354.

Kasch, J., Schutjens, V. A. J. M., Bootsma, M. C., Van Dam, F. W., Kirkels, A. F., van der Molen, M. K., Rimac, A., & Rebel, K. T. (2023). Distance and presence in interdisciplinary online learning. A challenge-based learning course on sustainable cities of the future. *Journal of Integrative Environmental Sciences*, *20*(1), 2185261. Advance online publication. doi:10.1080/1943815X.2023.2185261

Katukula, K., Set, B., & Nyambe, J. (2023). Language ideologies and the use of mother tongue as the medium of instruction and learning in junior primary schools: A case of parents and teachers in a Namibian Schools. *Ivannore Journal of Education, 11*(4), 70–79. doi:10.22159/ijoe.2023v11i4.48164

Katuu, S. (2015). The development of archives and records management education and training in Africa–challenges and opportunities. *Archives and Manuscripts, 43*(2), 96–119. doi:10.1080/01576895.2015.1050677

Keene, S. (2008). *Collections for People: the stored collections of UK museums as a public resource*. UCL Institute of Archaeology. Available at https://www.collectionslink.org.uk/assets/userfiles/use_my_collection/collections_for_people.pdf

Kemper, J. A., & Ballantine, P. W. (2019). Moral identity and green consumption. *Journal of Business Research, 95*, 236–247.

Kemper, J. A., Ballantine, P. W., & Hall, C. M. (2019). Combining the 'why' and 'how' of teaching sustainability: The case of the business school academics. *Environmental Education Research, 25*(12), 1751–1774. doi:10.1080/13504622.2019.1667959

Kendall, J. C., Duley, T. C., Little, T. S., Permaul, J. S., & Rubin, S. (1986). *Strengthening experiential education within institution*. National Society for Internship and Experiential Education.

Kestly, T. A. (2014). *The Interpersonal Neurobiology of Play: Brain-Building Interventions for Emotional Well-Being (Norton Series on Interpersonal Neurobiology)*. WW Norton & Company.

Kidd, C. V. (1992). *The Evolution of Sustainability*. doi:10.1007/BF01965413

Kilbourne, W. E., & Beckmann, S. C. (1998). Review and critical assessment of research on marketing and the environment. *Journal of Marketing Management, 14*(6), 513–532. doi:10.1362/026725798784867716

Kildiroglu, M. (2020). Geçmişten günümüze Türklerde nevruz. *Kafkas Üniversitesi Sosyal Bilimler Enstitüsü Dergisi, 25*, 315–338. doi:10.9775/kausbed.2020.018

Killen, R. (2015). Teaching strategies for quality teaching and learning. Cape Town: Juta and Company (Pty) Ltd.

Kim, E. (2023). Sustainable New Product Development for Ten Thousand Villages, a Fair-Trade Social Enterprise: Empowering Women and Economic Development through Problem-Based Service Learning. *Sustainability (Basel), 15*(8), 6452. Advance online publication. doi:10.3390/su15086452

Kim, E., & Coonan, T. (2023). Advancing Sustainability Education through a Cross-Disciplinary Online Course: Sustainability and Human Rights in the Business World. *Sustainability (Basel), 15*(6), 4759. doi:10.3390/su15064759

Kirchherr, J., Reike, D., & Hekkert, M.P. (2017). Conceptualizing the Circular Economy: An Analysis of 114 Definitions. SSRN *Electronic Journal, 127*. doi:10.2139/ssrn.3037579

Kızıltepe, Z. (2002). Good and effective teacher. *Education in Science, 27*, 10–14.

Klein, J. T. (2020). Sustainability and Collaboration: Cross-disciplinary and Cross-Sector horizons. *Sustainability (Basel), 12*(4), 1515. doi:10.3390/su12041515

Klein, N., Ramos, T., & Deutz, P. (2020). Circular Economy Practices and Strategies in Public Sector Organizations: An Integrative Review. *Sustainability (Basel), 12*(10), 4181. doi:10.3390/su12104181

Klein, N., Ramos, T., & Deutz, P. (2022). Factors and strategies for circularity implementation in the public sector: An organisational change management approach for sustainability. *Corporate Social Responsibility and Environmental Management, 29*(3), 509–523. Advance online publication. doi:10.1002/csr.2215

Kocabas, S., Ozfidan, B., & Burlbaw, L. (2020). American STEM Education in its global, National, and Linguistic Contexts. *EURASIA STEM Journal of Mathematics, Sciences and Technology and Technology Education, 16*(1), 1–23.

Koç, S. E. (2019). The relationship between emotional intelligence, self-directed learning readiness and achievement. *International Online Journal of Education & Teaching, 6*(3), 672–688.

Koehn, P. H., & Uitto, J. I. (2014). Evaluating sustainability education: Lessons from international development experience. *Higher Education, 67*(5), 621–635. doi:10.1007/s10734-013-9669-x

Kolb, D., Boyatzis, R. E., & Mainemelis, C. (1999). *Experiential learning theory: Previous research and new directions.* Available at https://learningfromexperience.com/media/2010/08/experiential- learning-theory.pdf

Kolb, A. Y., & Kolb, D. A. (2005). Learning styles and learning spaces: Enhancing experiential learning in higher education. *Academy of Management Learning & Education, 4*(2), 193–212. doi:10.5465/amle.2005.17268566

Kolb, A. Y., & Kolb, D. A. (2011). Experiential learning theory: A dynamic, holistic approach to management learning, education and development. In *The SAGE Handbook of Management Learning*. Education and Development. doi:10.4135/9780857021038.n3

Kolb, D. (1984). *Experiential learning theory: Experience as source of learning and development.* Prentice-Hall.

Kollmuss, A., & Agyeman, J. (2002). Mind the gap: Why do people act environmentally and what are the barriers to pro-environmental behavior? *Environmental Education Research, 8*(3), 239–260. doi:10.1080/13504620220145401

Kong, Y. (2021). The role of experiential learning on students' motivation and classroom engagement. *Frontiers in Psychology, 12*, 771272. doi:10.3389/fpsyg.2021.771272 PMID:34744950

Konietzko, J., Bocken, N., & Hultink, E. J. (2020, January 5). ATool to Analyze, Ideate and Develop Circular Innovation Ecosystems. *Sustainability (Basel), 12*(1), 417. doi:10.3390/su12010417

Kooli, C. (2023). Chatbots in education and research: A critical examination of ethical implications and solutions. *Sustainability (Basel), 15*(7), 5614. doi:10.3390/su15075614

Kootshabe, T. J., & Mnjama, N. (2014). Preservation of government records in Botswana. *Journal of the South African Society of Archivists, 47*, 26–49.

Köroğlu, Z. Ç., & Öz, G. (2023). Pre-Service English Teachers' Speaking Skills Assessment Literacy: Transformation to Digital Assessment. *The Reading Matrix: An International Online Journal, 23*(1).

Köroğlu, Z. Ç., Öz, G., & Yüce, E. (2023). Inclusive Education Practices During the COVID-19 Pandemic From the Eyes of Turkish EFL Student Teachers: A Qualitative Research. In Promoting Diversity, Equity, and Inclusion in Language Learning Environments (pp. 175-192). IGI Global.

Korthagen, F. A. (2004). In search of the essence of a good teacher: Towards a more holistic approach in teacher education. *Teaching and Teacher Education, 20*(1), 77–97. doi:10.1016/j.tate.2003.10.002

Kostadinova, E. (2016). *Sustainable consumer behaviour: A literature review.* Economic Alternatives.

Kotler, P., & Armstrong, G. (2017). *Principles of marketing*. Pearson.

Kreber, C. (2001). Learning experientially through case studies? A conceptual analysis. *Teaching in Higher Education, 6*(2), 217–228. doi:10.1080/13562510120045203

Kristiansen, P., & Rasmussen, R. (2014). *Building a Better Business Using the Lego Serious Play Method.* John Wiley & Sons.

Krtalic, M., & Hasenay, D. (2011). Exploring a framework for comprehensive and successful preservation management in libraries. *The Journal of Documentation*, *68*(3), 353–377. doi:10.1108/00220411211225584

Kumar, V., & Ghodeswar, B. M. (2015). Impact of corruption on firm's competitive performance: An analysis of Indian firms. *Paradigm*, *19*(1), 20–33.

Kumar, V., & Ghodeswar, B. M. (2015). Importance of corporate social responsibility to societies. *Journal of Human Values*, *21*(1), 63–75. doi:10.1177/0971685814558677

Kuncoro, T., Ichwanto, M. A., & Muhammad, D. F. (2023). VR-Based Learning Media of Earthquake-Resistant Construction for Civil Engineering Students. *Sustainability (Basel)*, *15*(5), 1–12. doi:10.3390/su15054282

Kustandi, C., Fadhillah, D., Situmorang, R., Prawiladilaga, D., & Hartati, S. (2020). VR Use in Online Learning for Higher Education in Indonesia. *International Journal of Interactive Mobile Technologies*, *14*(1), 31–47. doi:10.3991/ijim.v14i01.11337

Kyaw, B. M., Saxena, N., Posadzki, P., Vseteckova, J., Nikolaou, C. K., George, P. P., Divakar, U., Masiello, I., Kononowicz, A. A., Zary, N., & Car, L. T. (2019). Virtual reality for health professions education: Systematic review and meta-analysis by the digital health education collaboration. *Journal of Medical Internet Research*, *21*(1), 1–13. doi:10.2196/12959 PMID:30668519

Lachance, J. R., Honigsfeld, A., & Harrell, G. (2019). Equal educational opportunity for English learners: ESL teachers' conceptualizations on the importance of academic language. *TESOL Journal*, *10*(2), e00415. doi:10.1002/tesj.415

Lai, Y.-C., & Peng, L.-H. (2019). Effective Teaching and Activities of Excellent Teachers for the Sustainable Development of Higher Design Education. *Sustainability (Basel)*, *12*(1), 28. doi:10.3390/su12010028

Lam, S. (2012). Project-Based learning. In Springer eBooks (pp. 2707–2709). doi:10.1007/978-1-4419-1428-6_1095

Lantu, D. C., Suharto, Y., Fachira, I., Permatasari, A., & Anggadwita, G. (2022). Experiential learning model: Improving entrepreneurial values through internship program at start-ups. *Higher Education. Skills and Work-Based Learning*, *12*(1), 107–125. doi:10.1108/HESWBL-01-2021-0014

Laroche, M., Bergeron, J., & Barbaro-Forleo, G. (2001). Targeting consumers who are willing to pay more for environmentally friendly products. *Journal of Consumer Marketing*, *18*(6), 503–520. doi:10.1108/EUM0000000006155

Laurie, R., Nonoyama-Tarumi, Y., Mckeown, R., & Hopkins, C. (2016). Contributions of education for sustainable development (ESD) to quality education: A research synthesis. *Journal of Education for Sustainable Development*, *10*(2), 226–242. doi:10.1177/0973408216661442

Lau, W. S., & Shea, M. (2022). Empowering English learners in the classroom through culturally responsive social-emotional teaching practices. *Journal of Multilingual and Multicultural Development*, •••, 1–18. doi:10.1080/01434632.2022.2078337

Lawson, H. A. (2005). Empowering people, facilitating community development, and contributing to sustainable development: The social work of sport, exercise, and physical education programs. *Sport Education and Society*, *10*(1), 135–160. doi:10.1080/1357332052000308800

Lazareva, S. A., & Marchuk, T. L. (2019). STEM technology as a means of forming the engineering thinking of school children. *Perm Pedagogical Journal*, *10*, 76–79.

Leal Filho, W., Amaro, N., Avila, L. V., Brandli, L., Damke, L. I., Vasconcelos, C. R., ... Salvia, A. (2021). Mapping sustainability initiatives in higher education institutions in Latin America. *Journal of Cleaner Production, 315*, 128093. doi:10.3390/su141912289

Lee, C. L. (1984). *Outdoor Education Activities for Elementary School Students.* Academic Press.

Lee, E.A.L., Wong, K.W., & Fung, C.C. (2010). How does desktop virtual reality enhance learning outcomes? A structural equation modeling approach. *Computers & Education, 55*(4), 1424-1442.

Leedy, P. D., & Ormrod, J. E. (2014). *Practical research.* Pearson.

Lee, E. A. L., & Wong, K. W. (2014). Learning with desktop virtual reality: Low spatial ability learners are more positively affected. *Computers & Education, 79*, 49–58. doi:10.1016/j.compedu.2014.07.010

Lee, J. W. Y. (2008). A review of the theories of corporate social responsibility: Its evolutionary path and the road ahead. *International Journal of Management Reviews, 10*(1), 53–73. doi:10.1111/j.1468-2370.2007.00226.x

Lee, K.-H. (2009). Why and How to Adopt Green Management into Business Organizations? The Case Study of Korean SMEs in Manufacturing Industry. *Management Decision, 47*(7), 1101–1121. doi:10.1108/00251740910978322

Lee, K.-H., & Shin, D. (2010). Consumers' responses to CSR activities: The linkage between increased awareness and purchase intention. *Public Relations Review, 36*(2), 193–195. Advance online publication. doi:10.1016/j.pubrev.2009.10.014

Leeuw, A., Valois, P., Ajzen, I., & Schmidt, P. (2015). Using the theory of planned behavior to identify key beliefs underlying pro-environmental behavior in high-school students: Implications for educational interventions. *Journal of Environmental Psychology, 42*, 128–138. Advance online publication. doi:10.1016/j.jenvp.2015.03.005

LehmannDonald, R. (1972). *HulbertJames.* AreThree-PointScalesAlwaysGoodEnough. doi:10.1177/002224377200900416

Lehmann, M., Christensen, P., Du, X., & Thrane, M. (2008). Problem-oriented and project-based learning (POPBL) as an innovative learning strategy for sustainable development in engineering education. *European Journal of Engineering Education, 33*(3), 283–295. doi:10.1080/03043790802088566

Leiva, L. R., Miranda, L. P., & Riquelme-Sanderson, M. (2021). Social justice in the preparation of English language teachers. *Mextesol J., 45*, 1–14.

Lemmi, C., & Perez, G. (2023). Translanguaging in Elementary Science. *International Journal of Science Education*, 1–27. doi:10.1080/09500693.2023.2185115

León, J., Domínguez, E. G., Núñez, J. L., Pérez, A., & Martin-Albo, J. (2011). Translation and validation of the Spanish version of the Echelle de Satisfaction des Besoins Psychologiques in academic context. *Anales de Psicología, 27*(2), 405–411 Retrieved from http://digitum.um.es/jspui/handle/10201/26530

Le, T. N., Lai, M. H., & Wallen, J. (2009). Multiculturalism and subjective happiness as mediated by cultural and relational variables. *Cultural Diversity & Ethnic Minority Psychology, 15*(3), 303–313. doi:10.1037/a0015507 PMID:19594259

Levy, P., Aiyegbayo, O., & Little, S. (2009). Designing for inquiry-based learning with the Learning Activity Management System. *Journal of Computer Assisted Learning, 25*(3), 238–251. doi:10.1111/j.1365-2729.2008.00309.x

Lewis, L. H., & Williams, C. J. (1994). In L. Jackson & R. S. Caffarella (Eds.), *Experiential Learning: A New Approach* (pp. 5–16). Jossey-Bass.

Liao, Y. C., & Ringler, M. (2023). Backward design: Integrating active learning into undergraduate computer science courses. *Cogent Education, 10*(1), 2204055. doi:10.1080/2331186X.2023.2204055

Liedtka, J. (2015). Design Thinking. *Journal of Product Innovation Management, 32*(6), 925–938. doi:10.1111/jpim.12163

Lim, C. K., Haufiku, M. S., Tan, K. L., Ahmed, M. F., & Ng, T. F. (2022). Systematic Review of Education Sustainable Development in Higher Education Institutions. *Sustainability (Basel), 14*(20), 13241. doi:10.3390/su142013241

Lindenmayer, D. B. (2020). Improving restoration programs through more significant connection with ecological theory and better monitoring. *Frontiers in Ecology and Evolution*, *8*, 50. Advance online publication. doi:10.3389/fevo.2020.00050

Lind, J., & Lindström, J. (2023). Towards a framework for exploring the indirect value of tourist attractions in place branding: The case of Tom Tits Experiment Science Center. *Scandinavian Journal of Hospitality and Tourism*, 1–23. doi:10.1080/15022250.2023.2233486

Li, P., Fang, Z., & Jiang, T. (2022). Research Into improved Distance Learning Using VR Technology. *Frontiers in Education*, *7*, 757874. doi:10.3389/feduc.2022.757874

Lopez-Agudo, L. A., & Marcenaro-Gutierrez, O. D. (2017). Engaging children in lessons: The role of efficient and effective teachers. *School Effectiveness and School Improvement*, *28*(4), 650–669. doi:10.1080/09243453.2017.1364272

López-Bonilla, J. M. (2018). The effect of corruption on economic growth in Mexico. International *Journal of Economics, Commerce and Management*, *6*(10), 45–55.

Lucas, C., Williams, K., Tudball, J., & Walpola, R. L. (2018). Community, hospital and industry preceptor perceptions of their role in experiential placements - The need for standardization of preceptor responsibilities and evaluations on students. *Currents in Pharmacy Teaching & Learning*, *10*(11), 1447–1455. doi:10.1016/j.cptl.2018.08.002 PMID:30514534

Ludwig, V. U., Brown, K. W., & Brewer, J. A. (2020). Self-regulation without force: Can awareness leverage reward to drive behavior change? *Perspectives on Psychological Science*, *15*(6), 1382–1399. doi:10.1177/1745691620931460 PMID:32857672

Lugones, M. (2008). The coloniality of gender. *Worlds and Knowledge Otherwise*, (Spring), 1–17.

Luo, H., Li, G., Feng, Q., Yang, Y., & Zuo, M. (2021). Virtual reality in K-12 and higher education: A systematic review of the literature from 2000 to 2019. *Journal of Computer Assisted Learning*, *37*(3), 887–901. doi:10.1111/jcal.12538

Lutz. (1991). *The role of attitude theory in marketing*. Academic Press.

Macheridis, N., & Paulsson, A. (2021). Greening higher education? From responsibilization to accountabilization in the incorporation of sustainability in higher education. *International Journal of Sustainability in Higher Education*, *22*(8), 208–222. doi:10.1108/IJSHE-09-2020-0338

MacIntyre, P. D., & Gregersen, T. (2012). Affect: The role of language anxiety and other emotions in language learning. In S. Mercer, S. Ryan, & M. Williams (Eds.), *Language learning psychology: Research, theory and pedagogy* (pp. 103–118). Palgrave. doi:10.1057/9781137032829_8

MacIntyre, P. D., Ross, J., & Clément, R. (2019). Emotions are motivating. In M. Lamb (Ed.), *The Palgrave handbook of motivation for language learning* (pp. 183–202). Palgrave Macmillan. doi:10.1007/978-3-030-28380-3_9

Mahameed, M. A. (2022). The implementation of sustainability practices in Arab higher education institutions. *Journal of Financial Reporting and Accounting*. doi:10.1108/JFRA-11-2022-0415

Makransky, G., & Lilleholt, L. (2018). A structural equation modeling investigation of the emotional value of immersive virtual reality in education. *Educational Technology Research and Development*, *66*(5), 1141–1164. doi:10.1007/s11423-018-9581-2

Maniatis, P. (2015). Investigating factors influencing consumer decision-making while choosing green products. *Journal of Cleaner Production*, *132*, 215–228. Advance online publication. doi:10.1016/j.jclepro.2015.02.067

Mariani, A., & Ismail, Z. (2013). Pengaruh Kompetensi Guru Matematik Ke Atas Amalan Pengajaran Kreatif. In *2nd International Seminar on Quality and Affordable Education (ISQAE 2013)* (pp. 181-187). Academic Press.

Mariani, P. (2023). *Climate Change and International Cooperation — ALI Social Impact Review*. ALI Social Impact Review. https://www.sir.advancedleadership.harvard.edu/articles/climate-change-and-international-cooperation

Marks, B., & Thomas, J. (2022). Adoption of virtual reality technology in higher education: An evaluation of five teaching semesters in a purpose-designed laboratory. *Education and Information Technologies*, *27*(1), 1287–1305. doi:10.1007/s10639-021-10653-6 PMID:34257511

Martindale, L., Cannone, C., Niet, T., Hodgkins, R., Alexander, K., & Howells, M. (2023). Empowering Tomorrow's Problem Solvers: Nexus Thinking and CLEWs Modelling as a Pedagogical Approach to Wicked Problems. *Energies*, *16*(14), 5539. doi:10.3390/en16145539

Martin, R. (2007). How successful leaders think. *Harvard Business Review*, *85*(6), 60. PMID:17580648

Martin, S. C. (2003). The influence of outdoor schoolyard experiences on students' environmental knowledge, attitudes, behaviors, and comfort levels. *Journal of Elementary Science Education*, *15*(2), 51–63. doi:10.1007/BF03173843

Matsuo, M. (2014). Instructional skills for on-the-job training and experiential learning: An empirical study of Japanese firms. *International Journal of Training and Development*, *18*(4), 225–240. doi:10.1111/ijtd.12035

Matthews, J., & Wrigley, C. (2017). Design and Design Thinking in Business and Management Higher Education. *Journal of Learning Design*, *10*(1), 41–54. doi:10.5204/jld.v9i3.294

Maturén, R., & Pizzolante, F. A. (2020). Entrepreneurial intentions of university students in a developing country: The case of Venezuela. *Journal of Entrepreneurship in Emerging Economies*, *12*(1), 1–23. doi:10.1108/JEEE-04-2019-0038

Mburayi, L., & Wall, T. (2018). Sustainability in the professional accounting and finance curriculum: An exploration. *Higher Education. Skills and Work-Based Learning*, *8*(3), 291–311. doi:10.1108/HESWBL-03-2018-0036

McDonald, F. (1975). *Research on teaching and its implications for policymaking: Report on phase II of the beginning teacher evaluation study*. Educational Testing Service.

McGrath, O., Hoffman, C., & Dark, S. (2023). Future Prospects and Considerations for AR and VR in Higher Education Academic Technology, Emerging Technologies and Trends. *EDUCAUSE Review*. https://er.educause.edu/articles/2023/4/future-prospects-and-considerations-for-ar-and-vr-in-higher-education-academic-technology

McKeown, R., Hopkins, C. A., Rizi, R., & Chrystalbridge, M. (2002). *Education for Sustainable Development Toolkit*. Energy, Environment and Resources Center, University of Tennessee Knoxville. http://kpe-kardits.kar.sch.gr/Aiforia/esd_toolkit_v2.pdf

McLean, A. (2018). *"Tää oli kiva tunti!": A social and emotional learning intervention in an EFL classroom*. Academic Press.

McMillan, J. H. (1996). *Educational research: Fundamentals for the consumer*. HarperCollins College Publishers.

McNiff, J., & Whitehead, J. (2006). *All you need to know about Action Research*. Sage.

Melani, B. Z., Roberts, S., & Taylor, J. (2020). Social emotional learning practices in learning English as a second language. *Journal of English Learner Education*, *10*(1), 3.

Mendoza, B. (2016). Coloniality of Gender and Power: From Postcoloniality to Deconiality. In L. Disch & M. Hawkesworth (Eds.), *The Oxford Handbook of Feminist Theory* (pp. 1–24). Oxford University Press.

Mensa, J. (2019). Sustainable Development: Meaning, history, principles, and implications for action: Literature review. *Cogent Social Sciences*, *36*(1), 1–21. doi:10.1080/23311886.2019.1653531

Merchant, Z., Goetz, E. T., Cifuentes, L., Keeney-Kennicutt, W., & Davis, T. J. (2014). Effectiveness of virtual reality-based instruction on students' learning outcomes in K-12 and higher education: A meta-analysis. *Computers & Education, 70*, 29–40. doi:10.1016/j.compedu.2013.07.033

Mezirow, J. (1991). Transformative dimensions of adult learning. Jossey-Bass.

Michael, F. L., Sumilan, H., Bandar, N. F. A., Hamadi, H. A. N. A., Jonathan, V., & Nor, N. N. M. (2020). Sustainable development concept awareness among students in higher education: A preliminary study. *Journal of Sustainability Science and Management, 15*(7), 113–122. doi:10.46754/jssm.2020.10.011

Micheli, P., Wilner, S. J., Bhatti, S. H., Mura, M., & Beverland, M. B. (2019). Doing design thinking: Conceptual review, synthesis, and research agenda. *Journal of Product Innovation Management, 36*(2), 124–148. doi:10.1111/jpim.12466

Mikropoulos, T. A., & Natsis, A. (2011). Educational virtual environments: A ten-year review of empirical research (1999-2009). *Computers & Education, 56*(3), 769–780. doi:10.1016/j.compedu.2010.10.020

Miles, M. B., & Huberman, A. M. (1994). *Qualitative data analysis: An expanded sourcebook.* Sage Publications.

Minor, L. C., Onwuegbuzie, A. J., Witcher, A. E., & James, T. L. (2002). Preservice teachers' educational beliefs and their perceptions of characteristics of effective teachers. *The Journal of Educational Research, 96*(2), 116–127. doi:10.1080/00220670209598798

Mittelstaedt, J. D., Moore, T., & West, J. E. (2014). Sustainability and strategic management decision-making: The role of TBL accounting in translating sustainability concepts into business practices. *Journal of Business and Management, 20*(1), 26–45.

Moe, A., Jansen, P., & Pietsch, S. (2018). *Childhood preference for spatial toys: Gender differences and relationships with mental rotation in STEM and non-STEM students' learning and individual differences.* doi:10.1016/j.lindif.2018.10.003

Montesinos, L., Salinas-Navarro, D. E., & Santos-Diaz, A. (2023). Transdisciplinary experiential learning in biomedical engineering education for healthcare systems improvement. *BMC Medical Education, 23*(1), 207. doi:10.1186/s12909-023-04171-x PMID:37013525

Mori, L. (2022). Global Journal of Foreign Language Teaching. *Language Teaching, 12*(4), 267–276.

Mori, R. Junior, Fien, J., & Horne, R. (2019). Implementing the UN SDGs in universities: Challenges, opportunities, and lessons learned. Sustainability. *Sustainability (New Rochelle, N.Y.), 12*(2), 129–133. doi:10.1089/sus.2019.0004

Morshidi, S., & Wan, C. D. (2022). Higher Education in Malaysia. In L. P. Symaco & M. Hayden (Eds.), *International Handbook on Education in South East Asia* (pp. 1–23). Springer International Handbooks of Education. doi:10.1007/978-981-16-8136-3_14-1

Moshman, D. (2014). *Epistemic cognition and development: The psychology of justification and truth.* Psychology Press. doi:10.4324/9781315884684

Moura, M. M., Frankenberger, F., & Tortato, U. (2019). Sustainability in Brazilian HEI: Practices Overview. *International Journal of Sustainability in Higher Education, 20*(5), 832–841. Advance online publication. doi:10.1108/IJSHE-01-2019-0021

Msibi, N. M. (2015). *Preservation of public records and archives in Swaziland government ministries and the Department of Swaziland National Archives, Master of Information Studies* [Dissertation]. University of KwaZulu-Natal, Pietermaritzburg, South Africa.

Mufwene, S. (2006). Race, racialism, and the study of language evolution in America. In M. Picone & K. Davis (Eds.), *Language Variety in the South: Historical and Contemporary Perspectives* (pp. 1–49). University of Alabama Press.

Muller, C., Katz, S. R., & Dance, L. J. (1999). Investing in teaching and learning: Dynamics of the teacher-student relationship from each actor's perspective. *Urban Education*, *34*(3), 292–337. doi:10.1177/0042085999343003

Murray, D. E. (2020). The world of English language teaching: Creating equity or inequity? *Language Teaching Research*, *24*(1), 60–70. doi:10.1177/1362168818777529

Murray, H. G. (1983). Low-inference classroom teaching behaviours and student ratings of college teaching effectiveness. *Journal of Educational Psychology*, *75*(1), 138–149. doi:10.1037/0022-0663.75.1.138

Murray, K. (2002). *Preservation Education and Training for South African Library and Archive Professionals and Students, MBibl*. University of Cape Town.

Murray, K. (2006). Preservation education in South African library and archive degree programmes. *South African Journal of Library and Information Science*, *72*(02), 98–107. doi:10.7553/72-2-1130

Muzaffar, H., Nikolaus, C. J., & Nickols-Richardson, S. M. (2023). Students' reflections on an experiential learning rotation with the peer education about the weight steadiness club program. *Pedagogy in Health Promotion*, *9*(2), 138–149. doi:10.1177/2373379920960647

Nadanyiova, M., Gajanova, L., & Majerova, J. (2020). Green Marketing as a Part of the Socially Responsible Brand's Communication from the Aspect of Generational Stratification. *Sustainability (Basel)*, *12*(17), 7118. doi:10.3390/su12177118

Naim, A. (2022). Understanding the customer centric approach to add value to social ecrm (SECRM). American. *Journal of Technology and Applied Sciences*, *3*, 17–31.

National Research Council. (2011). *Successful K-12 STEM education: Identifying effective approaches in science, technology, engineering, and mathematics*. Washington, DC: National Academy Press. https://www.researchgate.net/publication/260517903_Introducing_STEM_Education_Implications_for_Educating_Our_Teachers_For_the_Age_of_Innovation

National University of Science and Technology (NUST). (2019). *Faculty of-Communication Information and Science Prospectus, Department of Records and Archives*. Retrieved from: https://www.nust.ac.zw/cis/images/documents/faculty-of-comm-info-science-prospectus-2019.pdf

Nesenbergs, K., Abolins, V., Ormanis, J., & Mednis, A. (2021). Use of Augmented and Virtual Reality in Remote Higher Education: A Systematic Umbrella Review. *Education Sciences*, *11*(1), 1–12.

Ngulube, P. (2007a). Preserving South Africa's paper trail and making public records available for present and future generations. *ESARBICA Journal*, *26*(1), 35–50. doi:10.4314/esarjo.v26i1.31014

Ngulube, P. (2007b). The Achilles Heel of the Preservation of Documentary Materials in Sub Saharan Africa: Knowledge and Skills or Funding? *Restaurator (Copenhagen)*, *28*(3), 158–168. doi:10.1515/REST.2007.159

Nguyen, H., Lobo, A., & Greenland, S. (2017). An empirical analysis of corporate social responsibility (CSR) disclosure in an emerging economy: The case of Vietnam. *Journal of Business Ethics*, *144*(2), 399–411. doi:10.1007/s10551-015-2826-1

Nhongo, R. (2013). A National Language Policy for Zimbabwe in the Twenty first Century:Myth or reality? *Journal of Language Teaching and Research*, *4*(6), 1208–1215. doi:10.4304/jltr.4.6.1208-1215

Nhongo, R., & Tshotsho, B. (2020). Is terminology the real problem in failure to implement STEM in African Languages? Translanguaging as an intevention strategy. *Multicultural Education*, *6*(2), 78–89.

Nielsen Media Research. (2015). https://www.nielsen.com/us/en/insights/reports/2015/the-sustainability-imperative.html

Niemi, K. (2020, December 15). *Niemi: CASEL is updating the most widely recognized definition of social-emotional learning. Here's why.* https://www.the74million.org/article/niemi-casel-is-updating-the-most-widely recognizeddefinition- of-social-emotional-learning-heres-why/

Nieves, L., & Morales, W. (2022). Una mirada a la brecha entre actitud y comportamiento del consumidor en la economía circular. *Tendencias*, *23*(1), 372–394. doi:10.22267/rtend.222301.194

Nikou, S., & Maslov, I. (2021). An analysis of students' perspectives on e-learning participation – the case of the COVID-19 pandemic. *Campus-Wide Information Systems*, *38*(3), 299–315. doi:10.1108/IJILT-12-2020-0220

Nir, E., & Musial, J. (2020). The power of experiential learning in emotional courtroom spaces. *Journal of Criminal Justice Education*, *31*(4), 542–562. doi:10.1080/10511253.2020.1817515

North Umbria University. (2023). *Preventive Conservation MA, Newcastle.* Available at https://www.northumbria.ac.uk/search/?q=conservation&trigger=

Northeast Document Conservation Center (NEDCC). (2021). *About Preservation Training at NEDCC.* Available at https://www.nedcc.org/about/overview

Nsibirwa, B. Z. (2012). *Preservation of, and access to, legal deposit materials in South Africa* [PhD Dissertation]. University of Kwazulu-Natal, South Africa.

Nsibirwa, Z. (2015). Staff education and the preservation of legal deposit materials in South Africa. Innovation. *Journal of Appropriate Librarianship and Information Work in Southern Africa*, *2015*(50), 1263–1145.

Nyemba, W. R., Mbohwa, C., & Carter, K. F. (2021). Academia and Industry Collaborations: A research and professional perspective. In EAI/Springer Innovations in Communication and Computing. doi:10.1007/978-3-030-70493-3_4

Nyilasy, G., Gangadharbatla, H., & Paladino, A. (2014). *Perceived greenwashing: the interactive*. Academic Press.

Nziramasanga, D. (1999). *Zimbabwe Report of the Presidential Commission of Inquiry into Education and Training.* Harare: Republic of Zimbabwe.

O'Brien, W., Doré, N., Campbell-Templeman, K., Lowcay, D., & Derakhti, M. (2021). Living labs as an opportunity for experiential learning in building engineering education. *Advanced Engineering Informatics*, *50*, 101440. doi:10.1016/j.aei.2021.101440

O'Shea, A. (2014). Models of WIL. In S. Ferns (Ed.), *Work integrated learning in the curriculum. Higher Education Research and Development Society of Australia guide* (pp. 7–14). Australia Collaboration Education Network Ltd.

Obrecht, M., Feodorova, Z., & Rosi, M. (2022). Assessment of environmental sustainability integration into higher education for future experts and leaders. *Journal of Environmental Management*, *316*, 115223. doi:10.1016/j.jenvman.2022.115223 PMID:35569358

OECD. (2019). *OECD Future of Education and Skills 2030.* OECD Learning Compass, 2030.

Office of Science and Technology Policy. (2021). *Progress Report on the Implementation of the Federal STEM education Strategic Plan.* Office of Science and technology.

Ogbodo, J., Otaguruagu, E., Ngonebu, C., & Ogenyi, L. (2015). *New Horizons in Communication Skills: A Use of English Course Text.* Nsukka: Afro-Orbis Publishing Ltd.

Ohly, S., Plückthun, L., & Kissel, D. (2017). Developing Students' Creative Self-Efficacy Based on Design-Thinking: Evaluation of an Elective University Course. *Psychology Learning & Teaching*, *16*(1), 125–132. doi:10.1177/1475725716681714

Onel, N., & Mukherjee, A. (2016). Corporate social responsibility, consumer evaluations, and purchase intentions in a developing country. *Journal of Business Research*, *69*(12), 5721–5729. doi:10.1016/j.jbusres.2016.05.016

Othman, N., & Amiruddin, M. H. (2010). Different perspectives of learning styles from the VARK model. *Procedia: Social and Behavioral Sciences*, *7*, 652–660. doi:10.1016/j.sbspro.2010.10.088

Öz, G., & Şahinkarakaş, Ş. (2023). The Relationship Between Turkish EFL Learners' Academic Resilience and English Language Achievement. *The Reading Matrix: An International Online Journal, 23*(1).

Özcan, S. (2015). Climate, seasons, weather events and public calender in Turkish proverbs. *The Journal of International Social Research*, *8*(36), 179–187.

Ozden, M. (2008). Environmental awareness and attitudes of student teachers: An empirical research. *International Research in Geographical and Environmental Education*, *17*(1), 40–55. doi:10.2167/irgee227.0

Öz, G. (2022). An Investigation into Turkish EFL Learners' Ambiguity Tolerance. *Turkuaz Uluslararası Türk Dünyası Bilimsel Araştırmalar Dergisi*, *3*(1), 63–70. doi:10.54970/turkuaz.1119010

Ozoemena, J., Ngwoke, F., & Nwoke, B. (2021). Prospects of mother tongue as a medium of instruction in Nigerian primary level education. *English Language Teaching, 14*(4), 1-8.

Ozoemena, T., Ngweke, F., & Nwoko, B. (2021). Prospects of mother tongue as a medium of instruction in Nigerian primary level education. *English Language Teaching*, *14*(4), 1–8. doi:10.5539/elt.v14n4p1

Painter-Morland, M., Sabet, E., Molthan-Hill, P., Goworek, H., & de Leeuw, S. (2016). Beyond the curriculum: Integrating sustainability into business schools. *Journal of Business Ethics*, *139*(4), 737–754. doi:10.1007/s10551-015-2896-6

Palacios-González, M., & Chamorro, A. (2021). Analysis of the predictive variables of socially responsible consumption. *Business Strategy & Development*, *5*(3), 187–196. Advance online publication. doi:10.1002/bsd2.189

Palavecinos, M., Amérigo, M., Ulloa, J. B., & Muñoz, J. (2016). Preocupación y conducta ecológica responsable en estudiantes universitarios: Estudio comparativo entre estudiantes chilenos y españoles. *Psychosocial Intervention*, *25*(3), 143–148. doi:10.1016/j.psi.2016.01.001

Palinkas, L. A., Horwitz, S. M., Green, C. A., Wisdom, J. P., Duan, N., & Hoagwood, K. (2015). Purposeful Sampling for Qualitative Data Collection and Analysis in Mixed Method Implementation Research. *Administration and Policy in Mental Health*, *42*(5), 533–544. doi:10.1007/s10488-013-0528-y PMID:24193818

Palmberg, I. E., & Kuru, J. (2000). Outdoor activities as a basis for environmental responsibility. *The Journal of Environmental Education*, *31*(4), 32–36. doi:10.1080/00958960009598649

Palmer, J. A., & Neal, P. (2003). *The handbook of environmental education*. Routledge.

Panayiotou, M., Humphrey, N., & Wigelsworth, M. (2019). An empirical basis for linking social and emotional learning to academic performance. *Contemporary Educational Psychology*, *56*, 193–204. doi:10.1016/j.cedpsych.2019.01.009

Panhwar, F., & Buriro, G. (2020). An overview of theories and approaches Code-Swtiching. *International Research Journal of Arts and Humanities*, *48*(48), 265–281.

Panke, S., & Harth, T. (2018). Design Thinking for Inclusive Community Design:(How) Does it Work? In EdMedia+ Innovate Learning (pp. 284–296). Association for the Advancement of Computing in Education (AACE).

Panke, S. (2019). Design thinking in education: Perspectives, opportunities and challenges. *Open Education Studies, 1*(1), 281–306. doi:10.1515/edu-2019-0022

Papert, S., & Harel, I. (1991). Situating constructionism. *Constructionism, 36*, 1–11.

Parong, J., & Mayer, R. E. (2018). Learning science in immersive virtual reality. *Journal of Educational Psychology, 110*(6), 785–797. doi:10.1037/edu0000241

Parrish, D. R. (2015). The relevance of emotional intelligence for leadership in a HE context. *Studies in HE, 40*(5), 821–837.

Parry, S., & Metzger, E. P. (2023). Barriers to learning for sustainability: A teacher perspective. *Sustainable Earth Reviews, 6*(1), 2. Advance online publication. doi:10.1186/s42055-022-00050-3

Patton, M. Q. (1990). *Qualitative evaluation and research methods*. SAGE Publications, Inc.

Paul, J., Modi, A., & Patel, J. (2016). Predicting green product consumption using theory of planned behavior and reasoned action. *Journal of Cleaner Production, 135*, 732–739. doi:10.1016/j.jclepro.2016.06.139

Payne, L., & O'Neil, J. K. (2019). Cultural sustainability in higher education. In Springer eBooks (pp. 1–7). doi:10.1007/978-3-319-63951-2_109-1

Pearson, C., Lyall, J., Sloggett, R., & Cook, I. (2011). Preserving the past: how to ensure the development of conservation profession in Australia is preserved for the future. *Australian Institute for the Conservation of Cultural Material, A.I.C.C.M., National Conference 2011*.

Pereira Heath, T., & Chatzidakis, A. (2012). A dialectical perspective on the social life of sustainable consumption: Exploring the dyadic tensions between individual and political expressions of sustainable consumption. *Journal of Business Ethics, 111*(2), 203–215.

Peresuh, M., & Masuku, J. (2002). The role of primary language in the bilingual and bicultural education in Zimbabwe. *Zambezia, 29*(1), 27–37.

Peterson, T. O. (2004). Are you considering trying problem-based learning?: Three critical success factors for implementation. *Journal of Management Education, 28*(5), 630–647. doi:10.1177/1052562904267543

Pick, H. L. Jr, & Saltzman, E. (2014). *Modes of perceiving and processing information*. Psychology Press. doi:10.4324/9781315802565

Pierson, A., Clark, D., & Brady, C. (2021). Scientific Modeling and Translanguaging: A multilingual and multimodal approach to support science learning and engagement. *Science Education, 150*(4), 776–813. doi:10.1002/sce.21622

Pisitsankkhakarn, R., & Vassanadumrongdee, S. (2020). Enhancing purchase intention in circular economy: An empirical evidence of remanufactured automotive product in Thailand. *Resources, Conservation and Recycling, 156*, 104702. doi:10.1016/j.resconrec.2020.104702

Pomering, A. (2017). Marketing for sustainability: Extending the conceptualisation of the marketing mix to drive value for individuals and society at large. *Australasian Marketing Journal, 25*(2), 157–165. doi:10.1016/j.ausmj.2017.04.011

Potkonjak, V., Gardner, M., Callaghan, V., Mattila, P., Guetl, C., Petrović, V. M., & Jovanović, K. (2016). Virtual laboratories for education in science, technology, and engineering: A review. *Computers & Education, 95*, 309–327. doi:10.1016/j.compedu.2016.02.002

Pozo-Munoz, C., Rebolloso-Pacheco, E., & Fernandez-Ramirez, B. (2000). The "Ideal Teacher". Implications for student evaluation of teacher effectiveness. *Assessment & Evaluation in Higher Education, 25*(3), 253–263. doi:10.1080/02602930050135121

PRIA. (2014). *A manual for participatory training methodology in development*. New Delhi: PRIA.

Primus, D. J., & Sonnenburg, S. (2018). Flow Experience in Design Thinking and Practical Synergies with Lego Serious Play. *Creativity Research Journal*, *30*(1), 104–112. doi:10.1080/10400419.2018.1411574

PRME. (2021). *Principles for responsible management education*. https://www.unprme.org/about

Purcell, W. M., Henriksen, H., & Spengler, J. (2019). Universities as the engine of transformational sustainability toward delivering the sustainable development goals. *International Journal of Sustainability in Higher Education*, *20*(8), 1343–1357. Advance online publication. doi:10.1108/IJSHE-02-2019-0103

Pye, E. (2008). The benefits of access through handling outweigh the risks. *Conservation and Access Preprints of the 22nd IIC Congress*, 162–165. 10.1179/sic.2008.53.Supplement-1.162

Pye, E. (2016). Challenges of conservation: Working objects. *Science Museum Group Journal*. Advance online publication. doi:10.15180/160608

Radianti, J., Majchrzak, T. A., Fromm, J., & Wohlgenannt, I. (2020). A systematic review of immersive virtual reality applications for higher education: Design elements, lessons learned, and research agenda. *Computers & Education*, *147*, 103778. doi:10.1016/j.compedu.2019.103778

Rahmawati, Y., Taylor, E., Taylor, P. C., Ridwan, A., & Mardiah, A. (2022). Students' engagement in education as sustainability: Implementing an ethical dilemma-STEAM teaching model in chemistry learning. *Sustainability (Basel)*, *14*(6), 3554. doi:10.3390/su14063554

Ralph, M., & Stubbs, W. (2013). Integrating environmental sustainability into universities. *Higher Education*. Advance online publication. doi:10.1007/s10734-013-9641-9

Ramirez, T., Brush, K., Raisch, N., Bailey, R., & Jones, S. M. (2021). Equity in social emotional learning programs: A content analysis of equitable practices in PreK-5 SEL Programs. *Frontiers in Education*.

Rangel-Lyne, L., Alberto Sánchez-Jaime, R., Cuellar-Hernández, L., & Morales-Ramírez, D. (2023). Socially Responsible Actions for a Sustainable Life Model: The Case of the BBVA Financial Group in Mexico in Responsible Consumption and Sustainability. In *Case Studies from Corporate Social Responsibility*. Social Marketing, and Behavioral Economics. Springer Business Cases.

Rauf, H. L., Shareef, S. S., & Othman, N. N. (2021). Innovation in Architecture Education: Collaborative Learning Method Through Virtual Reality. *Journal of Higher Education Theory and Practice*, *21*(16), 33–40.

Razzouk, R., & Shute, V. (2012). What is design thinking and why is it important? *Review of Educational Research*, *82*(3), 330–348. doi:10.3102/0034654312457429

Reeve, E. M. (2015). STEM thinking! *Technology and Engineering Teacher*, *74*(4), 8–16.

Repetto, R. (1992). Accounting for Environmental Assets. *Scientific American*, *266*(6), 94–101. doi:10.1038/scientificamerican0692-94 PMID:1585151

Resnick, M., & Robinson, K. (2017). *Lifelong kindergarten: Cultivating creativity through projects, passion, peers, and play*. MIT press. doi:10.7551/mitpress/11017.001.0001

Reyes-Rodriguez, J. F., & López-Bonilla, J. M. (2018). The relationship between corruption and economic growth: An empirical analysis. *Journal of Economics and Sustainable Development*, *9*(15), 15–22.

Ritchie, H., & Roser, M. (2020). *CO2 emissions*. Our World in Data.

Ritter, B. A., Small, E. E., Mortimer, J. W., & Doll, J. L. (2018). Designing management curriculum for workplace readiness: Developing students' soft skills. *Journal of Management Education, 42*(1), 80–103. doi:10.1177/1052562917703679

Rivière, E., Jaffrelot, M., Jouquan, J., & Chiniara, G. (2019). Debriefing for the Transfer of Learning: The Importance of Context. *Academic Medicine, 94*(6), 796–803. doi:10.1097/ACM.0000000000002612 PMID:30681450

Rizos, V., Behrens, A., van der Gaast, W., Hofman, E., Ioannou, A., Kafyeke, T., Flamos, A., Rinaldi, R., Papadelis, S., Hirschnitz-Garbers, M., & Topi, C. (2016). Implementation of Circular Economy Business Models by Small and Medium-Sized Enterprises (SMEs): Barriers and Enablers. *Sustainability (Basel), 8*(11), 1212. doi:10.3390/su8111212

Roberts, P., & Peters, M. A. (2019). Neoliberalism, HE and Research. Brill Sense.

RobertsC. (2021). *Conservation VS Restoration: What is the difference*. National Museum of the Pacific War. Retrieved from; https://www.pacificwarmuseum.org/about/news/conservation-vs-restoration-what-is-the-difference#

Roberts, J. (1996). Green Consumers in the 1990s: Profile and Implications for Advertising. *Journal of Business Research, 36*(3), 217–231. doi:10.1016/0148-2963(95)00150-6

Robinson, K., & Aronica, L. (2015). *Creative schools: Revolutionizing education from the ground up*. Penguin UK.

Rocha Beardall, T. (2023). Abolition as praxis and virtual community-based learning. *Teaching Sociology, 51*(1), 79–91. doi:10.1177/0092055X221096663

Rodd, P., & Sanders, K. (2018). The imperative of critical pedagogy in times of cultural austerity: A case study of the capacity to reimagine education as a tool for emancipation. *New Zealand Sociology, 33*(3), 33.

Rodríguez, B. C., Durán-Zuazo, V. H., Soriano, M., García-Tejero, I. F., Ruíz, B. G., & Tavira, S. C. (2022). Conservation Agriculture as a Sustainable System for Soil Health: A review. *Soil Systems, 6*(4), 87. doi:10.3390/soilsystems6040087

Roffey, S., & McCarthy, F. (2013). Circle solutions, a philosophy and pedagogy for learning positive relationships: What promotes and inhibits sustainable outcomes? *International Journal of Emotional Intelligence, 5*(1), 36–55.

Rogers, E. M. (1962). *Diffusion of Innovations*. Free Press.

Rogers, E. M. (1995) *Diffusion of Innovations (4th ed.)*. The Free Press.

Rojas-Berrio, S., & Rueda, F. (2020). Sustainable consumption: Associated factors and scale validation to measure appliance purchase intentions. *International Journal of Green Economics, 14*(3), 191. doi:10.1504/IJGE.2020.111469

Roos, J., & Victor, B. (1999). Towards a new model of strategy-making as serious play. *European Management Journal, 17*(4), 348–355. doi:10.1016/S0263-2373(99)00015-8

Ross, M. (1989). The relation of implicit theories to the construction of personal histories. *Psychological Review, 96*(2), 341–357. doi:10.1037/0033-295X.96.2.341

Rubin, D. C., Deffler, S. A., & Umanath, S. (2019). Scenes enable a sense of reliving: Implications for autobiographical memory. *Cognition, 183*, 44–56. doi:10.1016/j.cognition.2018.10.024 PMID:30412854

Ruddiman, W. F. (2007). The early anthropogenic hypothesis: Challenges and responses. *Reviews of Geophysics, 45*(4), 1–37. doi:10.1029/2006RG000207

Rushton, S., Morgan, J., & Richard, M. (2007). Teacher's Myers-Briggs personality profiles: Identifying effective teacher personality traits. *Teaching and Teacher Education, 23*(4), 432–441. doi:10.1016/j.tate.2006.12.011

Russo, R., Cacciatore, F., Lagazio, C., & Renga, F. (2019). Enhancing green consumption: A dynamic framework based on cognitive dissonance theory. *Journal of Cleaner Production, 225*, 234–243. doi:10.1016/j.jclepro.2019.03.042

Sakellariou, C., & Fang, Z. (2021). Self-efficacy and interest in STEM subjects as predictors of the STEM gender gap in the US: The role of unobserved heterogeneity. *International Journal of Educational Research, 109*, 101821. Advance online publication. doi:10.1016/j.ijer.2021.101821

Salas Zapata, W., & Ortiz Muñoz, S. (2019). Analysis of meanings of the concept of sustainability. *Sustainable Development (Bradford), 27*(1), 153–161. doi:10.1002/sd.1885

Saldana, J. (2013). *The coding manual for qualitative researchers* (2nd ed.). SAGE Publications Ltd.

Saleem, A., Aslam, S., Sang, G., Dare, P. S., & Zhang, T. (2023). Education for sustainable development and sustainability consciousness: Evidence from Malaysian universities. *International Journal of Sustainability in Higher Education, 24*(1), 193–211. doi:10.1108/IJSHE-05-2021-0198

Salovaara, J. J., Soini, K., & Pietikäinen, J. (2019). Sustainability science in education: Analysis of master's programmes' curricula. *Sustainability Science, 15*(3), 901–915. doi:10.1007/s11625-019-00745-1

Sam, C., & Van Der Sijde, P. (2014). Understanding the concept of the entrepreneurial university from the perspective of higher education models. *Higher Education, 68*(6), 891–908. doi:10.1007/s10734-014-9750-0

Sampietro, M. (2019). Executive learning and development. In Handbook of Vocational Education and Training: Developments in the Changing World of Work. Springer. doi:10.1007/978-3-319-94532-3

Sandoval, W. (2014). Science education's need for a theory of epistemological development. *Science Education, 98*(3), 383–387. doi:10.1002/sce.21107

Sarantakos, S. (2005). *Social Research* (3rd ed.). Palgrave Macmillan. doi:10.1007/978-0-230-20901-5

Schifter, D., & Ajzen, I. (1985). Intention, Perceived Control, and Weight Loss: An Application of the Theory of Planned Behavior. *Journal of Personality and Social Psychology, 49*(3), 843–851. doi:10.1037/0022-3514.49.3.843 PMID:4045706

Schön, D. A. (1983). *The reflective practitioner: How professionals think in action* (Vol. 5126). Basic books.

Schonert-Reichl, K. A., Kitil, M. F., & Hanson-Peterson, J. (2017). To reach the students, teach the teachers: A national scan of teacher preparation and social and emotional learning. A report prepared for the Collaborative for Academic, Social, and Emotional Learning (CASEL). University of British Columbia.

Schonert-Reichl, K. A., & Roeser, R. W. (2016). Mindfulness in education: Introduction and overview of the handbook. In *Handbook of mindfulness in education* (pp. 3–16). Springer. doi:10.1007/978-1-4939-3506-2_1

Schore, A. N. (2018). The right brain implicit self: A central mechanism of the psychotherapy change process. In *Unrepressed unconscious, implicit memory, and clinical work* (pp. 73–98). Routledge. doi:10.4324/9780429484629-4

Schulz, K. P., & Geithner, S. (2014). *Creative tools for collective creativity–the serious play method using LEGO bricks. In Learning and Collective Creativity: Activity-Theoretical and Sociocultural Studies*. Routledge.

Segaetsho, T., & Mnjama, N. M. (2019). Information and communication technology use in the preservation of paper-based materials in the Eastern and African Region. Academic Press.

Segaetsho, T. (2018). *Environmental Considerations- Preservation of library and archival papers*. LAP Lambert Publishers.

Segaetsho, T. (2020). *Transcending Theories of Preservation in Library and Archives: The Perspective in the ESARBICA, Cases on Electronic Record Management in the ESARBICA Region*. IGI Global. doi:10.4018/978-1-7998-2527-2.ch013

Segaetsho, T., & Moloi, J. (2019). Integration of Digital Preservation Knowledge, Skills and Competencies in the Teaching Curricula of Library and Information Studies at the University of Botswana. *Mousaion, 37*(1). Advance online publication. doi:10.25159/2663-659X/6436

Segura, E., & Zamar, M. D. (2021). Sustainable economic development in higher education institutions: A global analysis within the SDGs framework. *Journal of Cleaner Production, 294*, 126133. Advance online publication. doi:10.1016/j.jclepro.2021.126133

Sera, L., & Wheeler, E. (2017). Game on: The Gamification of the Pharmacy Classroom. *Currents in Pharmacy Teaching & Learning, 9*(1), 155–159. doi:10.1016/j.cptl.2016.08.046 PMID:29180148

Seyderhelm, A. J., Blackmore, K. L., & Nesbitt, K. (2019, November). Towards Cognitive Adaptive Serious Games: A Conceptual Framework. In *Joint International Conference on Entertainment Computing and Serious Games* (pp. 331-338). Springer. 10.1007/978-3-030-34644-7_27

Shadiev, R., Yang, L., & Huang, Y. M. (2022). A review of research on 360-degree video and its applications to education. *Journal of Research on Technology in Education, 54*(5), 784–799. doi:10.1080/15391523.2021.1928572

Shameenda, K. L., & Kanyengo, W. K. (2012). An Audit of Skills and Qualifications in Preservation and Conservation Techniques: The Case of the University of Zambia Libraries. *Journal of Archival Organization, 10*(02), 137–149. doi: 10.1080/15332748.2012.736769

Shan, F. (2018). Multilingual Education in China: Taking the situation of Guizhou minority area as examples. *Theory and Practice in Language Studies, 8*(2), 197–202. doi:10.17507/tpls.0802.03

Shanika, V. G., Jayawickrama, T. S., & Perera, B. A. K. S. (2023). The Adaptability of Sustainability Education in Quantity Surveying Degree Programs in Sri Lanka: The Industry Perception. *International Journal of Construction Education and Research, 19*(3), 299–320. doi:10.1080/15578771.2022.2100017

Shao, K. Q., Yu, W. H., & Ji, Z. M. (2013). An exploration of Chinese EFL students' emotional intelligence and foreign language anxiety. *Modern Language Journal, 97*(4), 917–929. doi:10.1111/j.1540-4781.2013.12042.x

Sharma, R., & Foropon, C. (2019). Blockchain technology in supply chain management: A review. *International Journal of Production Research, 57*(7), 2117–2139. doi:10.1080/00207543.2018.1533261

Sharma, U., & Stewart, B. (2022). Enhancing sustainability education in the accounting curriculum: An effective learning strategy. *Pacific Accounting Review, 34*(4), 614–633. doi:10.1108/PAR-02-2021-0029

Shaughnessy, J. M. (2013). Mathematics in a STEM Context. *Mathematics Teaching in the Middle School, 18*(6), 324. doi:10.5951/mathteacmiddscho.18.6.0324

Shrum, L. J., Lowrey, T. M., & McCarty, J. A. (1994). Buyer characteristics of the green consumer and their implications for advertising strategy. *Journal of Advertising, 23*(2), 21–31. doi:10.1080/00913367.1994.10673464

Shulman, L. (1987). Knowledge and teaching: Foundations of the new reform. *Harvard Educational Review, 57*(1), 1–23. doi:10.17763/haer.57.1.j463w79r56455411

Shulman, L. S. (2005). Signature pedagogies in the professions. *Daedalus, 134*(3), 52–59. doi:10.1162/0011526054622015

Shulman, L. S., & Sherin, M. G. (2004). Fostering communities of teachers as learners: Disciplinary perspectives. *Journal of Curriculum Studies, 36*(2), 135–140. doi:10.1080/0022027032000135049

Sidiropoulos, E. (2014). Education for sustainability in business education programs: A question of value. *Journal of Cleaner Production, 85*, 472–487. doi:10.1016/j.jclepro.2013.10.040

Siekmann, G. (2016). *What is STEM? The need for unpacking its definitions and applications*. NCVER.

Sinatra, G. M., Kienhues, D., & Hofer, B. K. (2014). Addressing challenges to public understanding of science: Epistemic cognition, motivated reasoning, and conceptual change. *Educational Psychologist, 49*(2), 123–138. doi:10.1080/00461520.2014.916216

Sklad, M., Diekstra, R., De Ritter, M., & Ben, J. (2012). Effectiveness of school-based universal social, emotional, and behavioral programs: Do they enhance students' development in the area of skill, behavior, and adjustment? *Psychology in the Schools, 49*(9), 892–909. doi:10.1002/pits.21641

Slavich, G. M., & Zimbardo, G. Z. (2012). Transformational teaching: Theoretical underpinnings, basic principles, and core methods. *Educational Psychology Review, 24*(4), 569–608. doi:10.1007/s10648-012-9199-6 PMID:23162369

Smith, K., Maynard, N., Berry, A., Stephenson, T., Spiteri, T., Corrigan, D., Mansfield, J., Ellerton, P., & Smith, T. C. (2022). Principles of Problem-Based Learning (PBL) in STEM Education: Using expert wisdom and research to frame educational practice. *Education Sciences, 12*(10), 728. doi:10.3390/educsci12100728

Smith, P. K. (1982). Does play matter? Functional and evolutionary aspects of animal and human play. *Behavioral and Brain Sciences, 5*(1), 139–155. doi:10.1017/S0140525X0001092X

Spronken-Smith, R., Bullard, J. O., Ray, W., Roberts, C., & Keiffer, A. (2013). Where might dunes be on Mars? Engaging students through inquiry-based learning in geography. In *Active Learning and Student Engagement* (pp. 72–87). Routledge.

Sreen, N., Purbey, S., & Sadarangani, P. (2017). Impact of culture, behavior and gender on green purchase intention. *Journal of Retailing and Consumer Services, 41*, 177–189. Advance online publication. doi:10.1016/j.jretconser.2017.12.002

Sroufe, R., & Ramos, D. P. (2015). Leveraging collaborative, thematic problem-based learning to integrate curricula. *Decision Sciences Journal of Innovative Education, 13*(2), 151–176. doi:10.1111/dsji.12063

Stake, R. E. (2005). Qualitative Case Studies. In N. K. Denzin & Y. S. Lincoln (Eds.), *The SAGE Handbook of Qualitative Research* (3rd ed., pp. 443–466). Sage Publications.

Stankevičienė, K., Mockevičienė, D., & Giedraitis, V. (2020). The role of digital marketing in the contemporary business environment. *The Engineering Economist, 31*(1), 103–114. doi:10.5755/j01.ee.31.1.21578

Steffen, W., Broadgate, W., Deutsch, L., Gaffney, O., & Ludwig, C. (2015). The trajectory of the Anthropocene: The Great Acceleration. *The Anthropocene Review, 2*(1), 81–98. doi:10.1177/2053019614564785

Steg, L., & Vlek, C. (2009). Encouraging pro-environmental behaviour: An integrative review and research agenda. *Journal of Environmental Psychology, 29*(3), 309–317. doi:10.1016/j.jenvp.2008.10.004

Stentoft, D. (2017). From saying to doing interdisciplinary learning: Is problem-based learning the answer? *Active Learning in Higher Education, 18*(1), 51–61. doi:10.1177/1469787417693510

Stern, P. C. (2000). Toward a coherent theory of environmentally significant behavior. *The Journal of Social Issues, 56*(3), 407–424. doi:10.1111/0022-4537.00175

Stokols, D., Misra, S., Moser, R. P., Hall, K. L., & Taylor, B. K. (2008). The ecology of team science: Understanding contextual influences on transdisciplinary collaboration. *American Journal of Preventive Medicine, 35*(2), S96–S115. doi:10.1016/j.amepre.2008.05.003 PMID:18619410

Study Portal. (2022). *Master's degrees in Library Science in Europe*. Available at: https://www.mastersportal.com/search/master/library-science/europe

Suárez-Eiroa, B., Fernández, E., Martínez, G., & Soto-Oñate, D. (2019). Operational principles of Circular Economy for Sustainable Development: Linking theory and practice. *Journal of Cleaner Production, 214*, 952–961. doi:10.1016/j.jclepro.2018.12.271

Sudbury-Riley, L., & Kohlbacher, F. (2015). Ethically minded consumer behavior: Scale review, development, and validation. *Journal of Business Research, 69*(8), 2697–2710. Advance online publication. doi:10.1016/j.jbusres.2015.11.005

Suganda, L. A., Petrus, I., & Zuraida, Z. (2021). EFL Teacher's Code Switching in the Social Emotional Learning Context. *Indonesian Research Journal in Education, 5*(2), 317-344.

Summers, M., Childs, A., & Corney, G. (2005). Education for sustainable development in initial teacher training: Issues for interdisciplinary collaboration. *Environmental Education Research, 11*(5), 623–647. doi:10.1080/13504620500169841

Surya, A., Retnawati, H., & Haryanto, H. (2023). Findings and Implications of Social Emotional Learning (SEL) in Paternalistic Culture in Elementary Schools: A Systematic Literature Review. *Pegem Journal of Education and Instruction, 13*(3), 151–158.

Sustainability Report. (2021). *YPF*. https://sustentabilidad.ypf.com/index_en.html

Sutton-Smith, B. (1966). Piaget on play: A critique. *Psychological Review, 73*(1), 104–110. doi:10.1037/h0022601 PMID:5906212

Su, Y. H., Feng, L. Y., Yang, C. C., & Chen, T. L. (2012). How teachers support university students' lifelong learning development for sustainable futures: The student's perspective. *Futures, 44*(2), 158–165. doi:10.1016/j.futures.2011.09.008

Symaco, L. P., & Wan, C. D. (2017). Development of Higher Education in Malaysia: Issues and Challenges. In M. Samuel, M. Y. Tee, & L. P. Symaco (Eds.), *Education in Malaysia, Education in the Asia-Pacific Region: Issues, Concerns and Prospects* (pp. 53–66). Springer Singapore Pte. Limited. doi:10.1007/978-981-10-4427-4_4

Taheri, M., Unterholzer, T., Hölzle, K., & Meinel, C. (2016). An educational perspective on design thinking learning outcomes. In *ISPIM Innovation Symposium* (p. 1). The International Society for Professional Innovation Management (ISPIM).

Talvio, M., Lonka, K., Komulainen, E., Kuusela, M., & Lintunen, T. (2013). *Revisiting Gordon´s teacher effectiveness training: An intervention study on teachers´ social and emotional learning*. Academic Press.

Taylor, S., & Todd, P. (1995). Decomposition and Crossover Effects in the Theory of Planned Behavior: A Study of Consumer Adoption Intentions. *International Journal of Research in Marketing, 12*(2), 137–155. doi:10.1016/0167-8116(94)00019-K

Tekgöz, G., Şahin, E., & Ertepinar, H. (2010). Çevre okuryazarlığı, öğretmen adayları ve sürdürülebilir bir gelecek. *Hacettepe Üniversitesi Eğitim Fakültesi Dergisi, 39*, 307–320.

Terlau, W., & Hirsch, S. (2015). Managing complexity in stakeholder networks: A study on the evolution of sustainability partnerships. *Journal of Cleaner Production, 108*, 1069–1078. doi:10.1016/j.jclepro.2015.07.002

TerraChoice. (2010). *The sins of greenwashing: home and family edition*. Retrieved from http://sinsofgreenwashing.org/findings/the-seven-sins/

Thaler, R. H., & Sunstein, C. R. (2017). *Un pequeño empujón: El impulso que necesitas para tomar mejores decisiones sobre salud, dinero y felicidad*. Taurus.

The Association of Canadian Archivist. (2022). *Graduate archival education in Canada*. Available at: https://archivists.ca/Archival-Education-Programs

The Talloires Declaration 10 Point Action Plan. (1990). Association of University Leaders for Sustainable Future.

The United Nations (UN). (2016). *Transforming our world: the 2030 Agenda for Sustainable Development* (A/RES/70/1). UN.

Theobald, K. A., & Ramsbotham, J. (2019). Inquiry-based learning and clinical reasoning scaffolds: An action research project to support undergraduate students' learning to 'think like a nurse'. *Nurse Education in Practice, 38*, 59–65. doi:10.1016/j.nepr.2019.05.018 PMID:31176910

Thompson, M., Kanmaz, T., Cheryl, C., & Tanzer, K. (2020). *Recommendations for Participation in Experiential Education Activities during a National, Regional or Local Crisis.* Available at https://www.acpe-accredit.org/wp-content/uploads/Experiential- Education-ACPE-Proposal-Response-Final.pdf

Thorndike, E. L. (1920). Intelligence and its uses. *Harper's Magazine.*

Thuneberg, H. M., Salmi, H. S., & Bogner, F. X. (2018). How creativity, autonomy and visual reasoning contribute to cognitive learning in a STEAM hands-on inquiry-based math module. *Thinking Skills and Creativity, 29*, 153–160. doi:10.1016/j.tsc.2018.07.003

Tietenberg, T., & Lewis, L. (2016). *Environmental and Natural Resource Economics* (10th ed.). Routledge. doi:10.4324/9781315523972

Tijsma, G., Horn, A., Urias, E., & Zweekhorst, M. B. M. (2023). Training students in inter- and transdisciplinary sustainability education: Nurturing cross-faculty staff commitment and continuous community collaboration. *International Journal of Sustainability in Higher Education, 24*(4), 765–787. doi:10.1108/IJSHE-02-2022-0049

Titus, P. A. (2000). Marketing and creative problem-solving process. *Journal of Marketing Education, 22*(3), 225–235. doi:10.1177/0273475300223006

Titus, P. A. (2007). Applied creativity: The creative marketing breakthrough model. *Journal of Marketing Education, 29*(3), 262–272. doi:10.1177/0273475307307600

Tollin, N., Küller, R., Hartig, T., & Sarlöv-Herlin, I. (2015). The restorative effects of sensory experiences in nature. *Environment and Behavior, 47*(2), 119–147.

Too Good To Go. (n.d.). https://www.toogoodtogo.com/es

Toom, A. (2015). Teaching Leo Vygotsky's Theory of Socialcultural Theory Development with the use of literature and Information Technology. In A. Toom (Ed.), *Academic research of SSaH 2015* (pp. 60–79). Czech Technical Republic.

Toom, A. (2015). Teaching Leo Vygotsky's theory of sociocultural development with the use of literature and information technology. In *Academic research of SSaH*. Czech Technical University.

Triana, N. (2022). The need for sustainability and CSR in undergraduate business education. *Journal of Undergraduate Research (Gainesville, Fla.), 24*. Advance online publication. doi:10.32473/ufjur.24.130792

Tse, M. M. Y., Kwan, T. S., & Lee, P. H. (2016). The development and psychometric evaluation of the Perception of Play Questionnaire for older adults. *Educational Gerontology, 42*(2), 79–88. doi:10.1080/03601277.2015.1071600

Turan, S. (1990). *Türk kültür tarihi.* Bilgi Yayınevi.

Turkish Republic Ministry of Culture and Tourism. (2019). *Hıdrellez traditions.* https://www.ktb.gov.tr/EN-98558/Hıdrellez-traditions.html

Uca, A. (2007). Türk toplumunda hidirellez. *İ. A.Ü. Türkiyat Araştırmaları Enstitüsü Dergisi, 34*, 113–138.

UGC. (2020). *Curriculum Framework and Credit Sustem for Four-Year Undergraduate Programme*. UGC.

Ulaga, W. (2018). The journey towards customer centricity and service growth in B2B: A commentary and research directions. *AMS Review, 8*(1-2), 80–83. doi:10.1007/s13162-018-0119-x

Uluocha, A. (2019). Assessment of preservation practices at Lagos State House of Assembly Library. In W. M. Olatokun, A. O. Aremu, & A. Adetimirin (Eds.), Research on contemporary issues in media resources and information and communication technology use (pp. 465-479). Department of Library, Archival and information studies, University of Ibadan.

UN. (2023). *Global Partnerships - United Nations Sustainable development*. United Nations Sustainable Development. https://www.un.org/sustainabledevelopment/globalpartnerships/

UN. (2023). *United Nations: The 17 goals*. Retrieved in August, 1, 2023 from https://sdgs.un.org/goals

UNEP FI. (2022). *Acerca de*. Retrieved from https://www.unepfi.org/about/

UNESCO. (2017). *Spring celebration, Hıdrellez*. https://ich.unesco.org/en/RL/spring-celebrationhdrellez-01284

UNESCO. (2022). *Transforming higher education institutions into lifelong learning institutions*. UIL POLICY BRIEF 14, UNESCO Institute for Lifelong Learning.

UNESCO. Director-General, 2009-2017 (Bokova, I.G.) (2014). *Estrategia de educación de la Unesco, 2014 - 2021*. https://unesdoc.unesco.org/ark:/48223/pf0000231288_spa

United Nations Educational Scientific and Cultural Organization (UNICEF). (2014). *Sustainable Development post 2015 begins with Education*. UNICEF.

United Nations Educational Scientific and Cultural Organizations (UNESCO). (2012). *Educational for Scientific and Cultural Development Source Book*. UNESCO.

United Nations. (2015). *Transforming our world: the 2030 Agenda for Sustainable Development*. Available at https://sustainabledevelopment.un.org/post2015/transformingourworld

United Nations. (2022). *Transforming Education Summit 2022 - Concept and Programme Outline, convened by the United Nations Secretary-General*. UN.

University for Continuing Eduacation. Krems. (2016). Interdisciplinary Methods in Graphic Art, Book and Document Conservation. *Auszug aus Mitteilungsblatt, 20*(98). Available at https://www.iiconservation.org/sites/default/files/news/attachments/7420-krems_university_course_details.pdf

University of the Sunshine Coast. (n.d.). Retrieved from usc.edu.in: https://www.usc.edu.au/about/structure/schools/school-of-law-and-society/sustainability-research-centre

Valentim, N. M. C., Silva, W., & Conte, T. (2017). The students' perspectives on applying design thinking for the design of mobile applications. In *Proceedings of the 39th International Conference on Software Engineering: Software Engineering and Education Track* (pp. 77–86). IEEE Press. 10.1109/ICSE-SEET.2017.10

Vanykina, G. V., & Sundukova, T. O. (2020). Educational robotics in educational institutions: An empirical study of motivation and STEM skills. *Pedagogika Informatiki, 3*, 1–10.

Varey, R. J. (2010). Sustainable marketing: Myth or reality? *Journal of Business Research, 63*(4), 111–126.

Villa Castaño, L. E., Perdomo-Ortiz, J., Durán León, W. F., Dueñas Ocampo, S., & Arredondo Trapero, F. G. (2018). Measuring socially responsible consumption: A study of Colombia–Mexico. *Academia (Caracas), 31*(3), 553–568. Advance online publication. doi:10.1108/ARLA-12-2016-0356

Villarroel, V., Bernavente, M., Chuecas, M. J., & Bruna, D. (2020). Experiential learning in higher education. A student-centred teaching method that improves perceived learning. *Journal of University Teaching & Learning Practice*, *17*(5), 1–16. doi:10.53761/1.17.5.8

Villegas, D. (2017). Colombia's Nationwide EFL policy and the construction of equity in policy documents. *App. J. App. Lang. Stu.*, *11*(4), 57–74. doi:10.17011/apples/urn.201708083437

von Thienen, J., Royalty, A., & Meinel, C. (2017). Design thinking in higher education: How students become dedicated creative problem solvers. In *Handbook of research on creative problem-solving skill development in higher education* (pp. 306–328). IGI Global. doi:10.4018/978-1-5225-0643-0.ch014

Vygotsky, L. (1978). Mindset in society: The Development of higher psychological processes. Havard University Press.

Wan, C. D., & Abdullah, D. (2021). Internationalisation of Malaysian Higher Education: Policies, Practices and the SDGs. *International Journal of Comparative Education and Development*, *23*(3), 212–226. doi:10.1108/IJCED-08-2020-0052

Wang, Y., Derakhshan, A., & Zhang, L. J. (2021). Researching and practicing positive psychology in second/foreign language learning and teaching: The past, current status and future directions. *Front. Psychol.*, *12*(731721). doi:10.3389/fpsyg.2021.731721

Ward, K., & Hartley, J. (2006). Using a virtual learning environment to address one problem with problem-based learning. *Nurse Education in Practice*, *6*(4), 185–191. doi:10.1016/j.nepr.2005.12.003 PMID:19040876

Watson, R. T., Boudreau, M. C., & Chen, A. J. (2010). Information systems and environmentally sustainable development: Energy informatics and new directions for the IS community. *Management Information Systems Quarterly*, *34*(1), 23–38. doi:10.2307/20721413

Webb, D., Mohr, L., & Harris, K. (2008). A re-examination of socially responsible consumption and its measurement. *Journal of Business Research*, *61*(2), 91–98. doi:10.1016/j.jbusres.2007.05.007

Webster, F. E. Jr. (1975). Determining the Characteristics of the Socially Conscious Consumer. *The Journal of Consumer Research*, *2*(3), 188–196. doi:10.1086/208631

Weißhuhn, P., Helming, K., & Ferretti, J. (2017). Research impact assessment in agriculture—A review of approaches and impact areas. *Research Evaluation*, *27*(1), 36–42. doi:10.1093/reseval/rvx034

Wei, X., Song, W., Shao, Y., & Cai, X. (2022). Progress of ecological restoration research based on bibliometric analysis. *International Journal of Environmental Research and Public Health*, *20*(1), 520. doi:10.3390/ijerph20010520 PMID:36612842

Wen, Y., Wu, L., He, S., Ng, N. H. E., Teo, B. C., Looi, C. K., & Cai, Y. (2023). Integrating augmented reality into inquiry-based learning approach in primary science classrooms. *Educational Technology Research and Development*, *71*(4), 1–21. doi:10.1007/s11423-023-10235-y PMID:37359487

Wersun, A., Dean, B. A., Mills, R., Perkiss, S., Acosta, P., Anastasiadis, S., Gibbons, B., Gonzalez-Perez, M. A., Heithaus, T., Jun, H., Mesicek, R. H., & Bayerlein, L. (2019). An exploration of student learning for sustainability through the WikiRate student engagement project. *International Journal of Management Education*, *17*(3), 100313. doi:10.1016/j.ijme.2019.100313

West Dean College. (2023). *Master of Arts Conservation Studies Programme Specification and Unit Descriptors.* West Dean College, University of Sussex. Available at https://assets.westdean.ac.uk/pdfs/original/60954-ma-conservation-studies-programme-specification-june23.pdf

Weybrecht, G. (2017). Greening the business of business education. *Corporate Knights, 16*(4), 32-35. https://www.jstor.org/stable/26789227

Wilhelmsson, M., Pelling, S., Ludvigsson, J., Hammar, M., Dahlgren, L. O., & Faresjö, T. (2009). Twenty years experience of interprofessional education in Linköping–ground-breaking and sustainable. *Journal of Interprofessional Care, 23*(2), 121–133. doi:10.1080/13561820902728984 PMID:19225972

Willimas, J. (2011). STEM education: Proceed with caution. *Design and Technology Education: An International Journal, 16*(1).

Wilson, P. (1992). *Library conservation course in Australia, Conservation DistList.* https://cool.culturalheritage.org/byform/mailing-lists/cdl/1992/0236.html

Wilson, F. R. (2010). *The hand: How its use shapes the brain, language, and human culture.* Vintage.

Wimberly, R., Faulkner, G., & Moxley, R. (1978). Dimensions of teacher effectiveness. *Teaching Sociology, 6*(1), 7–20. doi:10.2307/1317161

Winter, J., Zhai, J., & Cotton, D. R. E. (2022). Teaching environmental sustainability in China: Opportunities and challenges for business and economics faculty in higher education. *Environmental Education Research, 28*(2), 318–332. doi:10.1080/13504622.2021.2012560

Wiswall, M., Stiefel, L., Schwartz, A. E., & Boccardo, J. (2014). Does attending a STEM high school improve student performance? Evidence from New York City. Academic Press.

Wisz, M. S., Satterthwaite, E. V., Fudge, M., Fischer, M., Polejack, A., St John, M., Fletcher, S., & Rudd, M. A. (2020). 100 Opportunities for More Inclusive Ocean Research: Cross-Disciplinary Research Questions for Sustainable Ocean Governance and Management. Frontiers in Marine Science, 7. doi:10.3389/fmars.2020.00576

Witcher, A., & Onwuegbuzie, A. J. (1999). *Characteristics of effective teachers: perceptions of preservice teachers.* Paper presented at the Annual Meeting of the Mid-South Educational Research Association, Point Clear, AL. Retrieved from ERIC database. (ED438246).

Wolsko, C., Marino, E., Doherty, T. J., Fisher, S., Green, A. S., Goodwin, B., Reese, R., & Wirth, A. (2016). Systems of access: A multidisciplinary strategy for assessing the social dimensions of sustainability. Sustainability: Science. *Sustainability, 12*(1), 88–100. doi:10.1080/15487733.2016.11908156

World Commission on Economic Development (WCED). (1987). *From One Earth to One World: An Overview of Our Common Future: Report of the World Commission on Environment and Development.* Oxford University Press.

Wright, T. S. (2003). A Tenth Year Anniversary Retrospect: The Effect of the Halifax Declaration on Canadian Signatory Universities. *Canadian Journal of Environmental Education.*

Wrigley, C., & Straker, K. (2017). Design thinking pedagogy: The educational design ladder. *Innovations in Education and Teaching International, 54*(4), 374–385. doi:10.1080/14703297.2015.1108214

WTWCo. (2021). *How employee insights power Danone's sustainable growth agenda.* Retrieved from https://www.wtwco.com/en-CM/Insights/2021/09/how-employee-insights-power-danones-sustainable-growth-agenda

Wu, W.-Y., & Pham, T. (2020). Examining the Moderating Effects of Green Marketing and Green Psychological Benefits on Customers' Green Attitude, Value and Purchase Intention. *Sustainability (Basel), 12*(18), 7461. doi:10.3390/su12187461

Xie, F., & Derakhshan, A. (2021). A conceptual review of positive teacher interpersonal communication behaviours in the instructional context. *Frontiers in Psychology, 12*, 1–10. doi:10.3389/fpsyg.2021.708490 PMID:34335424

Xie, F., Liu, G., & Zhuang, T. (2021). A comprehensive review of urban regeneration governance for developing appropriate governance arrangements. *Land (Basel), 10*(5), 545. doi:10.3390/land10050545

Xiong, Z., & Li, H. (2019). Ecological deficit tax: A tax design and simulation of compensation for ecosystem service value based on ecological footprint in China. *Journal of Cleaner Production, 230*, 1128–1137. doi:10.1016/j.jclepro.2019.05.172

Yahiaoui, F., Aichouche, R., Chergui, K., Brika, S. K. M., Almezher, M., Musa, A. A., & Lamari, I. A. (2022). The Impact of e-Learning Systems on motivating students and enhancing their outcomes during COVID-19: A Mixed-Method Approach. *Frontiers in Psychology, 13*, 874181. Advance online publication. doi:10.3389/fpsyg.2022.874181 PMID:35967714

Yang, Y. T. C. (2012). Building virtual cities, inspiring intelligent citizens: Digital games for developing students' problem-solving and learning motivation. *Computers & Education, 59*(2), 365–377. doi:10.1016/j.compedu.2012.01.012

Yan, J., & She, Q. (2011). Developing a trichotomy model to measure socially responsible behavior in China. *International Journal of Market Research, 53*(2), 253–274. doi:10.2501/IJMR-53-2-253-274

Yarime, M., Trencher, G., Mino, T., Scholz, R., Olsson, L., Van Ness, B., Frantzeskaki, N., & Rotmans, J. (2012). Establishing sustainability science in higher education institutions: Towards integrating academic development, institutionalization, and stakeholder collaborations. *Sustainability Science, 7*(S1), 101–113. doi:10.1007/s11625-012-0157-5

Yazan, B., & Rudolph, N. (Eds.). (2018). *Criticality, Teacher Identity, and (In)equity in English Language Teaching: Issues and Implications*. Springer. doi:10.1007/978-3-319-72920-6

Yıldırım, A., & Şimşek, H. (2011). Sosyal bilimlerde nitel araştırma yöntemleri (8. Baskı) [Qualitative research methods in social sciences (8th Edition)]. Ankara: Seçkin Yayıncılık.

Yu, L., & Zin, Z. M. (2023, May). The critical thinking-oriented adaptations of problem-based learning models: a systematic review. In *Frontiers in Education* (Vol. 8, p. 1139987). Frontiers. doi:10.3389/feduc.2023.1139987

Yuniar, G. K. (2017). Story-Reading as a Teaching method for young Children: Teacher's Challenges. In *3rd International Conference on Early Childhood (ICECE-16)* (pp. 179-186). Atlantis Press.

Yu, X., Paudel, K. P., Li, D., Xiong, X., & Gong, Y. (2020). Sustainable Collaborative Innovation between Research Institutions and Seed Enterprises in China. *Sustainability (Basel), 12*(2), 624. doi:10.3390/su12020624

Zalat, M. M., Hamed, M., & Bolbol, S. A. (2021). The experiences, challenges, and acceptance of e-learning as a tool for teaching during the COVID-19 pandemic among university medical staff. *PLoS One, 16*(3), e0248758. doi:10.1371/journal.pone.0248758 PMID:33770079

Zervos, S. (2012). *Teaching Conservation and Preservation at the Department of Library Science and Information Systems of the TEI of Athens*. Restorator.

Zguir, M. F., Dubis, S., & Koç, M. (2021). Embedding education for Sustainable Development (ESD) and SDGs values in curriculum: A comparative review on Qatar, Singapore and New Zealand. *Journal of Cleaner Production, 319*, 128534. doi:10.1016/j.jclepro.2021.128534

Zhexembayeva, N. (2014). *Overfished Ocean Strategy: Powering Up Innovation for a Resource-Deprived World*. Berrett-Koehler Publishers.

Zins, J. E. (Ed.). (2004). *Building academic success on social and emotional learning: What does the research say?* Teachers College Press.

About the Contributors

Chai Lee Goi is a senior lecturer at the Department of Management, Marketing & Digital Business, Faculty of Business, Curtin University, Malaysia. His research interests include consumer behaviour, Internet marketing, sustainable development, big data and educational technology.

* * *

Usha Ajithkumar is Professor & Head, Department of Education, SASTRA University, Tamil Nadu, India. Her research interests are in the areas of ICT based innovative practices in education, teaching pedagogy, human rights, child rights, TVET and curriculum design and development. She is committed to promoting a culture of excellence, ethics, and engagement in her work.

Pinar Ayyıldız is an interpreter, sociologist, teacher trainer and holds MA and Doctor of Philosophy degrees in Educational Management as well as the Cambridge Diploma in Teaching English to Speakers of Other Languages (DELTA). She also acted as a DELTA trainer. She has worked as a head teacher, academic coordinator, dean of students and director of English preparatory school in higher education institutions. To date she has taught several courses at tertiary level. She concentrates on topics like Epistemology of Educational Sciences, Equity in Education and Educational Leadership. She has various academic memberships and is the editor of international journals like Participatory Educational Research (PER) and Philosophy and Theory in Higher Education Journal (PTHE). She is currently an associate professor of Management Information Systems Department and the coordinator of instructional processes, namely, teaching, learning, testing in Ankara Medipol University, Turkey.

Atul Bamrara received his PhD in Business Management from HNB Garhwal University. He has shared his expertise in Research Methodology, Project Management, Information Security, and Education Technology with students of Business Management, Computer Science and Education in India and abroad. He is currently working with Department of School Education, Government of Uttarakhand - India. His responsibilities include training to various in-service teachers in ICT Inclusion, Mathematics & Science Pedagogy, e-content development etc. He has received Young Scientist Award 2018, Commonwealth of Learning Award, Teacher Innovation Award and Nation Builder Award for his extraordinary performance. He has widely published his research papers in reputed journals of Information Security, Education Technology and Operations Management.

Poonam Bamrara received his Masters in Sociology and currently associated with Department of School Education, Government of Uttarakhand. Her research areas include Education Technology, Science Pedagogy and Language Learning. She shares her expertise with in-service as well as pre service teachers in these domains.

William Chakabwata holds a Master of Education Degree in Curriculum and Arts Education, which he earned from University of Zimbabwe. He has taught at Midlands State University in Namibia and also at Namibia University of Science and Technology(NUST). He is currently, studying for a PhD, in curriculum and Instruction with University of South Africa (UNISA.

Chorng Yuan Fung is a multidisciplinary academic and researcher. He is an accountant by profession and holds a PhD in educational psychology and a Master of Science in human resource development. His research interests include behavioural accounting, human resource development, educational psychology, educational technology and pedagogy. He publishes regularly in reputable indexed journals on these areas of interest. Fung also speaks regularly in seminars and workshops organised by corporates, universities and professional accounting bodies on topics related to his research interests.

Catherine Hayes is Professor of Health Professions Pedagogy and Scholarship at the University of Sunderland, UK. She is a UK National Teaching Fellow and Principal Fellow of the UK Higher Education Academy. As a graduate of Podiatric Medicine in 1992, Catherine was a Founding Fellow of the Faculty of Podiatric Medicine at the Royal College of Physicians and Surgeons (Glasgow) in 2012 and was awarded Fellowship of the Royal College of Podiatry (London) in 2010. She is currently Programme Leader of the University of Sunderland's Professional Doctorate pathways for the DBA, EdD, DPM and DProf.

Kiat Sing Heng joined Swinburne University of Technology Sarawak Campus as a Lecturer in 2007. She has served in various academic management and leadership roles, including Associate Dean (Accreditation and Curriculum Enhancement) and Program Coordinator for Undergraduate Business – Non-accounting and Postgraduate Business courses. In terms of teaching, she has taught economics subjects for foundation, diploma and undergraduate business courses. She was also an internal auditor, Malaysia Research Assessment (MyRA) Internal Audit, appointed by the Ministry of Higher Education Malaysia from 1 November 2017 to 31 October 2020.

Minna-Maarit Jaskari is a University Lecturer (PhD) at the University of Vaasa, Finland. Marketing, design thinking, and research methods are among her award-winning courses. Her current research interests include consumer experiences, sonic branding, virtual brand communities, temporality and spatiality of market spaces, and experiential and innovative pedagogies in higher education. She has published in number of academic journals, including Qualitative Market Research, the Journal of Marketing Education, and the Nordic Journal of Business. She also serves on the editorial board of the Journal of Marketing Education.

Ana Isabel Jiménez Zarco has a PhD in Economic and Business Sciences from the University of Castilla-La Mancha and a postgraduate diploma in the Construction of Models in Ecology and Natural Resource Management from the Universitat Politècnica de Catalunya. She received an award for her doctoral thesis from the Instituto de Estudios Económicos (IEE) in Madrid. Up until 2002, she conducted her teaching and research activities in the Marketing Department of the University of Castilla-La Mancha. From 2003 to 2006, she was co-director of the master's degree in Marketing and Communication Management and, from 2006 to 2012, director of the second-cycle degree in Business Administration at the UOC. She has also made a number of research visits to the University of East Anglia and the University of Brighton in the UK and the Universidad Autónoma Chapingo in Mexico. As a researcher, her main areas of interest are entrepreneurship, value co-creation processes, business and social innovation, brand image and identity and the analysis of technology adoption and dissemination processes. Since 2015, she has been an evaluator for the European Union's Marie Skłodowska-Curie Actions – Innovative Training Networks programme (H2020-MSCA-ITN) and, since 2012, also an evaluator for Spain's National Evaluation and Forward Planning Agency (ANEP), in the field of business and social innovation. She has authored more than 70 articles for national and international journals, 25 of which have appeared in international JCR-SSCI-indexed journals, as well as numerous book chapters. She has acted as associate editor of Frontiers in Psychology, and as a reviewer for Computers in Human Behavior, Journal of Business Venturing, European Journal of Marketing and Interactive Learning Environments.

Jorge Lara Guillén is a graduate in Mechanical Engineering from the University of Zaragoza, master in Innovation and Digital Transformation from the Universitat Oberta de Catalunya (UOC) and PhD student in Business and Management from the UOC. His career has developed in the home appliance industry . He currently works as Business Excellence Manager for BSH Home Appliances' headquarters in Munich, Germany.

M. Dolores Méndez-Aparicio is a Doctor in Legal, Economic and Social Sciences (University of Burgos, Spain). Certification in Customer Experience by the Association for the Development of Customer Experience, DEC and regular collaborator of the same. Master in Information and Knowledge Society (UOC, Barcelona, Spain). Degree in Advertising and Public Relations (UOC, Barcelona, Spain). Management Computer Engineering (Polytechnic University of Madrid, Spain). Head of Applications at Mutua Madrileña (Spain). Line of research in marketing 5.0, he has published in various congresses, magazines, reports and blogs. Her latest publications in 2021 have been in Frontiers in Psychology, IGI Global, Nova Science Publisher, Springer Oikonomics and Empresa and Humanismo Magazine, to 2023. As a regular contributor to the Association for the Development of Customer Experience (DEC), she has participated in the definition of the Employee Wave methodology and in various annual reports presented to the Association since 2019. Regular reviewer of publications in ACEDE, Aemark, Frontiers and others.

Gülşah Öz has been working as a Research Assistant at the Department of Foreign Languages Education, Aksaray University, Turkey. Her academic background mainly consists of English Language Teaching. She graduated from Çukurova University and got her MA degree from Çağ University. Then, she started her PhD at Çukurova University in 2020 and she is still a PhD student. She is especially interested in multiculturalism, social justice and equity in education, language teachers' education, foreign language teaching.

Thatayaone Segaetsho is a Lecturer at the Department of Library and Information Studies, University of Botswana. Previously he worked as a Conservator (April 2008 – 2021) at the University of Botswana Library Services. He holds Doctor of Philosophy (PhD) in Library and Information Studies obtained at

University of Botswana in 2017, BSc (Chemistry)-University of Botswana (UB), Post Graduate Diploma (Conservation) and Masters of Arts (Conservation) both obtained at Camberwell College of Arts in the UK. Dr Segaetsho is currently the Secretariat of Africa China Research Group (ACRG) (June 2015-Current) at University of Botswana. He is the Chairperson of the Disaster Management Committee of the University of Botswana Library, and Secretary for University of Botswana Sustainable Development Goals (UB SDG HUB) Committee. He also served as an Assistant Secretariat (January 2013-December 2013) for The Review Committee of the Reorganization of the UB Administration, University of Botswana. He was a member of The Organizing Committee for the IFLA/RISS satellite meeting, The International Federation of Library Associations and Institutions (IFLA): Reference and Information Services Section (RISS) 2015 satellite meeting Gaborone.

Şenol Şen is currently an associated professor in the Department of Mathematics and Science Education at Hacettepe University, Ankara, Türkiye. He received B.Sc., M.Sc., and Ph.D. degree from the Faculty of Education at Hacettepe University in 2009, 2011 and 2015. His research interests include self-regulated learning, chemistry education, inquiry-based learning, problem-based learning, and technology integration.

Chia Hua Sim (Janice) is currently the Associate Dean (Academic Practice) of the Faculty of Business, Design and Arts. She has more than 10 years of experience in academia. In her current role, she works with Head of School to facilitate staff participation in professional development activities. This includes facilitating the development of new faculty members and the enculturation of these staff members into Swinburne Sarawak. Given support from Melbourne counterpart, she works with Digital Learning Technologies at the Sarawak campus to promote new educational initiatives. As a member of the Academic Practice Committee, she also works closely with faculty/school representatives to identify and prioritise skill enhancement areas, and to facilitate continuous development of academic and professional skills of faculty members.

Ranjit Singha is a Doctorate Research Fellow at Christ (Deemed to be University) and a distinguished American Psychological Association (APA) member. His expertise lies in research and development across various domains, including Mindfulness, Addiction Psychology, Women Empowerment, UN Sustainable Development Goals, and Data Science. He has earned certifications from renowned institutions, including IBM and The University of Oxford Mindfulness Centre, UK, in Mindfulness. Additionally, he holds certifications as a Microsoft Innovative Educator, Licensed Yoga Professional, Certified Mindfulness Teacher, and CBCT Teachers Training from Emory University, USA. Mr Ranjit's educational qualifications include PGDBA (GM), MBA (IB), MSc in Counseling Psychology, and completion of a Senior Diploma in Tabla (Musical Instrumentation). His dedication to continuous learning is evident through his involvement in the SEE Learning® (Social, Emotional, and Ethical) Learning program. As a committed researcher and educator, Mr Ranjit focuses on mindfulness and compassion-based interventions. He has an impressive publication record, having authored twenty-three research papers, ten chapters, four books, and five edited books. His research interests encompass various aspects of mindfulness, such as assessment, benefits of mindfulness-based programs, change mechanisms, professional training, mindfulness ethics, cognitive and neuropsychology, and studies related to high-risk behaviours. Apart from his research endeavours, Mr Ranjit has extensive teaching experience, instructing courses in diverse subjects like Forensic Psychology, Positive Psychology, Organizational Planning, Strategic

Management, Psycho Metric Tests, Counseling Skills, Disaster Management, Basic Computer Science, Business Planning, Business Law, and Auditing. He has mentored numerous Postgraduate and undergraduate research projects, demonstrating his commitment to nurturing young minds in psychology. Furthermore, Mr Ranjit actively provides personal counselling services, showcasing his genuine concern for his students' well-being and academic success. His unwavering dedication to research and education has solidified his position as a valuable contributor to psychology.

Surjit Singha is a dedicated researcher actively involved in Research & Development (R&D) and publishing on a wide range of topics, including UN Sustainable Development Goals, Organizational Climate, Workforce Diversity, Organizational Culture, HRM, Marketing, Finance, IB, Global Business, Business, AI, Women Studies, Cultural Studies, and Data Science. Currently serving as a faculty member in the Department of Commerce at Kristu Jayanti College (Autonomous), Dr. Surjit holds the position of Editor, reviewer, and author for various esteemed Global Publications and Journals. With over 13 years of experience in Administration, Teaching, and Research in both Industry and Higher Education Institutions (HEIs), Dr. Surjit is committed to imparting knowledge and guiding students in their research pursuits. As a research mentor for students, Dr. Surjit plays a vital role in nurturing young minds and fostering their academic growth. Additionally, Dr. Surjit heads the website management, promotion, and social media publicity, leading a team of approximately 35 faculty members. Dr. Surjit's dedication to continuous learning is evident through certifications as a Life Skill Trainer by RGNYID (An Institution of National Importance introduced by the Act of Parliament), a Microsoft Innovative Educator by Microsoft, and a Yoga Volunteer certified by The Yoga Certification Board, Ministry of AYUSH, Gov. of India, and evaluated by Morarji Desai National Institute of Yoga. As a dedicated leader in academia, Dr. Surjit has guided and mentored 21 student projects in the academic year 2022-23, resulting in all 21 students publishing papers under Dr. Surjit's astute guidance. As NSS Faculty Mentor, Dr. Surjit has successfully mentored 65 students in the academic year 2022-23, ensuring all students achieve credits and qualifications. Dr. Surjit has an impressive track record with over 65 publications in SCOPUS-indexed articles, including chapters in books, edited books, and various textbooks. Granted 2 US Copyright and received two fully funded minor research projects (Completed and Published) from Kristu Jayanti College (Autonomous). With an unwavering commitment to excellence in education and research, Dr. Surjit Singha continues to impact various domains profoundly, elevating the institution's status and fostering an environment of knowledge dissemination and growth.

Senar Temel is a professor in Hacettepe University, Faculty of Education, Department of Mathematics and Science Education, Division of Chemistry Education. She has published many articles in national and international peer-reviewed journals. Her research interests include self-regulated learning, chemistry education, inquiry-based learning, and problem-based learning.

Adem Yilmaz is an academic, analyst, educational researcher, and a measurement and evaluation specialist. He holds a PhD in Measurement and Evaluation. His fields of expertise are as follows: teaching and learning of science, science education, scale development, measurement and evaluation, SPSS, LISREL, AMOS, EFA/DFA analyses. In addition, he concentrates on topics like teacher training, quality in higher education, and accreditation. Dr. Yilmaz is a Measurement, Learning & Evaluation (MLE) specialist and is the editor of several international books on Measurement and Evaluation. He is currently a faculty member and is working as an associate professor of education in Kastamonu University, Turkey. He is also a member of several committees in the same university.

Index

F

G

H

I

L

M

N

O

P

R

S

T

U

V

Z

* 9 7 8 1 6 6 8 4 9 8 5 9 0 *